Conservation Science and Advocacy for a Planet in Peril

Conservation Science and Advocacy for a Planet in Peril

Speaking Truth to Power

Edited by

Dominick A. DellaSala

Wild Heritage, A Project of Earth Island Institute, Berkeley, CA, United States

Elsevier
Radarweg 29, PO Box 211, 1000 AE Amsterdam, Netherlands
The Boulevard, Langford Lane, Kidlington, Oxford OX5 1GB, United Kingdom
50 Hampshire Street, 5th Floor, Cambridge, MA 02139, United States

British Library Cataloguing-in-Publication Data
A catalogue record for this book is available from the British Library

Library of Congress Cataloging-in-Publication Data
A catalog record for this book is available from the Library of Congress

ISBN: 978-0-12-812988-3

For Information on all Elsevier publications
visit our website at https://www.elsevier.com/books-and-journals

Publisher: Candice Janco
Acquisitions Editor: Marisa LaFleur
Editorial Project Manager: Alice Grant
Production Project Manager: Vignesh Tamil
Cover Designer: Christian Bilbow

Typeset by MPS Limited, Chennai, India
Transferred to Digital Printing 2022

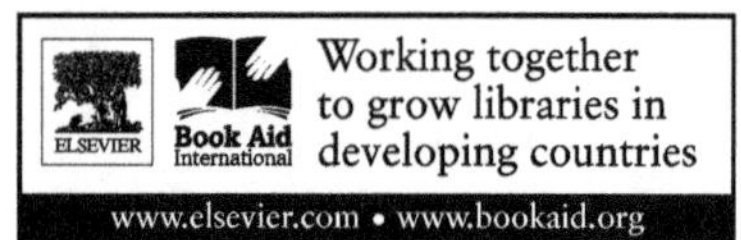

Contents

List of contributors

Franz Baumann New York University, Graduate School of Arts and Science, Program in International Relations, New York, NY, United States

Hal Beecher Washington Department of Fish and Wildlife (Retired), Olympia, WA, United States

Monica L. Bond Department of Evolutionary Biology and Environmental Studies, University of Zurich, Zurich, Switzerland

Bill Bradbury Former Oregon Secretary of State, Bandon, OR, United States

Jim Burroughs Oklahoma Department of Wildlife Conservation, Oklahoma City, OK, United States

Joel Clement Harvard Belfer Center for Science and International Affairs, Cambridge, MA, United States

Dominick A. DellaSala Wild Heritage, A Project of Earth Island Institute, Berkeley, CA, United States

Angus Duncan Natural Resources Defense Council—Consultant, Former Chair, Oregon Global Warming Commission, Salem, OR, United States; Former Chair, Northwest Power Planning Council, Portland, OR, United States

Peter Goldman Washington Forest Law Center

Noah Greenwald Center for Biological Diversity, Portland, OR, United States

Michael Halpern Formerly with the Union of Concerned Scientists, Cambridge, MA, United States

Chad Hanson John Muir Project of the Earth Island Institute, Berkeley, CA, United States

Robert M. Hughes Amnis Opes Institute, Corvallis, OR, United States; Oregon State University, Corvallis, OR, United States

David Johns School of Government, Portland State University, Portland, OR, United States

Jeremy T. Kerr University of Ottawa, Ottawa, ON, Canada

Arthur C. Knutson, Jr. California Department of Fish and Wildlife (Retired), Sacramento, CA, United States

Beverly E. Law Oregon State University (Professor Emeritus), Corvallis, OR, United States

Derek E. Lee Department of Biology, Pennsylvania State University, University Park, Pennsylvania, United States

Jennifer Mamola John Muir Project of the Earth Island Institute, Washington, DC, United States

Richard McIntyre The Community Governance Partnership (CGP), Sacramento, CA, United States; Cannabis Removal on Public Lands Project (CROP Project), Oakland, CA, United States

J. Hal Michael, Jr. Ecologists Without Borders, Olympia, WA, United States

Randi Spivak Public Lands Program, Center for Biological Diversity, Tucson, AZ, United States

Robert L. Vadas, Jr. Participating in his own capacity

Kara A. Whittaker Washington Forest Law Center

Augusta C.F. Wilson Climate Science Legal Defense Fund, New York, NY, United States

Biographies

Franz Baumann, PhD, started his career at the European Parliament in Luxembourg before transferring to the European Commission in Brussels, joining Siemens in Munich and, in 1980 the United Nations, where he served in four cities on three continents in a variety of functions. In 2009, he was appointed as an Assistant Secretary-General. His last assignment was Special Adviser on Environment and Peace Operations. He is currently a visiting research professor at New York University, a senior fellow and a member of the Board of Trustees of the Hertie School, Berlin, a Vice President of the Academic Council on the United Nations System, and a member of the Board of Advisers of the Centre for United Nations Studies at the University of Buckingham, England.

Hal Beecher, PhD, worked for The Nature Conservancy in Oregon and Washington before 36 years as an instream flow biologist with the Washington Department of Fish and Wildlife, where he researched salmonid hydroecology and worked with water users and managers to protect instream habitat while accommodating water use. He represented the agency on the Instream Flow Council (www.instreamflowcouncil.org), served as its president 2006–08, and was a co-author of two IFC books. He retired in 2016.

Monica Bond, PhD, is a wildlife biologist and biodiversity advocate with the Wild Nature Institute and a research associate with the University of Zurich. She has published 45 peer-reviewed scientific journal articles and book chapters, has worked as a field organizer for Green Corps, an Endangered Species grassroots organizer for the National Wildlife Federation, and a staff biologist for the Center for Biological Diversity. She spent the past two decades studying Spotted Owls and served on the Dry Forest Landscapes Working Group for the Northern Spotted Owl Recovery Plan. She travels around the world researching and advocating for the conservation of imperiled wildlife and habitats.

Bill Bradbury is a retired Oregon public official whose career includes both salmon recovery and climate change. He was first elected to the Oregon House of Representatives in 1980 and introduced the bill creating the Salmon and Trout Enhancement Program. Since its enactment, Oregon has benefitted from over 3 million volunteer hours restoring salmon. In 2006, Bill was one of the first 50 people trained by the former Vice President Al Gore to present the now well-known slide show ("An Inconvenient Truth") about global (and local) impacts of a changing climate. In 2017, Bill received "the Green Ring Award" from Mr. Gore "in recognition of outstanding work towards solving the climate crisis."

Jim Burroughs, MS, has been working for over 30 years at the Oklahoma Department of Wildlife Conservation. His primary responsibilities include managing both lakes and rivers and presently employed as the state stream supervisor representing the agency on environmental flow/environmental flow policy and water quality standards. His research experience ranges from braided prairie streams throughout the stream spectrum to Ozark spring-fed streams.

Joel Clement is a senior fellow at the Harvard Kennedy School's Belfer Center for Science and International Affairs. With a background in resilience, climate change adaptation, and Arctic social-ecological systems, he is conducting research and working with partners to improve the knowledge and tools necessary to reduce climate risk and increase resilience in frontline areas such as the Arctic region. Joel served as an executive for 7 years at the US Department of the Interior. In July 2017, he became the first public whistleblower of the Trump Administration. Since then, he has received multiple awards for ethics, courage, and his dedication to the role of science in public policy. Joel has been featured and interviewed on CNN, MSNBC, PBS, ABC, NBC, CBS, and Democracy Now and has been published by the Washington Post, Denver Post, The Guardian, NBC Think, and the Georgetown Journal of International Affairs.

Dominick A. DellaSala PhD, is Chief Scientist at Wild Heritage (www.wildheritage.org), and former President of the Society for Conservation Biology, North America Section (http://www.conbio.org). He is an internationally renowned author of over 200 science papers on forest and fire ecology, conservation biology, endangered species management, and landscape ecology. Dominick has given plenary and keynote talks ranging from academic conferences to the United Nations Earth Summit. He has appeared in National Geographic, Science Digest, Science Magazine, Scientific American, Time Magazine, Audubon Magazine, National Wildlife Magazine, High Country News, Terrain Magazine, NY Times, LA Times, USA Today, Jim Lehrer News Hour, CNN, MSNBC, "Living on Earth (NPR)," several PBS shows and documentaries and even Fox News! Dominick has served on numerous committees, including White House Council task forces on forests and the Oregon's Global Warming Commission carbon task force reporting to the governor. He is editor of numerous scientific journals and encyclopedias and has published award-winning books on climate change, forest policy, and scientific integrity. Dominick is motivated by his work to leave a living planet for his daughters, grandkids and all those that follow.

Angus Duncan was the founding President of the Bonneville Environmental Foundation, which supports renewable energy development and watershed restoration in the Pacific Northwest. He has worked in private sector renewable energy project development, in state and local government, as a Member and Chair of the Northwest Power Planning Council, and as Director of Energy

Policy, US Department of Transportation. In 2004, Angus chaired the Drafting Committee that wrote Oregon's greenhouse gas reduction goals and climate strategy since adopted by the Governor and Legislature. He served as Chair of Oregon's Global Warming Commission 2008–20 and consults with NRDC on energy and climate policies in the Pacific Northwest.

Peter Goldman has been the Director and Managing Attorney of the Washington Forest Law Center since its inception in 1997. After clerking for Washington State Supreme Court Justice James M. Dolliver, he joined the criminal division of the King County Prosecutor's office, where he spent 11 years as a trial lawyer in all of the office's divisions. During his last 5 years there, he was a senior deputy prosecuting attorney in the Appellate Division, where he handled over 200 complex criminal appeals. Peter has litigated and advocated in the public interest to protect the Pacific Northwest's private and state-owned forest lands.

Noah Greenwald, MS, is the endangered species director at the Center for Biological Diversity, where he has spent more than two decades working to protect hundreds of species as threatened or endangered under the Endangered Species Act. He also works to educate the public about the importance of protecting biodiversity and the multitude of threats to the survival of North American wildlife. Before he joined the Center in 1997, Noah worked as a field biologist, surveying Northern Spotted Owls, Marbled Murrelets, and banding Hawaiian songbirds.

Michael Halpern, PhD, spent 17 years at the Union of Concerned Scientists where he helped build its Scientific Integrity Program and Center for Science and Democracy. He has extensive experience advocating for solutions that ensure government decisions are fully informed by scientific information, that the public understands the scientific basis for those decisions, and scientists are able to more effectively engage the public. Michael has led major projects to defend scientists from harassment and create conditions that make science and scientists more resilient to political, industry, and ideological influence. He has testified before Congress and numerous federal and state advisory panels about the use of science in policymaking and has appeared in national and international media outlets, including the Associated Press, CNN, National Public Radio, NBC, *The New York Times*, *The Boston Globe*, *Rolling Stone*, and *The Washington Post*. He wrote this chapter in a personal capacity.

Chad Hanson, PhD, is a research ecologist and the director of the John Muir Project of Earth Island Institute, located in Big Bear City, California. His research is focused on fire ecology in conifer forest ecosystems, and he is the coeditor and co-author of the 2015 book, "*The Ecological Importance of Mixed-Severity Fires: Nature's Phoenix*" (Elsevier, Inc.). He has published dozens of scientific studies and articles in peer-reviewed journals pertaining to forest and fire ecology, and recently finished a second book, focusing on forest protection to mitigate climate change and the myths about wildland fire that are impeding progress. His research includes natural postfire forest regrowth and carbon sequestration, carbon flux in wildland fires, current forest fire patterns and trends, fire history, habitat selection of rare wildlife species associated with habitat created by high-intensity fire, and adverse impacts to wildlife caused by logging.

Robert Hughes, PhD, for 32 years was a contracted researcher with the Environmental Protection Agency focusing on regional and national aquatic ecosystem studies and assessments. He currently works part-time for the Amnis Opes Institute on biological assessments of streams, lakes, and rivers in Europe, Brazil, and China. Bob is a past-president of the American Fisheries Society (AFS). He is an AFS and Society for Freshwater Science Fellow and Life Member, Oregon AFS Lifetime Achievement Awardee (2017), Best Paper Awardee *Lake and Reservoir Management* (2014), AFS Distinguished Service Awardee (2013), Oregon AFS Fisheries Worker of the Year Awardee (2011), Fulbright Scholar (2010, 2007), AFS Western Division Special Recognition Awardee (2010), and Best Paper Awardee *Transactions of the American Fisheries Society* (2008). Robert has authored or co-authored 237 peer-reviewed publications and given 77 invited international presentations on five continents in 15 nations.

David Johns, PhD, has advocated for large-scale conservation for decades. A co-founder of the Wildlands Network, Yellowstone to Yukon Conservation Initiative, and Conservation Biology Institute, he currently serves as Chair of the Marine Conservation Institute which is home to the Blue Parks Initiative. He has worked on conservation projects throughout the Americas, in the Russian Far East, Australia, Europe, southern Africa, and the global ocean. An activist, attorney, and conservation strategist, he is the author of *A New Conservation Politics* (2009), a manual on effective conservation advocacy, and *Conservation Politics: The Last Anti-Colonial Battle* (2019), about overcoming the root causes of ecological decline instead of treating symptoms. David taught politics and law at Portland State University as an adjunct for 40 years.

Jeremy Kerr, PhD, holds the University Research Chair in Macroecology and Conservation and is Full Professor and Chair of Biology at the University of Ottawa. His primary research has led to discoveries with broad impact, most recently around how and why climate change has affected species survival, with several publications in *Science* and elsewhere. Jeremy is strongly engaged in public science, working at the forefront of efforts to defend and recover scientific integrity, defend species at risk protection, and propose and implement policies on equity, diversity, and inclusion, which have helped create profound change in the structure of research funding in Canada. He is a Past-President of the Canadian Society for Ecology and Evolution, and an alumnus of the Global Young Academy.

Arthur Charles Knutson, Jr., MS, is a fisheries biologist that has worked on steelhead ecology in California. He also has been Peace Corps Volunteer high school science teacher trainer in Bihar, India; marine and fisheries biologist with the California Department of Fish and Game; and has 14 publications and 15 awards.

Beverly E. Law, PhD, is Professor Emeritus of Global Change Biology & Terrestrial Systems Science at the Oregon State University. She is a fellow of the American Geophysical Union and the Leopold Leadership Program. Beverly has over 250 journal articles and reports on the effects of climate, wildfire, and management on forest carbon processes from ecosystems to regions, carbon accounting, and land use strategies for mitigating climate change that benefit biodiversity.

Derek E. Lee, PhD, is principal scientist at the Wild Nature Institute and associate research professor at Pennsylvania State University. Derek has published more than 50 peer-reviewed scientific journal articles and book chapters on population biology, fire ecology, animal migration, animal coloration, and climate effects on marine predators.

Jennifer Mamola, BS, is the DC-based forest protection advocate for the John Muir Project of the Earth Island Institute. Prior to joining the project, she spent time as a lobbyist fighting for the health, safety, and security of Peace Corps Volunteers on Capitol Hill.

J. Hal Michael, Jr., MS, is a retired state biologist currently volunteering with Ecologist's Without Borders working globally on projects to improve ecological sustainability for people and resources. He was employed for more than 30 years by the Washington Departments of Game, Fisheries, and Fish and Wildlife. His work revolved around anadromous and resident salmonids and consisted of research, frontline management, environmental management of hatcheries, and developing projects to restore threatened and endangered fish stocks.

Richard McIntyre, a senior consultant on natural resource and political issues, has a long history in the field of major conservation campaigns, community organizing, and electoral politics. He is the Director of the CROP Project (Cannabis Removal on Public Lands), a groundbreaking effort to solve the seemingly intractable problem of black market grows on California public lands. He is the founder and Executive Director of Leadership For Jobs and a New Economy (LJANE/501c-4), and The Community Governance Partnership (501c-3), established in 2013. He is also the founder of The California Organizing Academy, a community organizer training program that has over 450 graduates, with 57 now holding public office or managing nonprofits. He has served on Presidential campaign staffs, ran a US Senate campaign in Oregon, and ran the California Assembly Special Election in 2016 in Assembly District 31.

Richard has a long association with the Midwest Academy, internationally noted trainers in community organizing.

Randi Spivak has over 25 years of experience as a policy and conservation advocate. She joined the Center for Biological Diversity in 2013 where she directs the Public Lands Program with the goal of strengthening management for wildlife and climate. She played a lead role in spearheading and leading the campaign to end fossil fuel leasing on federal public lands, which accounts for 25% of the United States' carbon pollution. Randi is championing a campaign to end logging of mature- and old-growth federal forests, the conservation of greater sage-grouse, hard rock mining reform, and securing permanent protection for 30% of the U S lands and waters. Prior to joining the Center for Biological Diversity, she was a vice president for government affairs at the Geos Institute and executive director of American Lands Alliance, a grassroots umbrella group that advocated for federal forest and wildlife protections.

Robert L. Vadas, Jr., PhD, is a research scientist, addressing biophysical issues with a focus on riparian/wetland, instream-flow, dam, and marine-hydrokinetic topics for Washington. He has worked for two consulting companies in Alberta and has completed two postdoctoral projects in British Columbia and California on fish habitat and other issues, with a collective focus on impacts to the floodplain, freshwater, and estuarine ecosystems. Bob has worked as a state agency biologist on estuarine fish habitat (especially salmonids) and exotic wetland plant issues, his contributions to this book are in his personal capacity.

Kara A. Whittaker, PhD, has been the Senior Scientist and Policy Analyst at the Washington Forest Law Center since 2008. At the center, she performs spatial and technical analyses on topics ranging from endangered species management to forest certification and geomorphology in support of legal cases and environmental policymaking. The emphasis of her work is identifying sustainable forest practices intended to minimize negative impacts on fish, wildlife, and people.

Augusta C.F. Wilson, JD, has been an attorney with the Climate Science Legal Defense Fund since 2018. The primary focus of her work is offering legal support and advice to scientists who are targeted because of the work they do. She has also practiced environmental law as an attorney with the Clean Air Council where she primarily worked on litigation involving pipelines carrying natural gas and fracking byproducts. Augusta was a fellow at the NYU School of Law's Guarini Center in Environmental, Energy, and Land Use Law, publishing on environmental and energy law.

Foreword: uncensored science is crucial for global conservation

Editor's Note: This essay by esteemed scientist James Hansen is a hybrid of the books' foreword and an independent treatise on the accelerated warming of the planet.

Science is needed today more than ever

We must follow the science to save our home planet, but what does the science tell us? Essays in this extraordinary work will help scientists communicate their findings to the public and policymakers, which is one objective of conservation scientist Dominick DellaSala, the book's mastermind. The lessons provided—in personal integrity, preparation, accessibility (the elevator speech), alliance-building, and political acumen—have wide applicability.

Science was a guiding force in the explosions of knowledge and political revolutions of the 17th and 18th centuries—the Age of Science and Reason, also called the Enlightenment. Rationalism was spurred by Galileo's telescopic observations. Science dispelled myths, such as the belief that the Sun orbited Earth. Medieval worldviews began to change.

However, there was no sudden global epiphany. Galileo, for his daughter's sake and his own sake, was forced to "confess" his heresy, comforted by the realization that history would provide fair assessment and judgment. A delay in understanding was not harmful to the world.

Today science still competes with beliefs. Yet the need for rationalism in understanding—of both our planet and our political systems—has never been greater. And we do not have the luxury of ample time that Galileo enjoyed.

Philosophers of the Enlightenment were mainly European, but the American Revolution and Constitution were the most important political products that emerged on the world scene. Concepts of freedom, equality, individual rights, and celebration of diversity were at the heart of this first democratic constitutional republican form of government, characterized by the rule of law with the consent of the governed, and by checks and balances among competing interests.

I was born in 1941—the year the United States entered World War II—and grew up in the post-war era, when, unlike the period after World War I, the United States provided global leadership, supported the reconstruction of war-torn regions, led the formation of the United Nations, and promoted the Universal Declaration of Human Rights. Global cooperation and commerce increased. Standards of living improved in nations adopting constitutional governments with individual rights, including nations defeated in World War II. The United States took the lead in establishing the international organizations that facilitated economic growth and security. Cooperation lifted all boats; it was not a zero-sum game.

When I was a kid, we were taught that America was a shining city on a hill. "Truth, justice and the American way" seemed almost synonymous when the comic hero Superman first uttered that phrase. Science provided a way to discover the truth. The objective scientific method is designed to uncover the unvarnished truth, independent of our preferences.

Presidents Harry Truman and Dwight Eisenhower respected science. Eisenhower had a Science Advisory Committee and he invested in our educational system. Not long before his death, Eisenhower said to James Killian, his former Science Adviser: "You know, Jim, this bunch of scientists was one of the few groups that I encountered in Washington who seemed to be there to help the country and not to help themselves" (Killian, 1977). Eisenhower wanted the truth and the country benefitted from it.

Eisenhower was concerned as he left office in 1960. He saw the growing power of special interests. His farewell address focused on the military-industrial complex, which grew with a perceived threat by the Soviet Union. That threat receded as NASA, formed in 1958, beat the Soviets to the moon, arms control treaties were negotiated, and the Berlin wall was torn down. But, like a cancer, the role of special interests and money in our government continued to grow.

Truth is the enemy of special interests

Gradually, the truth became malleable to politicians. They became elite and addicted to the money of special interests. They justified taking money as being required for their campaigns, but it also supported their lifestyles. Their first priority became their own reelection, not the best interests of the public.

The other party became the focus, the enemy. Negative campaigning worked. The next campaign began the morning after the last election. Bipartisanship waned. Governance and policies suffered. Wealth disparity grew. Opportunity was not equal. Unjustified military adventures abroad drained lives, treasure, and spirit. We did not seem to have a government working to achieve a more perfect union. Frustration of the public brought out the worst from fringe groups, including scapegoating and hatreds. Home and abroad, the public saw that America's professed idealism—of a shining city on a hill—was becoming a sham. Yet politicians attempted to prove their patriotism by the number of flags in their photo-ops.

We still live in a democracy with enormous potential, but we must work to make it work. The shock of recent events—angry, destructive protests—may be a godsend, if it invigorates the people who believe that we can still achieve a more perfect union. Founders of the American democracy foresaw the sort of deterioration that we have witnessed—corruption, really—and they believed that every so often a revolution may be required to restore government integrity. Not a shooting revolution—they hoped—but a revival of the spirit of public service.

We have reached such a time. I am optimistic that we can find a path out of our present dangerous partisanship. I believe that truth and science can help us find that path.

My perspective derives from a long career in science, including efforts to communicate implications of climate science to the public and politicians. Indeed, the chapters in this book and the world's precarious circumstance—on the cusp of previously only imagined global change—forces me to ponder: where did we go wrong? How could we scientists do a better job of informing the public and policymakers?

My first foray into the world of policy was innocent. I wrote a paper describing research carried out by six other young atmospheric physicists and me at the NASA Goddard Institute for Space Studies (Hansen et al., 1981). We warned that continued business-as-usual fossil fuel use could result in global warming as great as 5°C by the end of the 21st century. Such warming, we noted, might result in disintegration of the West Antarctic ice sheet and sea level rise of as much as several meters by the end of this century, as well as extreme regional climate consequences.

Furthermore, we dared to point out obvious policy implications of our study, writing:

> *Political and economic forces affecting energy use and fuel choice make it unlikely that the CO_2 issue will have a major impact on energy policies until convincing observations of the global warming are in hand. In light of historical evidence that it takes several decades to complete a major change in fuel use, this makes large climate change almost inevitable. However, the degree of warming will depend strongly on the energy growth rate and choice of fuels for the next century. Thus, CO_2 effects on climate may make full exploitation of coal resources undesirable. An appropriate strategy may be to encourage energy conservation and develop alternative energy sources, while using fossil fuels as necessary during the next few decades.*

Funding for our CO_2 research was promptly terminated by the U.S. Department of Energy. The impact was sobering and stressful, as I had to inform individuals that we had lost support for five young scientists. It was clear already in 1981 that special interests had inordinate sway in our government—and no special interest was more powerful than the fossil fuel industry. The funding blow added to pressures to close our Institute in New York and move remaining scientists to Greenbelt, Maryland. We survived in New York thanks to the help of two angels, one at Columbia University and one at Goddard Space Flight Center (Hansen, 2022a).

Funding constraints did not terminate our climate research. In Chapter 3 of this book, Sounding the Alarm, Former Assistant Secretary General of the United Nations Franz Baumann points to my 1988 testimony before the Senate Committee on Energy and Natural Resources (Hansen, 1988), when I concluded that "earth is warming by an amount that is too large to be a chance fluctuation and in my opinion the greenhouse effect. . .is changing our climate now." My conclusion was hardly universal then (Kerr, 1989), but I could state it with confidence based on the combination of paleoclimate evidence, global climate models, and ongoing observations of climate change. Altogether, it was clear that a basic change in the world's energy strategy was needed.

Remarkably, within a few years the UN Framework Convention on Climate Change was agreed upon in Rio in 1992 and would be signed by almost all nations. The stated objective of the Framework Convention is "stabilization of greenhouse gas concentrations in the atmosphere at a level that would prevent dangerous anthropogenic interference with the climate system. Such a level should be achieved within a timeframe sufficient to allow ecosystems to adapt naturally to climate change, to ensure that food production is not threatened and to enable economic development to proceed in a sustainable manner" (United Nations Framework Convention on Climate Change, 1992).

The Framework Convention was a political triumph, achieved at a time when the scientific community had barely begun to recognize that greenhouse-driven climate change was underway—as evidenced by the reactions to my Congressional testimony (Kerr, 1989). Yet even three decades later, the Framework Convention has had almost no effect in stemming the growth of atmospheric greenhouse gases. Indeed, after the 1997 Kyoto Protocol, global fossil fuel emissions of CO_2 (Fig. F.1), the principal drive of global warming, accelerated faster!

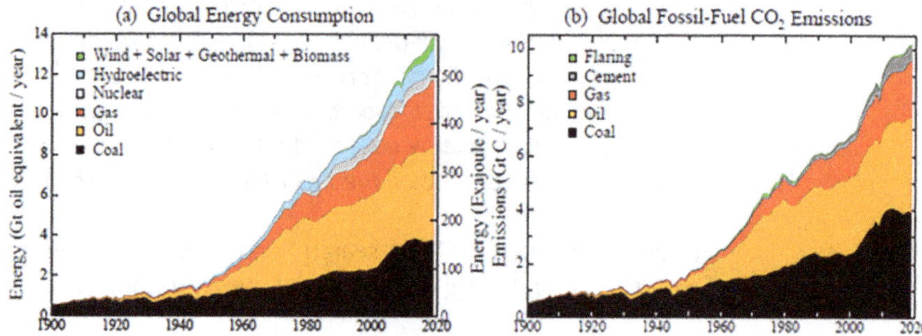

Figure F.1 Global energy consumption (A) and fossil fuel CO_2 emissions (B) from 1900 through 2019.

Censorship, as Dominick DellaSala realized in choosing chapters for this book, is a problem for conservation

Blatant censorship can be addressed by public objection, but institutional and personal costs discourage such revelation. Moreover, there are also more subtle constraints on communication, which are more difficult to address and may be more dangerous.

In my testimony to Congress I had no intent to be a whistleblower—I just reported science as I saw it. However, when NASA preferred to have someone from Headquarters testify in my stead in 1988, I did not acquiesce because his testimony would differ from mine (Hansen, 2022b). And in 1989 I informed the Senate committee that my written testimony had been altered, over my objection, by the White House Office of Management and Budget (Hansen, 2022c).

After a third incident—I submitted a paper (Hansen et al., 1990) on the need for small satellite observations that were not included in NASA's Earth Observing System—I was reprimanded for "fighting NASA for a third time." I objected, arguing that I acted under allegiance to scientific accuracy and the taxpayers. I took the issue to high levels—the NASA Administrator and the White House (Vice President Al Gore)—but without effect (Hansen, 2022d).

That's not surprising. The problems emanate from the highest levels. The NASA troops—and government employees in other agencies that I interacted with—are hard-working competent people. The problem is that they are constrained to work in an increasingly bureaucratic, inefficient system. Neither political party makes a serious effort to reduce bureaucracy and increase government vitality. On the contrary, both parties have increased the politicization and inefficiency of government agencies.

The political party controlling the executive branch installs political appointees to head Offices of Public Information at science agencies, which thus become Offices of Propaganda that attempt to make the incumbent Administration look good. Both parties allow their Office of Management and Budget to alter scientific testimony. These are fixable problems. The public can affect this situation via political parties, their platforms, and elections—as I discuss below.

Regarding my specific disagreements with NASA, I had to accept the punishments, which included a reduction of resources for the Institute that I headed. I was in love with science and uncomfortable with the hullabaloo that accompanied my testimonies, so I had already decided to retreat into scientific research. I continued to advocate for small satellite measurements but otherwise tried to focus on science and avoid controversy.

By 2004, I felt compelled to speak out again because of growing evidence that we were moving toward dangerous climate change but had no effective policies. Reactions to my talks exposed both continued government censorship and a censorship self-imposed by scientists.

Government censorship was so routine that NASA thought nothing of assigning a minder to screen my interactions with media. When censorship reached the level of "prior restraint"—I was required to tell NASA Public Affairs of interviews beforehand and let them replace me—I informed the New York Times. Prior restraint blatantly violates the Constitutional right of free speech. After a Public Affairs employee confirmed the censorship, the Times published a story that put an end to this specific censorship. NASA pretended that the censorship had been the work of a 24-year-old maverick, but Bowen (2008) in *Censoring Science* found that instructions came from the White House and the highest levels at NASA Headquarters.

When I was asked to testify to Congress again, the Director of Goddard Space Flight Center—whom I greatly respected—gently suggested that I would be wise to stick to climate science and not discuss policy. I could not have agreed less. Why should scientists not connect dots all the way to defining the actions needed to avoid dangerous climate change? If scientists do not speak up, policies will continue on the disastrous course defined by special interests.

Scientific reticence can amount to self-censorship

Indeed, damage from excessive reticence can exceed that from ham-handed government strictures.

In public lectures (Hansen, 2004, 2005a), I argued that the dangerous level of warming is lower than implied in UN Intergovernmental Panel on Climate Change (IPCC) reports. The IPCC "burning embers" method was used to calculate a 50% chance of exceeding the dangerous threshold if global warming reached 2.85°C relative to late 20th century climate or 3.45°C relative to 1880–1920 (Schneider and Mastrandrea, 2005). That comported with common sense: 2°C–3°C warming did not seem to be disastrous.

But paleoclimate data give pause. When Earth was last 3°C warmer—in the Pliocene—sea level reached at least 10–15 m (33–50 feet) higher (Dwyer and Chandler, 2009; Dumitru et al., 2019). IPCC relied on ice sheet models that needed millennia to yield large ice sheet change. The then-current IPCC (2001) report estimated sea level rise of only 40–45 cm by 2100, with 30 cm from thermal expansion of ocean water, 10–15 cm from alpine glaciers, and little change from the ice sheets—for the heavily studied IS92a scenario, which has 715 ppm CO_2 in 2100.

Field geologists who worked on the ice sheets—Konrad Steffen, Eric Rignot, and Jay Zwally—doubted those models. They expected more rapid ice sheet disintegration. So did I. In an editorial essay (Hansen, 2005b), I argued that a warmer ocean and marine-abutting ice sheets could yield a sea level rise of several meters in a century. Human-made climate forcing—imposed perturbation of Earth's energy balance—in IPCC scenarios for this century is larger and much more rapid than natural climate forcings. It seemed nonsensical to think it would take a millennium to achieve a large response.

Scientific reticence (Hansen, 2007) arises partly from the reward structure. A scientist crying danger is rebuked—by fellow scientists and funders—as we learned in the 1980s. But there is no penalty for "fiddling while Rome burns." Indeed, a scientist who lards his conclusions with excessive caveats and uncertainties is rewarded. Caution has merits, but in the case of ice sheets and sea level rise, we may rue reticence, if it locks in future disasters.

Something was wrong with ice sheet models

In fairness to the modelers, ice sheets are complex with processes occurring on a wide range of spatial scales, so modeling ice sheets is hard. However, it is easy to find instances in the paleoclimate record when sea level rose several meters in a century (Fairbanks, 1989; Deschamps et al., 2012). Ice sheet models did not capture such rapid change.

What could we do in the absence of good ice sheet models? Known cases of sea level rising several meters in a century imply exponential ice sheet disintegration, a process characterized by amplifying feedbacks that lead to collapse of a vulnerable portion of an ice sheet. Such a process can be characterized approximately by a doubling time for the rate of ice sheet mass loss.

I decided to do a climate modeling experiment in which—instead of using an ice sheet model—we used the observed rate of ice melt and let it grow exponentially. We would try alternatives—10 years and 20 years, for example—for the doubling time. Precise measurements of ice sheet mass were beginning to be made, so if our concept was right and high emissions continued, we would eventually get an empirical measure of the doubling time.

In October 2006, Reto Ruedy, Makiko Sato, and I— made a model run with meltwater injection from Antarctica and Greenland. The initial ice melt rate was from observations; it then increased with a 10-year doubling time up to a sea level rise of 5 m. Most of that water could be provided by West Antarctic ice, which rests on bedrock below sea level (Fig. F.2). Deep valley outlets on East Antarctica (Greenbaum et al., 2015) and Greenland (Catania et al., 2020) would expose additional ice to contact with ocean water.

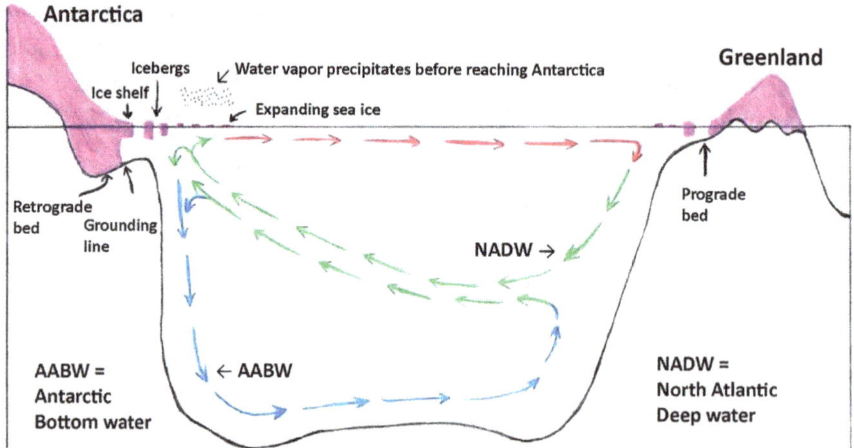

Figure F.2 Schematic diagram of the surface origin of two water masses that fill most of the world ocean: North Atlantic Deep Water (NADW) and Antarctic Bottom Water (AABW). The upper circulation cell is the Atlantic Meridional Overturning Circulation (AMOC) and the deeper circulation is the Southern Meridional Overturning Circulation (SMOC). Physical processes occurring near the ice sheets are discussed later in this foreword and this illustration is adapted from Hansen et al. (2016).

I was stunned by the model result

Within several decades the North Atlantic and Southern Ocean Overturning Circulations (dubbed AMOC and SMOC) had shut down. In a hot, warming world, sea ice around Antarctica held steady and then expanded northward!

Wally Broecker had long asserted that ice melt could shut down the AMOC and cause cooling in the North Atlantic and Europe. However, climate modelers did not confirm his expectations. Wally was not a climate modeler. He grumpily acceded to the modelers, but he retained a healthy skepticism of climate models (Hansen, 2022e). I shared Wally's skepticism.

Why did our result differ? Our climate model included an ocean model designed by Gary Russell with special attention to proper conservation of physical quantities. We also included the cooling effect of icebergs in the freshwater discharge from ice sheets. However, our model had coarse resolution compared with other models. We were certain to be hammered by other scientists if we presented new results without a good explanation for why they differed.

Now we were challenging both ice sheet models and ocean models! We had neither a glaciologist nor an oceanographer on our little team. We did not have the heft, nomenclature, or detailed understanding needed to challenge the leaders in those fields. So, we had a lot of work to do. Moreover, I was committed to protests against government inaction on climate, and I was involved in issues in energy science and economics about how to phase out carbon emissions.

In 2007, I read papers of geologist Paul Hearty and initiated correspondence with him

Hearty explored sites worldwide, focusing on places with minimal vertical movement of Earth's crust from tectonic uplift or crustal rebound caused by ice sheet melt. Hearty's papers (e.g., Hearty and Neumann, 2001; Hearty et al., 2007) are full of photos, maps, and descriptions that allow readers to almost feel that they are accompanying a classical geologist who skillfully reads the record of Earth's climate and sea level imprinted in the rocks.

My interest was in Hearty's conclusion that rapid sea level rise of at least a few meters occurred late in the Eemian interglacial period, raising sea level to +6–9 m (20–30 feet) relative to today. The Eemian is the most recent interglacial period prior to the present (Holocene) interglacial period during which civilization developed. Global temperature during the Eemian was 1°C–2°C warmer than the preindustrial Holocene, thus providing an indication of what may be in store as a consequence of human-made global warming.

Hearty was not alone in concluding that rapid sea level rise occurred in the Eemian. Rohling et al. (2008), via innovative analysis of Red Sea sediment cores, found evidence of sea level changes during the Eemian period. If these oscillations were real sea level change, they implied an average Eemian sea level change rate of more than 1 m per century. His group (Grant et al., 2012) also found that sea level changes lagged Antarctic climate changes by only 100–400 years and lagged Greenland climate changes by 200–400 years.

The paleoclimate sea level changes were in response to climate forcings—imposed changes of Earth's energy imbalance—that were weaker and changed much more slowly than the human-made climate forcings. Yet the paleoclimate forcings produced frequent large, rapid sea level change. The models IPCC relied on failed to produce such realistic, rapid change.

Something was wrong with those models. Of that, I was certain.

Providentially, I was invited to give the Bjerknes lecture at the American Geophysical Union meeting in 2008

In my talk, *Climate Threat to the Planet: Implications for Energy Policy and Intergenerational Justice* (Hansen, 2008), I had 1 hour to describe the climate situation and policy implications.

I had a suspicion about a problem in ocean models. When we doubled atmospheric CO_2, we found that global surface temperature after 100 years had only achieved 60% of its final warming. Could mixing of heat into the ocean really slow down the surface response that much? Such a long delay was not expected by the legendary Jule Charney (Hansen, 2022f).

Prior to my talk, I requested model results from three of the most prominent climate groups, and they generously cooperated. The global temperature response of these models was even slower than in our model! I emphasized this topic in my AGU lecture, hoping to encourage reporting of response functions—how fast global surface temperature responds to a forcing—for all models.

Response function information might spur more focus on ocean mixing and on observations to test the reality of ocean mixing in all models. Such a focus on the key (real world and model) physics is analogous to how Jule Charney focused his famous investigation of climate sensitivity (Charney et al., 1979). Charney would have jumped eagerly on the issue of ocean mixing and climate response time, but he died young, in 1981. We lesser scientists were on our own.

The climate modeling community did not jump on the ocean mixing issue, and I was focused on policy matters, as summarized in communications on my website (Columbia, 2021). But then, in 2010 I saw a paper by Karina von Schuckmann, a German post-doc working in a French oceanographic laboratory. She had the data that I had been waiting for.

Karina von Schuckmann analyzes data from thousands of Argo floats that were distributed around the world ocean during the first decade of this century

Argo floats (Argo, 2021) dive to a 2-km depth, rise to the surface while making measurements, and radio the data to a satellite. Precise ocean temperatures measured by the Argo floats were the data needed to define Earth's energy imbalance. That imbalance is important: it defines how much additional global warming is in the pipeline and it thus informs us about actions needed to stop further global warming.

Accurate determination of Earth's energy imbalance meant that we had two major "knowns" about the climate system, the other being observed global warming in the past century. There are three major unknowns: climate sensitivity to a forcing, the net climate forcing, and the delay of surface temperature change caused by ocean mixing of heat.

Climate sensitivity is constrained by paleoclimate data, which implies a sensitivity near 3°C for doubled CO_2 climate forcing. If we assume that sensitivity, we are left with two knowns and two unknowns—a solvable problem. Much of the climate forcing—that due to greenhouse gases, solar irradiance, and volcanic aerosols—is well known, but the human-made aerosol forcing is unknown. So, the two unknowns are aerosol forcing and ocean mixing.

With Karina's data for 2005–10, we concluded (Hansen et al., 2011) that the aerosol forcing was $-1.6 \pm 0.3\ W/m^2$. This large forcing is not a surprise to aerosol scientists; it is in the middle of the range that they estimated in IPCC reports. Most of their aerosol climate forcing is the "indirect" effect of aerosols on cloud cover and cloud brightness.

Our paper also confirmed the suspicion that ocean models mixed heat into the deep ocean too efficiently, but it did not tell us why. Did models create an artificial diffusion of heat via their finite differencing approximation of the equations of motion? Did the approximations used to represent mixing on scales smaller than the model's grid cause too much mixing? Did the coarse vertical resolution of ocean models cause excessive downward mixing?

Whatever the reason(s), excessive mixing makes it difficult to maintain a low-density ocean surface layer fed by meltwater. Therefore, SMOC and AMOC shut down more readily in the real world than in models. SMOC is more important than AMOC because SMOC shutdown accelerates Antarctic ice melt and sea level rise. The high sensitivity of SMOC implies that sea level rise could begin to run out of control within the next few decades.

How could we make the sea level threat and its implications clearer?

A persuasive case should include an explanation of the rapid Eemian sea level rise. The Eemian interglacial period—just slightly warmer than today—provides the closest real-world example of our likely near-term future. So—on our 40th wedding anniversary in January 2011—Anniek and I spent 3 days on the island Eleuthera, the Bahamas, where we could examine some of the field sites that Paul Hearty had described.

Hearty provided instructions for us to find the field sites. Giant boulders on a ridge as high as 20 m above today's sea level provided the most spectacular evidence. Fig. F.3 shows Anniek standing by a boulder dubbed "the bull" by Hearty. The boulders are "hammer-ringing hard" limestone of age at least several hundred thousand years, but they rest on younger, Eemian-age, soil—the Eemian period lasted from about 130,000 years ago to 118,000 years ago.

The boulders must have been washed up the ridge by powerful storm-driven waves—some even rolled down the opposite side of the ridge—providing evidence of the strength of Eemian storms. Quantitative implications for Eemian

Figure F.3 Anniek (height 1.6 m) stands beside one of the boulders that were washed to the top of a coastal ridge of North Eleuthera Island, Bahamas, by waves driven by powerful storms during the Eemian interglacial period.

storms are discussed in Chapter 48 of Sophie's Planet. Here we focus on the more important issue: the rapid rise of global sea level.

Hearty et al. (2007) review data from 15 places around the world for sea level during the Eemian. They conclude that sea level rose early in the Eemian to a level a few meters higher than today. Then late in the Eemian there was rapid additional sea level rise to a level as much as 6–9 m above the present level.

Hearty's geologic evidence indicated that the late-Eemian sea level rise required at most a few centuries, possibly less. Independently, Blanchon et al. (2009) used coral reef "back-stepping" on the Yucatan Peninsula to establish an even tighter time scale. As sea level rises, coral moves their reef building shoreward. From rapid back-stepping of coral in the late Eemian, Blanchon et al. inferred that 2–3 m of sea level rise occurred within several decades.

These data all fit together and made sense. We understood a lot about the causes of glacial–interglacial climate change, as discussed in Chapter 25 of Sophie's Planet (Hansen, 2022g). Milutin Milankovitch, building on hypotheses of 19th-century scientists, proposed in the 1920s that glacial–interglacial climate oscillations are caused by small changes of Earth's orbit and the tilt of Earth's spin axis. James Hays and colleagues confirmed the essence of this orbital theory by showing that climate-driven periodicities in ocean sediment cores matched the periodicities of Earth's orbital changes (Hays et al., 1976).

The facts pointed to a clear conviction: the West Antarctic ice sheet collapsed in the late Eemian. Understanding how the natural climate forcings caused rapid sea level rise has strong implications for our future and needed policies. However, before getting to that climactic story, we need to recognize an age-old sort of censorship that scientists have imposed upon scientists.

Resistance by scientists to scientific discovery is widely acknowledged, even though it clashes with the vision of science as open-minded and unbiased

Barber (1961) notes famous scientists who chaffed bitterly at this resistance and is disappointed that they offer only vague explanations, such as "scientists are human" or "fear of novelty."

Feynman (1986) described resistance that embarrasses physicists. Robert Millikan measured the electron charge in an experiment in which he observed the motion of a charged oil drop in the air. The value he reported was not quite right, and it took years to correct it. Why? When an experimenter's results differed too much from Millikan's, the experimenter would look for reasons that some data points were wrong and eliminate those, thus reporting a result not too different than Millikan's. Thus, only slowly did they inch themselves to the true value.

I believe that the IPCC-led climate research community is slowly—too slowly—inching toward conclusions about climate change that are needed for policy purposes. In my opinion, one of the reasons for this excessive slowness is an unusual form of resistance and censorship imposed by scientists who are respected authorities.

It is not unusual for authorities to disagree with a new conclusion. An example is reaction to my testimony in 1988, when I asserted that human-made global warming was underway and significant. The community did not agree. In a 1-week conference my views were almost universally criticized, as reported by Kerr (1989). Kerr—one of the top science writers in the world—provided insight when he quoted one of the experts as saying "if there were a secret ballot at this meeting on the question, most people would say the greenhouse warming is probably there" and another as saying "what bothers a lot of us is that we have a scientist telling Congress things that we are reluctant to say ourselves."

These differences were open and well reported. There was also time to resolve the differences. Nature would soon provide a clearer picture. There was enough time for governments to change energy policies. I was happy to withdraw from that debate.

Here I describe—via a relevant example—a different sort of resistance and censorship imposed anonymously by senior scientists. I raise this issue because—unlike the Galileo and Millikan cases—delay now does great harm. We can lock in large sea level rise if we do not understand the time scale of the relevant physical processes, the actions that are needed to avoid unacceptable consequences, and the policies needed to achieve good results. Large sea level rise would be practically irreversible on any time scale people care about. Also, if the AMOC shuts down totally, it will take centuries to recover (Hofmann and Rahmstorf, 2009).

Blackballing by grand poohbahs includes both resistance to discovery and censorship

To blackball is to ostracize. Blackballing may not be widespread, but it is relevant to the climate story and many of the chapters of this book. I use our paper Ice Melt, sea level rise, and superstorms (Hansen et al., 2016), hereinafter abbreviated as Ice Melt, as a case in point because it brings out the physics and the poohbahs.

Ice Melt is the paper I wanted to write after we ran the "freshwater" climate simulations in 2006. We had reframed the sea level rise problem, as described by Hansen et al. (2016) and in Chapter 48 of Sophie's Planet (Hansen, 2022h). Reframing seemed natural to me and its merits were demonstrated by the master Henk van de Hulst and related by his protégé Joop Hovenier. The idea is to look at an old problem in a new way, preferably a simpler way that provides physical insight. Hovenier said about van de Hulst's propensity to attack a well-worked problem from scratch "it takes a lot of guts!" (Hansen, 2022i).

The old way in the ice melt problem relied on ice sheet models. The models do not yield much ice melt in a century, although they might inch up from one IPCC report to another. Such small ice melt did not have much effect on overturning ocean circulation in climate models.

Reframing was based on real-world data

Paleoclimate data reveal frequent cases of sea level rise of several meters in a century. When an ice sheet, or part of it, becomes vulnerable because of climate change, the ice sheet contraction is often via rapid ice disintegration. Geologists call these "meltwater pulses." Some meltwater pulses may result from slow ice melt that builds up a lake trapped by the ice sheet until the lake suddenly bursts through. Such large lakes and outbursts occur in the geologic record, but they account for only about 1% of paleoclimate sea level rise (Harrison et al., 2019).

Sea level rise of several meters in a century implies exponential growth of the injection rate of freshwater onto the ocean. Exponential growth can be characterized by a doubling time for the period of rapid disintegration, which lasts until the vulnerable ice begins to be exhausted. Paleoclimate examples of sea level rise of several meters in a century imply a doubling time of no more than 10–20 years, at least for the last few doublings.

Our task is to find the doubling time for the West Antarctic ice sheet if greenhouse gases continue to grow rapidly. We can try alternative 10 and 20-year doubling times and cut off freshwater injection when sea level rise reaches 5 m, thus allowing examination of how the ocean and climate recover from the perturbation. This approach also reveals the freshwater injection rates that yield shutdown of the AMOC and SMOC. This procedure—as opposed to step function meltwater injection—mimics real-world ice sheet disintegration, as a given melt rate is preceded by a slower melt rate.

A climate model study that employs a doubling time for freshwater injection is appropriate because it resembles the real world. A common—but unrealistic—alternative is to compare a control run with no freshwater injection to an experiment with a fixed freshwater injection rate or a specified linearly increasing rate. Ice sheet disintegration is inherently exponential; thus, the appropriate way to determine the ice melt rate required to shut down the AMOC, for example, is with exponential freshwater growth.

A large block of time was needed to write Ice Melt. It had to wait until after our 2-year saga to publish a paper (Hansen et al., 2013) needed for a lawsuit against the government. In early 2014, we reran climate simulations with our latest climate model; results were similar to those in 2006. Still, progress in writing Ice Melt was slow. I had retired from the government to start a program at Columbia University—Climate Science, Awareness and Solutions (CSAS)—but it was difficult and time-consuming to obtain funding to cover the three people working with me.

Late in 2014—during the holidays—I received a message from an angel (Douglas Durst)

Douglas provided a gift to CSAS that—with the one-third match that Jeremy Grantham provided for all donations—would cover our costs for more than 2 years. I could work full time on Ice Melt, with the help of Reto Ruedy, Makiko Sato, and other co-authors. I camped out in my study that winter, with about 25 growing piles of papers on relevant subtopics stacked around the floor.

On June 11, 2015, I submitted Ice Melt to Atmospheric Chemistry and Physics (ACP). I chose ACP because the paper could be published promptly as a discussion paper, if it was accepted by the editors after their cursory review. There the paper would be freely available worldwide while it underwent peer review prior to publication in the print journal ACP. I wanted the paper to be available prior to the Paris COP (Conference of the Parties for the United Nations Framework Convention on Climate Change), which would begin on November 30, 2015.

Our paper outlined overall Eemian climate change. The Eemian interglacial period was initiated by large positive insolation anomalies at the latitude of Northern Hemisphere ice sheets—indeed, the largest Milankovitch (Earth orbital) anomaly of the past 400,000 years—which was sufficient to melt the ice sheets on North America and Eurasia and reduce the size of the Greenland ice sheet such that sea level was a few meters higher than today. By the latter part of the Eemian period, insolation anomalies were negative in the Northern Hemisphere but positive in the Southern Hemisphere. Before Northern Hemisphere ice sheets could grow, the small positive insolation anomaly in the Southern Hemisphere caused the West Antarctic ice sheet to collapse and sea level to rise rapidly.

After Ice Melt was published as a discussion paper on July 23, 2015, the Dursts arranged publicity, including an interview by Fareed Zakaria on Global Public Square on CNN. It was a good opportunity to discuss the threat of global sea level rise and policies needed to avoid that. It was clear that governments had no intent to take effective action, even if it made economic sense. I hoped to add pressure for more meaningful policies, such as carbon fee and dividend.

Referee responses to Ice Melt varied

Referee #1 described it as a "masterwork of scholarly synthesis, modeling virtuosity, and insight, with profound implications." Referee #2—an IPCC contributing author—seemed intent on preventing publication. Referee #3 fell somewhere between #1 and #2. Fortunately, the editor secured a Referee #4, who recommended publication and noted that we made several predictions that could be evaluated later.

The final paper was published in ACP on March 22, 2016. Durst's publicist sent it to the Associated Press writer, Seth Borenstein, who replied: "I sent the paper to a large number of the top climate scientists whose names you would recognize. The responses were near universal in their criticism of it as exaggerated and problematic." The Associated Press did not report on our paper. I thought nothing of it then, because others reported on it (Columbia, 2016).

I saw Borenstein in 2018, after it was clear that researchers in relevant disciplines ignored our paper. It was cited by people concerned about climate change but not by researchers in glaciology, oceanography, or paleoclimatology. Seth explained that the scientists he contacted were the leading scientists and five of the six warned him not to write about our paper.

When the grand poohbahs blackball a paper, others in their fields take the cue. Even when new data support our predictions or other scientists reach conclusions that we already published, that fact is not mentioned. I don't need the citations—that's not the problem. The problem, rather, is that our predicted climate change is vastly different than that of IPCC.

IPCC has turned the world on its head. All their reports claim that sea level rise of several meters in a century is highly unlikely, even with CO_2 reaching more than 700 ppm. We find the opposite. With IPCC business-as-usual scenarios, it is practically certain that we would have a devastating sea level rise this century. People need to understand this situation as soon as possible, while we still have a chance to adopt policies that are essential for conservation.

Prior analyses of ocean circulation focused on AMOC

That focus is understandable. Shutdown of AMOC yields large climate change in the North Atlantic with a downstream impact on Europe. The reduced northward ocean heat transport also warms the Southern Ocean—an interhemispheric "seesaw" effect (Stocker, 1998). However, the research community and IPCC concluded that AMOC would not shut down this century; it would only slow down somewhat more than it has already (IPCC, 2019).

Our conclusions differed dramatically. We found that SMOC is more important than AMOC because of its effect on future sea level rise. For business-as-usual scenarios used by IPCC, we found that SMOC would shut down by mid-century (Fig. 4). AMOC would also shut down this century and would not recover for centuries. Our approach to the problem also differed greatly from

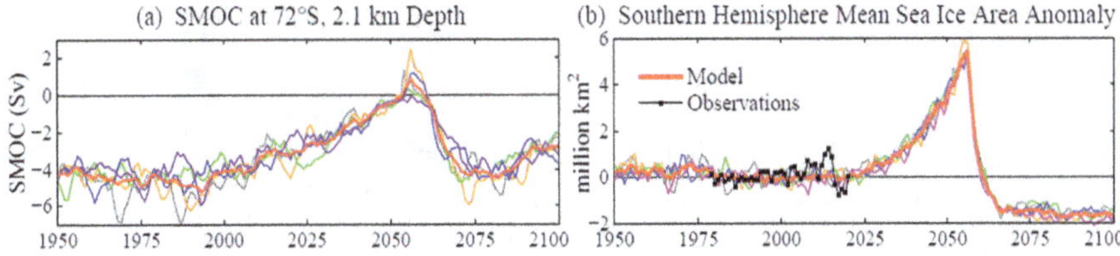

Figure F.4 (A) SMOC mean circulation at 72°S (excludes eddy-induced transport), and (B) annual sea ice area anomaly (10^6 km^2) relative to 1979–2000 in five model runs and observations. One Sverdrup (Sv) is 10^6 m^3/s or $\sim 3 \times 10^4$ Gt/year. A Gt is a billion tons. *SMOC*, Southern Meridional Overturning Circulation. *Figure adapted from Hansen et al. (2016)*

that of IPCC. While IPCC relies heavily on ice sheet models, our approach was based on empirical information from the real world.

Our climate simulations begin with observed rates of ice melt. Numerous paleoclimate cases of sea level rise by several meters per century imply that the collapse of an ice sheet—once climate change makes it vulnerable—is exponential with a doubling time of at most 1–2 decades. We used a 10-year doubling time for future melt rates in most of the simulations for Ice Melt.

We can obtain an empirical measure of doubling times from continuing observed changes in the masses of the Antarctic and Greenland ice sheets. Data for ice sheet mass loss through April 2021 yield best-fit values for the doubling time of 12, 10, and 7.5 years for Greenland, Antarctica as a whole, and West Antarctica, respectively (Hansen, 2022h). The doubling time for West Antarctica is crucial, because it is expected to be the shortest, in which case it will be the dominant source of global sea level rise later this century.

In our Ice Melt paper—with 10-year doubling times for Greenland and Antarctica—the AMOC shuts down this century (Fig. F.5). The high sensitivity of AMOC to freshwater injection that we find in our global climate model is supported by paleoclimate data showing that AMOC shutdowns occurred during interglacial periods when potential freshwater sources were no larger than today (Galaasen et al., 2020).

We sharply terminated freshwater injection onto the ocean when sea level rise reached 5 m in our climate simulations. The purpose was to see how fast AMOC and SMOC would recover once freshwater forcing was removed. We found, in agreement with the expected "hysteresis" behavior of AMOC (Stommel, 1961; Rahmstorf et al., 2005), that AMOC does not fully recover

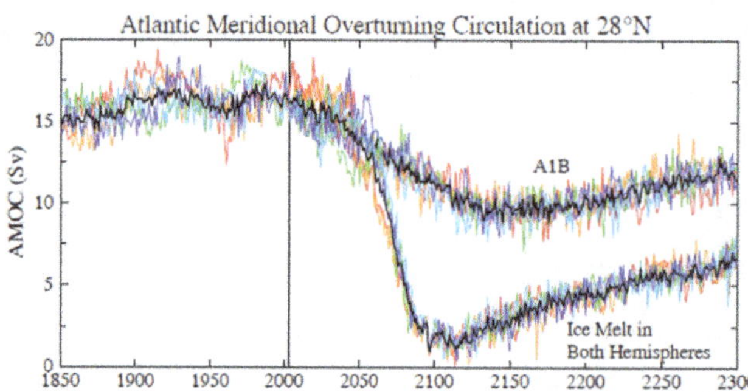

Figure F.5 AMOC strength at 28°N in five simulations and their mean (black line) for IPCC A1B scenario and ice melt in both hemispheres, two-thirds of it from Antarctica. *AMOC*, Atlantic Meridional Overturning Circulation. *Figure adapted from Hansen et al. (2016).*

even in 200 years (Fig. F.5). Furthermore, if ice sheet collapse and multimeter sea level rise occur, freshwater injection is likely to continue for centuries.

How can real-world ice melt be so much faster than in the ice sheet models that IPCC relies on? Ice sheet modeling is hard. Ice sheet processes occur on spatial scales ranging from microscale freeze–thaw effects that cause pot-holes in our streets to continental-scale "rivers" of ice that discharge icebergs to the ocean. However, as argued in my "slippery slope" paper (Hansen, 2005b), the crucial amplifying feedbacks are probably interactions between ice sheets and oceans abutting against them. Our global climate model results in Ice Melt revealed such specific amplifying feedbacks.

Shutdown of SMOC is a powerful feedback

The shutdown can spur the disintegration of the West Antarctic ice sheet (Fig. F.2). Our climate model correctly locates deep water formation along the Antarctic coast at places such as the Weddell Sea coast (Section 3.8.5 in Ice Melt), which supports the use of our model to study the SMOC feedback. That capability is absent in many Climate Model Intercomparison Project models used in the IPCC assessment (Heuze et al., 2015).

SMOC already slowed in our climate simulations by the late 20th century (Fig. F.4, which is Fig. 32 in Ice Melt) due to growing freshwater injection from Antarctica. Ocean current measurements are too sparse to accurately monitor SMOC, but sufficient for Purkey and Johnson (2012) to conclude that the real-world SMOC did slow during that period.

SMOC is an escape valve for ocean heat. As relatively warm water reaches the surface near Antarctica (see Fig. F.2), heat escapes to the air and space—especially in winter. The salty water cools there to high density and sinks, but as increasing light meltwater is added, the rate of sinking water decreases. As this surface escape valve for heat closes, that heat warms the deeper ocean, with maximum warming at 1–2 km depth. That is the depth of ice shelf grounding lines, the part of the ice shelf that exerts the strongest restraining force on landward ice [Fig. 14 of Jenkins and Doake (1991)]. West Antarctic ice shelves thus have begun to melt more rapidly (Rignot and Jacobs, 2002), and the ice streams feeding them have accelerated (Rignot, 2008).

Menviel et al. (2010) used a simplified Earth system model to show that the collapse of the West Antarctic ice sheet would cause expansion of sea ice on the Southern Ocean, suppression of Antarctic Bottom Water formation, and warming of the Southern Ocean at depth. Fogwill et al. (2015) used a high-resolution atmosphere-ocean model to investigate the effects of increasing freshwater flux from West Antarctica today, finding that increased ocean stratification reduced bottom water formation and increased ocean temperature at depth. Fogwill et al. (2015) submitted their paper on almost the same date in 2015 that we submitted our paper. They concluded, however, that they saw no significant atmospheric response to the freshwater injection. We found a significant accompanying atmospheric feedback.

Precipitation feedback is also important

Precipitation provides an amplifying feedback for sea level rise in our model, but a diminishing feedback in the climate models that IPCC has reported and relied on. Their models yield a large reduction of sea ice around Antarctica and increasing snowfall over the continent as Earth warms. This increased snowfall causes sea level to fall, thus at least partially offsetting sea level rise from ice sheet dynamical mass loss (Fig. F.2).

In our climate model described in Ice Melt, increasing meltwater cools the Southern Ocean surface enough to offset greenhouse gas warming. Indeed, the sea surface in the western portion of the Southern Ocean, where two-thirds of increased freshwater injection is occurring (Rignot et al., 2013), already has cooled while the rest of the planet has warmed (Fig. 31 in Ice Melt).

If high fossil fuel emissions continue, SMOC will shut down during the next few decades and sea ice in the Southern Ocean will expand several million square kilometers, according to our climate simulations (Fig. F.4B). These effects should begin to emerge this decade from the "noise" level of unforced and unpredictable climate variability.

Mother nature threw a curve ball

Before the ink had dried on our Ice Melt paper, the Antarctic sea ice cover plummeted (Fig. F.4B) to its lowest level in 40 years of satellite data (Parkinson, 2019). Antarctic Bottom Water (AABW) formation—the engine of SMOC—increased (Silvano et al., 2020). So, was the slowdown of SMOC over the prior few decades only temporary? Will Antarctic sea ice decrease now like Arctic sea ice, as predicted by IPCC models?

No, surely not. On the contrary, data that have accumulated since we submitted our paper in 2015 allow improved assessment of the basic time scales of the climate change problem. These time scales are central to our reframing of the ice melt problem and they are at the heart of our disagreement with the conclusions of IPCC. One merit of our approach is the role of empirical data, which will allow continual, easily understandable, evaluations as climate response unfolds.

Scientists agree that the greenhouse gas amounts in 2100 for business-as-usual assumptions would yield an eventual sea level rise that would be the demise of the world's coastal cities. The disagreement concerns the time scale on which sea level will rise. The most crucial time scale is the characteristic response time for the West Antarctic ice sheet, because West Antarctica contains enough ice to raise sea level a few meters by itself, and its disintegration would be accompanied by substantial contributions from East Antarctica and Greenland.

In Ice Melt we concluded that ice sheets disintegrate faster in the real world than in ice sheet models. In Chapter 48 of Sophie's Planet (Hansen, 2022h) I show that information arising since 2015 supports the suspicion that we voiced in Ice Melt: the real world is more sensitive to freshwater injection than even our climate model suggested.

Those conclusions do not mean that the climate problem is unsolvable. On the contrary, solution of the climate problem makes economic and practical sense—and conservation of nature can be one of many benefits. We are running out of time, however. We cannot afford to waste time on the ineffectual wishful thinking that has characterized past policy efforts.

Let us consider the main threats of climate change, the implications for policy, and the benefits that will accrue from positive action

Sea level rise sets the lowest bar on acceptable global warming. Global temperature by the mid-20th century reached approximately the maximum in the Holocene (Hansen et al., 2017), the current interglacial period in which civilization developed on stable shorelines. We must go back to a global climate no warmer than that of the mid-20th century, perhaps a bit cooler.

Restoration of a moderate global temperature will have many benefits besides saving our shorelines. Global warming has increased the severity of extreme climate events (Hansen et al., 2012), and there is more warming "in the pipeline" without additional increase of greenhouse gases. Climate zones are shifting poleward at a rate much faster than any time in Earth's history that we are aware of. If global warming continues at this rate, much of the low latitudes will become uncomfortable, if not intolerable for human habitation (Raymond et al., 2020).

Emigration pressures arising from climate change are already a global problem, illustrated in recent years by the effects of extended drought in Syria (Wendle, 2015; Kelley et al., 2015) and unprecedented tropical storms in Central America (Kitroeff and Volpe, 2021). Yet these cases pale compared with potential emigration pressures from large sea level rise. A scaling back of global warming is needed to solve these problems.

Extermination of species is another major irreversible effect of uncontrolled climate change. Restoration of mid-20th century climate will alleviate the extinction pressure caused by rapidly shifting climate zones. Other human-caused pressures on biodiversity need to be reduced by means of an increased set-aside of land for nature as covered by DellaSala in the closing chapter 16 of this book. The concept of a contiguous "nature's corridor" stretching through all climate zones of the Americas is described in Sophie's Planet.

We live on a spectacular planet, unrivaled by any other in the universe that we know. With a little more effort in our schools, we can help young people appreciate Earth's wonders and understand that climate change is not something to fear or worry about. Instead, we—and they—have an opportunity and challenge to take actions that will preserve both nature and human-made structures, including our great coastal cities.

We cannot eliminate weather and climate variability, but we can return to a condition in which historic 100-year floods occur only once a century, on average; one in which superstorms and firestorms in populated areas are rare; a planet whose low-latitude regions are not only livable, but able to support the abundant life that was historically adapted to those climate zones.

The challenge is great, but the rewards will be commensurate. We can achieve the goal of restoring and preserving nature's bounty, but only if we are honest about what is required. We must be guided by realistic scientific analysis, not by wishful thinking.

The United States and China must cooperate

The governments of China and the United States are beginning to appreciate the existential threat posed by accelerating global warming. Once they both realize that the climate problem must receive first priority and that solution requires their cooperation, the world can at last begin to address the matter seriously.

There is no point in casting blame, but a quantitative understanding of the cause of climate change is informative. China has the largest fossil fuel emissions now (left side of Fig. F.6) and China's energy future will have the greatest impact on climate. However, global warming is proportional to cumulative emissions (Hansen et al., 2007; Matthews et al., 2009), for which the United States is most responsible (right side of Fig. F.6). On a per-capita basis, the nations that industrialized early—such as the United Kingdom, Germany, and the United States—will always be far more responsible for climate change than China.

Nations of the West burned fossil fuels to raise their standards of living. There is plenty of fossil fuel in the ground for China, India, Indonesia, and the rest of

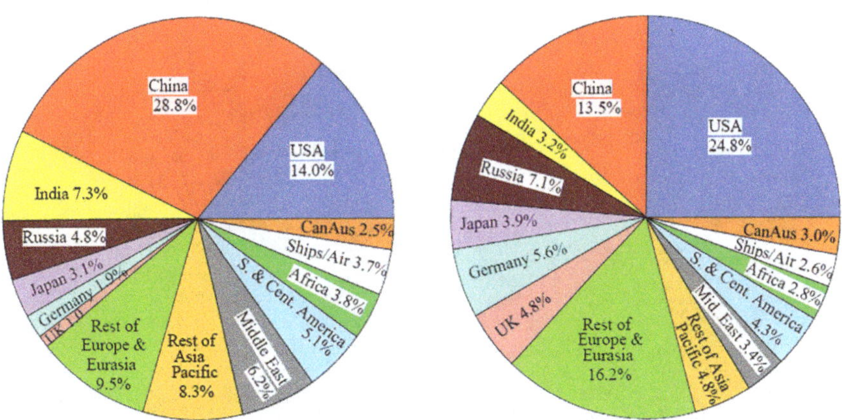

Figure F.6 Fossil fuel emissions in 2018 (left) and cumulative 1751–2018 emissions.

the world to rely on for that same purpose, but such a course would assure mutual destruction. We must find a better way.

Robert Daly—Director of the Kissinger Institute on China and the United States—invited me to join him and Ambassador Stapleton Roy in Beijing in 2014 for a Symposium on a New Type of Major Power Relationship (Hansen, 2022j). Topics were climate change and infectious disease. I gave the climate talk, which was blunt, and I provided all of my charts to our Chinese hosts.

My conclusion was that there are two major requirements for solving the climate problem. First is the need for a simple rising carbon (oil, gas, and coal) fee or tax. China and the United States would collect their own fee at their domestic mines and ports of entry. Although each nation would decide how to use the funds, my suggestion was to distribute the money uniformly to citizens, which helps address growing wealth disparities in most nations.

As the dominant economic powers in the world, these two nations can make a carbon fee near-global by imposing a border duty on products from countries without a carbon fee and by rebating the fee to domestic manufacturers on products sold to countries without a fee. Most countries would agree to impose a carbon fee, so that they can collect the money themselves.

The second requirement is a carbon-free alternative to fossil fuels for baseload dispatchable electric power that is as cheap or cheaper than fossil fuels. Such an energy source is a vital complement to renewable energies, even if the latter is used to maximum practical potential.

The final chart in my presentation, updated here as Fig. F.7, revealed the sorry state of global efforts to decarbonize energy use. France and Sweden made good progress by using nuclear power for a large portion of their electricity,

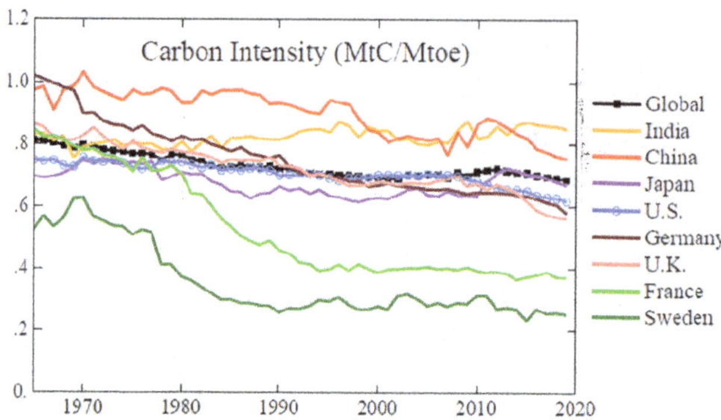

Figure F.7 Carbon intensity (carbon per unit energy) of global and national energies. MtC is megatons of carbon. Mtoe is megatons of oil equivalent.

but they stopped short of achieving carbon-free energy for transportation and some industrial processes.

The importance and urgency of the need for inexpensive, carbon-free, baseload electric power cannot be overstated. China and India obtain most of their energy from coal. They will not agree to an equivalent carbon fee until they have a viable alternative to coal, nor should we expect otherwise—all nations will strive to raise their living standards, as the West has done.

Follow the science, not popularism

Science and engineering can help us solve the climate problem, but we are in a race that we must win before either the physical climate system or global governance pass irreversible tipping points. Here I draw attention to "popularism," which is a bias of technical evaluations toward the answer that the audience wants to hear. I give two examples of why this bias is dangerous.

My colleague Pushker Kharecha and I spent years developing our understanding of energy choices and their effects on climate and the environment, including the organization of workshops at the East-West Center in Hawaii and in Washington, as described in Sophie's Planet. Participants included engineers and managers charged with making electrical grids safe and reliable. There was agreement that it is necessary to complement intermittent renewables with reliable, dispatchable power as can be provided by nuclear or fossil energies (Kharecha et al., 2010; Clack et al., 2017; Jenkins et al., 2018).

Nuclear power is the clear preference for climate, and modern nuclear power has the smallest environmental footprint of all energy sources. Even 1970's nuclear technology provided the safest energy during the past half-century in the United States, but there were serious nuclear accidents at Chernobyl, U.S.S.

R. and Fukushima, Japan. Neither of these accidents would have occurred with modern passively safe reactors that shut down in case of an anomaly without need for human intervention or need for external power to keep the nuclear fuel cool.

Modern reactors have the potential to be cheaper than fossil fuels, based on the cost of nuclear fuel and the material required to construct the power plants. However, achievement of that goal requires the same kind of support provided to renewable energies. Utility managers agree that we could already be on the verge of carbon-free electricity if we had adopted clean energy portfolio standards decades ago, rather than renewable portfolio standards.

As a political independent and an agnostic about nuclear power, but concerned about the world we leave for young people, I was distressed to see unscientific origins of bias against nuclear power. As a resident of Pennsylvania, I have been bombarded by specious antinuclear ads paid for and mailed by the American Petroleum Institute (Meyer, 2020). Recent ads focused on nuclear energy subsidies, while in fact, fossil fuels collect the largest absolute subsidies and renewables collect the most on a per unit energy basis. Earlier disinformation focused on nuclear waste. Nuclear waste is small in volume, can be safely stored, and even provide fuel for advanced reactors (Till and Chang, 2011). Fossil fuel waste is dumped in the atmosphere, resulting in air pollution that kills millions of people per year (Kharecha and Hansen, 2013).

Yet there is a danger that fossil fuels will continue to be the main complement to renewable energy. A small number of scientists assert that renewables can soon provide all the world's energy. Liberal media and "big green" environmental organizations (Hansen, 2022k) promote this disinformation, even though authoritative studies conclude that achievement of deep decarbonization of the world's energy by mid-century requires substantial contributions from nuclear power and/or carbon capture and storage. "Renewables can provide 100 percent of our energy" is a message that the public, most politicians, and the liberal media want to hear. Most informed scientists recognize this as wishful thinking that prolongs the reign of fossil fuels.

Let me give a second example of popularism

We all know that geoengineering is a terrible idea. We do not understand nature well enough that it makes any sense to mess with nature. Oops! We are geoengineering the dickens out of the planet right now! Because of increased greenhouse gases, Earth is out of energy balance, more energy coming in than going out. We are pouring energy into the ocean equal to the energy of 600,000 Hiroshima atomic bombs per day, every day of the year. That heat is ominously melting ice shelves around Antarctica.

The best measure of our geoengineering is Earth's energy imbalance. Changes in the ocean's heat content define about 90% of Earth's energy imbalance.

The main cause of Earth's energy imbalance is the human-made increase of greenhouse gases, with CO_2 from fossil fuel burning being the largest contributor. So, a popular position is that we must stop burning fossil fuels as rapidly as we can, even if it means that we need to stop eating red meat, severely reduce flying, and take other drastic measures.

For sure, we should phase down global fossil fuel emissions rapidly, while we also work to raise living standards globally to reduce poverty. We must find a realistic approach to address the broad needs of global society and conservation.

However, even with maximum effort, atmospheric CO_2 and other greenhouse gases will increase and climate impacts will grow in the near-term. Governments should soon get serious about reducing emissions, but past delay means that more actions are needed. We will also need a good understanding of how we can reduce our geoengineering of the planet.

One widely discussed way to reduce geoengineering is solar radiation management (NAS, 2021). It involves temporary reflection of a small share of solar energy hitting Earth so as to restore Earth's energy balance while the work to eliminate fossil fuel emissions and draw down excess atmospheric greenhouse gases is ongoing.

In October 2018 at the first joint meeting of the American Geophysical Union and the Chinese Academy of Sciences in Xi'an, China, I presented climate simulations in which aerosols were added to the atmosphere with alternative geographical distributions (Hansen, 2018). With aerosols over the Southern Ocean the effect is the opposite of what has been happening in the real world with increasing CO_2. Instead, Antarctic Bottom Water formation around the coast of Antarctica is invigorated, the Southern Oceans cools at the depths of the ice shelves, the ocean surface layer warms, and Earth as a whole begins to cool off.

Research to better understand the climate system and find ways to reduce geoengineering of Earth is warranted, in my opinion. We are likely to reach a point when—despite global efforts to phase out fossil fuel emissions—it is clear that we are headed toward large sea level rise and loss of our coastal cities. Humanity may then wish to consider options such as spraying tiny droplets and cloud condensation nuclei into the air from autonomous floats on the Southern Ocean, with the material being sprayed extracted from the ocean itself. Such aerosols as a tool for conservation are about as natural an approach as one can imagine. They may be capable of restoring Earth's energy balance while we work on reducing atmospheric greenhouse gases.

Yet the popular first reaction to such proposed research is to condemn it, perhaps because it seems unnatural. However, in our efforts to support nature we cannot afford to condemn such research any more than we can simply dismiss the potential and need for advanced nuclear power. In light of the global climate and environmental crises, we instead need to soberly evaluate and weigh the likelihood and range of risks, as we reconstruct a viable future.

President Biden has invigorated the climate issue in the United States

He has assembled experienced and committed advisers, including the United States Special Presidential Envoy for Climate John Kerry and White House National Climate Advisor Gina McCarthy. Their early work is substantial, with President Biden issuing executive orders to reverse the prior Administration's harmful actions, ordering a realistic assessment of the social cost of carbon, and pronouncing accelerated decarbonization targets ahead of the COP 26 meeting (26th Conference of the Parties to the UNFCCC) slated for Glasgow, Scotland in November.

However, the Biden team, in my view, still needs a fair and efficient centerpiece federal climate policy to drive decarbonization across all key U.S. economic sectors. To that end, in January 2021, along with Dan Galpern, my long-time legal and policy adviser, I urged President Biden to impose upstream carbon fees on the major fossil fuel producers, with revenues returned as dividends to citizens. Relying on the scholarship of E. Donald Elliott, formerly the General Counsel of EPA, we argued that the imposition of such fees could be done under existing executive authority (Hansen and Galpern, 2021). If adopted in conjunction with complementary policies under consideration by the Biden team, the imposition of carbon fees would rebuild U.S. international credibility and give us a fighting chance with respect to the climate crisis.

Speaking truth to power: closing thoughts

Science and technology have been a boon for humanity. Standards of living have improved dramatically over the last few centuries for much of the world. Yet as capabilities of our species increased, we also introduced problems, including global climate change.

Science and technology can help solve these problems and allow us to live more in harmony with nature. However, for science to work well it needs to be unfettered by censorship and ideology.

Climate facts are clear. We have passed the dangerous level of atmospheric greenhouse gases. Climate effects of the gases are already detectable and have the potential to cause global chaos if we do not rapidly phase down emissions and take other actions to minimize disruption of nature.

Policy implications are clear. There are three fundamental requirements.

First, we cannot continue to use the atmosphere as a free dumping ground for pollutants from fossil fuel burning. That means that we must have a rising fee or tax on carbon emissions. Such a fee can be readily imposed by the major powers on a practically global basis via border duties on products from countries that do not have an equivalent domestic carbon fee.

Second, clean-energy alternatives cheaper than fossil fuels must be available, including dispatchable (baseload) electric power. Modern passively safe nuclear power appears to be the best option in the near- and medium-term for baseload clean power, but a rising carbon price will allow alternatives such as fusion, carbon capture, and renewables plus energy storage to compete.

Third, we must use the power of ecosystems to sequester and store carbon. Potential carbon drawdown via improved agricultural and forestry practices is substantial, albeit requiring better quantification (Smith et al., 2014: Griscom et al., 2017). We need to protect primary forests, continue growing secondary forests, and reduce emissions from bioenergy (Kun et al., 2020).

So far, the world's nations have taken only baby steps toward the solution of the climate problem, with agreements that amount to wishful thinking, as emissions continue to grow, not decline. Words and goals amount to little, as long as the three fundamental requirements are not met.

When the United States and China realize that they are in the same boat, they may be able to use their combined strengths to move rapidly toward the achievement of the fundamental requirements. Until then, it is good that we have many friends, colleagues, and former students in China. We can cooperate on research that lays the groundwork for future international cooperation (Hansen, 2014).

As chronicled in the chapters of this book, we have very little time to change course. Scientists working with advocates need to stand up and be the voice for the planet.

James E. Hansen[1]

[1]Climate Science, Awareness and Solutions Program Earth Institute, Columbia University, New York City, NY, United States

References

Argo floats program. https://argo.ucsd.edu/

Barber, B., 1961. Resistance by scientists to scientific discovery. Science 134, 596–602.

Blanchon, P., Eisenhauer, A., Fietzke, J., Liebetrau, V., 2009. Rapid sea-level rise and reef back-stepping at the close of the last interglacial highstand. Nature 458, 881–884.

Bowen, M., 2008. Censoring Science: Inside the Political Attack on Dr. James Hansen and the Truth of Global Warming. Dutton, New York.

Catania, G.A., Stearns, L.A., Moon, T.A., Enderlin, E.M., Jackson, R.H., 2020. Future evolution of Greenland's marine-terminating outlet glaciers. J. Geophys. Res.: Earth Surf. 125, e2018JF004873.

Charney, J., Arakawa, A., Baker, D., Bolin, B., Dickinson, R., et al., 1979. Carbon Dioxide and Climate: A Scientific Assessment. The National Academies Press, Washington, DC, p. 33.

Clack, C.T.M., Qvist, S.A., Apt, J., Bazilian, M., Brandt, A.R., et al., 2017. Evaluation of a proposal for reliable low-cost grid power with 100% wind, water, and solar. Proc. Natl. Acad. Sci. USA 114, 6722–6727.

Communications., 2016. Columbia University website. http://www.columbia.edu/~jeh1/mailings/2016/20160322_IceMeltPaper.Abbreviation.pdf.

Communications., 2021. Columbia University website. http://www.columbia.edu/~jeh1/mailings/index.shtml.

Deschamps, P., Durand, N., Bard, E., Hamelin, B., Camoin, G., et al., 2012. Ice-sheet collapse and sea-level rise at the Bolling warming 14,6000 years ago. Nature 483, 559–564.

Dumitru, O.A., Austermann, J., Polyak, V.I., Fornos, J.J., Asmerom, Y., et al., 2019. Constraints on global mean sea level during Pliocene warmth. Nature 574, 233–236.

Dwyer, G.S., Chandler, M.A., 2009. Mid-Pliocene sea level and continental ice volume based on coupled benthic Mg/Ca palaeotemperatures and oxygen isotopes. Philos. Trans. R. Soc. A 367, 157–168.

Fairbanks, R.G., 1989. A 17,000-year glacio-eustatic sea-level record-influence of glacial melting rates on the younger rates on the Younger Dryas event and deep-ocean circulation. Nature 342, 637–642.

Feynman, R.P., 1986. Surely you're joking Mr. Feynman!. Bantam Books, New York, p. 322.

Fogwill, C.J., Phipps, S.J., Turney, C.S.M., Golledge, N.R., 2015. Sensitivity of the Southern Ocean to enhanced regional Antarctic ice sheet meltwater input. Earth's Future 3, 317–329.

Galaasen, E.V., Ninnemann, U.S., Kessler, A., Irvali, N., Rosenthal, Y., et al., 2020. Interglacial instability of North Atlantic Deep Water ventilation. Science 367, 1485–1489.

Grant, K.M., Rohling, E.J., Bar-Matthews, M., Ayalon, A., Medina-Elizalde, M., et al., 2012. Rapid coupling between ice volume and polar temperature over the past 150,000 years. Nature 491, 744–747.

Greenbaum, J.S., Blankenship, D.D., Young, D.A., Richter, T.G., Roberts, J.L., et al., 2015. Ocean access to a cavity beneath Totten Glacier in East Antarctica. Nat. Geosci. 8, 294–298.

Griscom, B.W., Adams, J., Ellis, P.W., Houghton, R.A., Lomax, G., et al., 2017. Natural climate solutions. Proc. Natl. Acad. Sci. U. S. A. 114, 11645–11650.

Hansen, J., Johnson, D., Lacis, A., Lebedeff, S., Lee, P., Rind, D., et al., 1981. Climate impact of increasing atmospheric carbon dioxide. Science 213, 957–966.

Hansen, J., 1988. Testimony to U.S. Senate on 23 June 1988. https://www.sealevel.info/1988_Hansen_Senate_Testimony.html.

Hansen, J., Rossow, W., Fung, I., 1990. The missing data on global climate change. Issues Sci. Tech. 7, 62–69.

Hansen, J., 2004. Dangerous Anthropogenic Interference, Distinguished Lecture. University of Iowa.

Hansen, J., 2005a. A Tribute to Charles David Keeling, AGU meeting, San Francisco, California.

Hansen, J.E., 2005b. A slippery slope: How much global warming constitutes "dangerous anthropogenic interference"? An editorial essay. Clim. Change 68, 269–279.

Hansen, J.E., 2007. Scientific reticence and sea level rise. Environ. Res. Lett. 2, 024002.

Hansen, J., Sato, M., Ruedy, R., Kharecha, P., Lacis, A., et al., 2007. Dangerous human-made interference with climate: A GISS modelE study. Atmos. Chem. Phys. 7, 2287–2312.

Hansen, J., 2008. Climate Threat to the Planet: Implications for Energy Policy and Intergenerational Justice. American Geophysical Union, San Francisco, California.

Hansen, J., Sato, M., Kharecha, P., von Schuckmann, K., 2011. Earth's energy imbalance and implications. Atmos. Chem. Phys. 11, 13421–13449.

Hansen, J., Sato, M., Ruedy, R., 2012. Perception of climate change. Proc. Natl. Acad. Sci. U. S. A. 109, 14726–14727.

Hansen, J., Kharecha, P., Sato, M., Masson-Delmotte, V., Ackerman, F., et al., 2013. Assessing "dangerous climate change": required reduction of carbon emissions to protect young people, future generations and nature. PLoS One 8, e81468.

Hansen, J., 2014. Symposium on a New Type of Major Power Relationship, Beijing. http://www.columbia.edu/~jeh1/mailings/2014/20140224_Beijing35.pdf.

Hansen, J., Sato, M., Hearty, P., Ruedy, R., Kelley, M., et al., 2016. Ice melt, sea level rise and superstorms: Evidence from paleoclimate data, climate modeling, and modern observations that 2 C global warming could be dangerous. Atmos. Chem. Phys. 16, 3761–3812.

Hansen, J., Sato, M., Kharecha, P., von Schuckmann, K., Beerling, D.J., et al., 2017. Young people's burden: requirement of negative CO_2 emissions. Earth Syst. Dynam 8, 577–616.

Hansen, J., 2018. Aerosol effects on climate and human health. http://www.columbia.edu/~jeh1/2018/China_Charts_Handout_withNotes.pdf, 17 October 2018.

Hansen, J., Galpern, D., 2021. Letter to President Biden, 25 January 2021. https://cprclimate.org/letter-to-president-biden-hansen-and-galpern/.

Hansen, J., 2022a. Chapter 24: Deep Water and Two Angels, in *Sophie's Planet*, Bloomsbury.

Hansen, J., 2022b. Chapter 28: Mother Earth Speaks, in Sophie's Planet, Bloomsbury.

Hansen, J., 2022c. Chapter 29: 1989: The White House Effect, in *Sophie's Planet*, Bloomsbury.

Hansen, J., 2022d. Chapter 33: Climsat: A Proposal, in *Sophie's Planet*, Bloomsbury.

Hansen, J., 2022e. Chapter 39: The Freshwater Experiment, in *Sophie's Planet*, Bloomsbury.

Hansen, J., 2022f. Chapter 17: Charney's Puzzle: Is Earth sensitive? in *Sophie's Planet*, Bloomsbury.

Hansen, J., 2022g. Chapter 25: Paleoclimate and "Slow" Feedbacks, in *Sophie's Planet*, Bloomsbury.

Hansen, J., 2022h. Chapter 48: Ice Melt, Sea Level Rise and Superstorms, in *Sophie's Planet*, Bloomsbury.

Hansen, J., 2022i. Chapter 8: The Netherlands & Anniek Dekkers, in *Sophie's Planet*, Bloomsbury.

Hansen, J., 2022j. Chapter 47: China and the Global Solution, in *Sophie's Planet*, Bloomsbury.

Hansen, J., 2022k. Chapter 45: Energy and World Peace, in *Sophie's Planet*, Bloomsbury.

Harrison, S., Smith, D.E., Glasser, N.F., 2019. Late Quaternary meltwater pulses and sea level change. J. Quatern. Sci. 34 (1), 1–15.

Hays, J.D., Imbrie, J., Shackleton, N.J., 1976. Variations in the Earth's orbit: Pacemaker of the ice ages. Science 194, 1121–1132.

Hearty, P.J., Neumann, A.C., 2001. Rapid sea level and climate change at the close of the Last Interglaciation (MIS 5e): Evidence from the Bahama Islands. Quatern. Sci. Rev. 20, 1881–1895.

Hearty, P.J., Hollin, J.T., Neumann, A.C., O'Leary, M.J., McCulloch, M., 2007. Global sea level fluctuations during the Last Interglaciation (MIS 5e). Quatern. Sci. Rev. 26, 2090–2112.

Heuze, C., Heywood, K.J., Stevens, D.P., Ridley, J.K., 2015. Changes in global ocean bottom properties and volume transports in CMIP5 models under climate change scenarios. J. Clim. 28, 2917–2944.

Hofmann, M., Rahmstorf, S., 2009. On the stability of the Atlantic meridional overturning circulation. Proc. Natl. Acad. Sci. U. S. A. 106, 20584–20589.

Intergovernmental Panel on Climate Change (IPCC), 2001. In: Houghton, J.T., et al., (Eds.), Climate Change 2001: The Scientific Basis. Cambridge University Press, Cambridge, U.K.

IPCC, 2019. Technical Summary. In: Portner, H.O., Roberts, D.C., Masson-Delmotte, V., Zhai, P., Tigor, M., et al., IPCC Special Report on the Ocean and Cryosphere in a Changing Climate.

Jenkins, A., Doake, C.S.M., 1991. Ice-ocean interaction on Ronee Ice Shelf, Antarctica. J. Geophys. Res. 96, 791–813.
Jenkins, J.D., Luke, M., Thernstrom, S., 2018. Getting to zero carbon emissions in the electric power sector. Joule 2 (12).
Kelley, C.P., Mohtadi, S., Cane, M.A., Seager, R., Kushnir, Y., 2015. Climate change in the Fertile Crescent and implications of the recent Syrian drought. Proc. Natl. Acad. Sci. U. S. A. 112, 3241–3246.
Kerr, R.A., 1989. Hansen vs. the World on the Greenhouse Threat. Science 244, 1041–1043.
Kharecha, P.A., Kutscher, C.F., Hansen, J.E., Mazria, E., 2010. Options for near-term phaseout of CO_2 emissions from coal use in the United States. Environ. Sci. Technol. 44, 4050–4062.
Kharecha, P.A., Hansen, J.E., 2013. Prevented mortality and greenhouse gas emissions from historical and projected nuclear power. Environ. Sci. Technol. 47, 4889–4895.
Killian, J.R., 1977. Sputnik, Scientists, and Eisenhower. MIT Press, Cambridge, MA.
Kitroeff, N., Volpe, D., 2021. Stay or Go? Storms were a Tipping Point for Many Hondurans. New York Times.
Kun, Z., DellaSala, D., Keith, H., Kormos, C., Mercer, B., Moomaw, W.R., et al., 2020. Recognizing the importance of unmanaged forests to mitigate climate change. GCB Bioenergy 12, 1034–1035.
Matthews, H.D., Gillett, N.P., Stott, P.A., Zickfeld, K., 2009. The proportionality of global warming to cumulative carbon emission. Nature 459, 829–832.
Menviel, L., Timmermann, A., Timm, O.E., Mouchet, A., 2010. Climate and biogeochemical response to a rapid melting of the West Antarctic Ice Sheet during interglacials and implications for future climate. Paleoceanography 25, PA4231.
Meyer, R., 2020. The Oil Industry Is Quietly Winning Local Climate Fights, The Atlantic.
NAS, National Academies of Sciences, Engineering, and Medicine, 2021. Reflecting Sunlight: Recommendations for Solar Geoengineering Research and Research Governance. National Academies Press, Washington, DC.
Parkinson, C.L., 2019. A 40-y record reveals gradual Antarctic sea ice increases followed by decreases at rates far exceeding the rates seen in the Arctic. Proc. Natl. Acad. Sci. U. S. A. 116, 14414–14423.
Purkey, S.G., Johnson, G.C., 2012. Global contraction of Antarctic bottom water between the 1980s and 2000s. J. Clim. 25, 5830–5844.
Rahmstorf, S., Crucifix, M., Ganopolski, A., Goosse, H., Kamenkovich, I., et al., 2005. Thermohaline circulation hysteresis: a model intercomparison. Geophys. Res. Lett. 32, L23605.
Raymond, C., Matthews, T., Horton, R.M., 2020. The emergence of heat and humidity too severe for human tolerance. Sci. Adv. 6, eaaw1838.
Rignot, E., Jacobs, S.S., 2002. Rapid bottom melting widespread near Antarctic ice sheet grounding lines. Science 296, 2020–2023.
Rignot, E., 2008. Changes in West Antarctica ice stream dynamics observed with ALOS PALSAR data. Geophys. Res. Lett. 35, L12505.
Rignot, E., Jacobs, S., Mouginot, J., Scheuchl, B., 2013. Ice shelf melting around Antarctica. Science 341, 266–270.
Rohling, E.J., Grant, K., Hemleben, Ch, Siddall, M., Hoogakker, B.A.A., Bolshaw, M., et al., 2008. High rates of sea-level rise during the last interglacial period. Nat. Geosci. 1, 38–42.
Schneider, S.H., Mastrandrea, M.D., 2005. Probabilistic assessment of "dangerous" climate change and emissions pathways. Proc. Natl. Acad. Sci. U. S. A. 102, 15728–15735.
Silvano, A., Foppert, A., Rintoul, S.R., Holland, P.R., Tamura, T. a, et al., 2020. Recent recovery of Antarctic Bottom Water formation in the Ross Sea driven by climate anomalies. Nat. Geosci. 13, 780–786.

Smith, P., Bustamante, M., Ahammad, H., Clark, H., Dong, H., et al., 2014. Agriculture, forestry and other land use (AFOLU). In: Edenhofer, O., Pichs-Madruga, R., Sokona, Y., Farahani, E., Kadner, S., et al.,Climate Change 2014: Mitigation of Climate Change. Contribution of Working Group III to the Fifth Assessment Report of the Intergovernmental Panel on Climate Change. Cambridge University Press, Cambridge, United Kingdom and New York, NY, USA.

Stocker, T.F., 1998. The seesaw effect. Science 282 (5386), 61–62.

Stommel, H., 1961. Thermohaline convection with two stable regimes of flow. Tellus 13 (2), 224–230.

Till, C.E., Chang, Y.I., 2011. ISBN: 978-1466384606 Plentiful Energy. CreateSpace, p. 400. United Nations Framework Convention on Climate Change, 1992. http://www.unfccc.int.

Wendle, J., 2015. Syria's climate refugees. Sci. Amer.

Further reading

Hansen, J., 2009. ISBN 978-1-60819-502-2 Storms of My Grandchildren. Bloomsbury, New York.

Preface

I grew up in the concrete jungle of Brooklyn, New York, an inner-city kid with a developing love affair with Nature. My cityscape featured sidewalk fire hydrants that capped pressurized water stored in the event of a fire emergency. Regularly spaced street "manhole" covers marked distances for single, double, triple, and home run in neighborhood stick-ball games.

Each summer, an adventurous teen would "monkey-wrench" the fire hydrant valve, liberating an exploding fountain of aquatic mayhem. Sidewalk trees shaded the expansive urban heat island. In July, my family would retreat to a farmhouse in upstate New York's Catskill Mountains where I discovered the hidden world of salamanders.

All science begins with a question informed by observations/data. As a city kid exploring the wide-open spaces of the Catskill Mountains, I was curious about the odd behavior of orange-colored salamanders with iridescent red spots (the red eft or land stage of the eastern red-spotted newt, *Notophthalmus viridescens*). After heavy rain or summer hailstorms, the salamanders would emerge from their tiny bunkers concealed in roadside swales to march off toward the banks of the East Branch of the Delaware River where they would complete their life cycle as aquatic adults. I wanted to know why something seemingly so vulnerable to barnyard cats would flash bright colors in defiance of the apparent risks and why they came out only after storms. I began collecting them in empty coffee cans for travel back to my city haunts in their new home-made backyard aquarium (I do not recommend this today because of the potential for escaped pets to become invasive!).

Years later, as an undergraduate, I would study warning coloration as a defense mechanism in amphibians. That summer (as a 21-year-old), I returned to the Catskills to find roads paved and the banks of the river permanently altered, not by beavers or salamanders, but levees for flood control. I never saw those red efts again.

The East Branch salamanders would launch my career as a field biologist in the 1980s–90s. As a graduate student at Wayne State University (Detroit, Michigan), I set up translucent "mist-nets" to capture, mark, and release breeding birds of all shapes, sizes, and colors in habitat-relations studies. I counted plants in 10-m-radius circular plots centered on nesting sites and cataloged insects captured with hand-held "sweep" nets for foraging studies. Years later, as a PhD student at the University of Michigan, I climbed the tallest trees in northern Michigan's hardwood forests to determine survival rates of newly hatched chicks in forests adjacent to oil and gas drilling pads. My fieldwork continued as a postdoc (Oregon State University and University of Wyoming) investigating dietary preferences of the federally threatened Northern Spotted Owl (*Strix occidentalis caurina*) in Oregon's old-growth forests, wintering habitat of the threatened Bald Eagle (*Haliaeetus leucocephalus*) in Oregon's Cascade Mountains, and dietary preferences of the endangered Least Tern (*Sternula antillarum*) on the Platte River in Nebraska. While the studies were aimed at helping imperiled species recover from a habitat-declining world, they prepped me for my first real job out of academia and into the hotly contested political arena of forestry.

In the early 1990s, I worked for a large international private consulting firm that secured a research grant from the USDA Forest Service to study the effects of clear-cut logging and other forestry alterations on breeding and overwintering birds, an endemic subspecies of wolf (Alexander Archipelago wolf, *Canis lupus ligoni*), and an endemic subspecies of deer (Sitka black-tail deer, *Odocoileus hemionus sitkensis*) in Alaska's Tongass rainforest. I followed marked census routes in logged versus unlogged rainforests at the crack of the Alaskan summer dawn (3 a.m.). Before each survey, I would suit-up in head-to-toe raingear with seams neatly duct-taped to protect against intrusive, biting insects. Calf-high rainboots lined with spiked soles ("Calk Logger Boots") were essential for safe footing in steep, treacherous terrain. Insect repellant sprayed copiously on clothing acted as field "cologne," while mosquito head-netting offered protection from the kamikaze-like rage of biting insect swarms so dense that the songs of birds were muffled by the insects' high-pitched wing vibrations. The Alaska dawn carried a musk-like scent of freshly deposited dew on spongy-rainforest carpeting, as I spied through binoculars with X-ray vision for concealed bird nests.

One day while counting birds along the census route, I stumbled on a pack of wolves unaware of my down-wind scent as they surrounded a young deer for the kill. I watched in awe as my carnivorous neighbors tore into exposed flesh. But I became overwhelmed with grief that my primal experience would soon be erased from the Nature's bountiful annals like the salamanders of East Branch. A scheduled clear-cut would soon sever the intricate predator–prey web. Thoughts of protest, tying myself to trees, and shouting to the Universe—this is just wrong—consumed my day.

Years of salamander, wolf, and bird studies inspired me to take action as I accepted a job at the World Wildlife Fund (WWF) to run the domestic forest conservation program. WWF was like a mini-United Nations in the 1990s, working globally to "save life on Earth." As conservation doctors, we were schooled in the science of treating a sick planetary patient in need of intensive care and resuscitation.

Armed with a scientific pedigree and a Brooklyn tough-kid attitude, I was set loose upon the mosh-pit of Washington DC politics. I would soon find out how nothing in field biology or the halls of academia prepared me for the food fight of congressional hearings where problem-solving and scientific knowledge are routinely thrown out the window by anti-environmental forces.

My first close-encounter-of-the-congressional kind came in 1994 when I testified during a "forest health" hearing called for by the House Natural Resources Committee, which was in the process of renaming itself the "Resources Committee" for exploitative interests. The Committee was proposing legislation aimed at logging forests deemed "sick and dying" from natural causes (insects, wildfires, and storms). I was the lone "minority" witness (Republicans were in control of the House and Democrats were allowed only one witness on the panel) with the job of backstopping environmental protections. My job was as a witness was to counter the Orwellian doctrine that "healthy forests" were only those where trees were planted in neatly spaced, dense rows resembling cornfields, competing nonconifer vegetation killed by herbicides, and old-growth "decadent" trees removed with chain saws. This sort of tree euthanasia was imported from Europe over a century ago and it has since left an indelible mark on forests nationwide.

In forestry, money grows on trees. The faster the rate of tree growth, the quicker the return on investment. In the timber industry world, environmental regulations are a debit against interest payments from cutting trees and endangered species mute the bottom line.

At the hearing, I listened in disbelief for hours as a parade of timber and corporate interests testified about how natural forests needed a chain-saw cure otherwise blocked by the Endangered Species Act and other environmental laws (like the Clean Water Act, National Environmental Policy Act). As the last witness of the day, it was my turn to enter the congressional theater of witness intimidation and abject denial of science. While other PhD witnesses were called respectfully by the Committee Chair as doctors, I was introduced as Mr. DellaSala, despite also having a PhD. Not a good start, to say the least.

The podium at congressional hearings uses a 5-minute traffic-light (green, yellow, and red) timer to ensure concise testimonials. As I thanked the Chair for inviting me, the green light went off, signaling the countdown had begun. Three minutes in, the yellow light warned of little time remaining. During my wrap up, the red light signaled, hard stop.

In what felt like a New York minute, I spoke about how forests, insects, and wildfires were natural elements of "healthy forests" that replenished soil nutrients, jump-started forest succession, and provided habitat for scores of wildlife. These natural elements were not the harbingers of forest death nor did they constitute a "forest health emergency" but, in reality, were part of the checks and balances in Nature's prodigious circle-of-life. The sickest forests were actually those infested with a metastasized network of roads and clear-cuts.

During the witness questioning period, I was accused of being an "elitist scientist" willing to put the needs of endangered species ahead of logging-dependent, low-income families. Representative Helen Chenoweth (R-ID), known at the time for her rambunctious statements about how salmon were not endangered by dams because they can be purchased in a can at the supermarket, kicked off the rebuttals. We soon became locked in a verbal sparring match so bizarre it was featured in *The New Republic* magazine. Chenoweth was the acting Committee attack dog that opened with a framed picture of an elk killed by wildfire in her home state of Idaho. Her booming voice proclaimed, "what is this, what is this?!" My response—"ah, a dead elk in a burned forest!" Chenoweth then pivoted to the off-topic question—"if this were a burning building instead of a forest fire would you leave people trapped inside?"

I paused for a moment as thoughts ran through my head of how nothing in the concrete jungle of Brooklyn, my exploration into the world of birds, salamander, and wolves, or the rigors of academia, prepared me for anything this harebrained. From my witness seat down-sight of Chenoweth's intimidating podium posture, I squinted at the photo she was holding as I prepared a counter punch that went something like this:

> *Congressman Chenoweth (she preferred congressman by the way), respectfully, I disagree with your position on forest health and here's why. The framed picture you are holding about the forest being unhealthy because of wildfire – do you see the background in that picture? Maintaining a defensive arms-crossed position, she affirmatively nodded. That backdrop, which you earlier stated prevented people from having a picnic in the forest because it's a fire-created "moonscape," in reality, they couldn't go there anyway because of the expansive damage caused by clearcutting the hillside. If the concern is forest health, the cause of the problem is not fire per se but the logging and road building that took place before the fire and now you are wanting to log the forest again after fire. That's adding insult to injury.*

Chenoweth's turned to the burnt elk photo and asked me how could I let this animal die in a fire?

I tried desperately to simplify population biology that I learned at my alma mater, which went something like this.

> *Individual animals die in forest fires all the time but what matters most is the viability of entire populations, particularly endangered ones, that lose habitat whenever watersheds are damaged by intensive developments like what you see in that hillside photo.*

My rebuttal was aimed at defense of the Endangered Species Act, the National Environmental Policy Act, and the Clean Water Act that the Committee was trying to tear down to speed up logging (referred to at the time by conservation groups as "lawless logging").

After an hour of point-counterpoint jousting that resembled a fencing match between arch rivals, I was finally released from the hearing room that consumed my entire day to provide a mere five-minute dose of sanity to the Committee. To my surprise, some of the industry witnesses approached me in the corridor and, with a sense of fair play, expressed empathy for the disrespectful treatment I had experienced.

The following year, I boarded a flight to Boise, Idaho where by happenstance so did Helen Chenoweth, who sat in the seat right next to me (I'm not making this up!). Not a word was spoken between us despite the hour-long flight. Instead, we each flashed the cover of the books we were reading—mine, "The Celestine Prophecy" (how events presumably happen for a reason), hers, "Can America Survive" (a right-wing conspiracy book about the "evil left" taking over the nation). I told this story at a tribal reservation in Montana later that week and won a reservation t-shirt and applause for the most bizarre account. It turned out that Chenoweth was not well liked by the tribe either!

In the years following, I have routinely testified in the House and Senate (state and federal) in defense of forests and endangered species (with many similar and shocking stories), shook hands with presidents and cabinet secretaries in victory laps (national monuments, roadless protections), and did my part to bridge the expansive divide between science and politics. I did this despite being labeled to this day an "advocate," "fake scientist," "nonscientist," "alarmist," "elitist," "fringe scientist," "minority opinion," and accused of having "confirmation bias." Despite the personal attacks, I believe whole-heartedly that speaking truth to power can be done in a respectful way that maintains scientific integrity, which is why I reached out to my colleagues in writing this book.

As scientists and advocates, it is our badge of honor to speak out. Let the record show that our nation's landmark environmental laws were won through hard-fought battles where scientists and advocates took on powerful and well-funded interests despite seemingly impossible odds, personal attacks, scant funding, and countless hours invested and then re-invested when the political landscape changed (e.g., the push–pull between pro- and anti-environmental presidential administrations). Often, we are chastised, disrespected, and even criticized by colleagues that think scientists should be seen and not heard, lest we tarnish our reputations.

But this is changing, as more scientists realize what's at stake if we are complicit. Witness the thousands that marched at rallies for scientific integrity in Washington DC soon after Donald Trump was elected president (January 2017). Many others also sounded the alarm about the global biodiversity and climate change emergencies (e.g., Alliance for World Scientists, https://scientistswarning.forestry.oregonstate.edu/). The youth movement, inspired by activist Greta Thunberg, works with leading climatologists like NASA scientist James Hansen and, in Europe, with *Scientists for Future* (https://www.scientists4future.org/stellungnahme/statement-text/), raising awareness about climate chaos. In sum, we are a force to be reckoned with and humanity's best chance for a sustainable future.

This book offers a collection of case studies of scientists and *advocates* willing to speak truth to power. The Oxford dictionary defines *advocacy* as "any action that speaks in favor of, recommends, argues for a cause, supports or defends, or pleads on behalf of others." Here, we substitute "on behalf of others" with "on behalf of the planet and future generations" and on "behalf of scientific integrity and speaking truth to power."

The antithesis of scientific integrity and a prime reason for scientists to be advocates was on full display by Donald Trump during the Covid-19 pandemic. His daily virus missives have put the lives of millions at risk by routinely ignoring science in favor of his gut feelings, election politics, and the offering of "red meat" conspiracy theories to his political base, which includes religious zealots that defy science, especially environmental science deemed as devil worship (https://www.nytimes.com/2020/03/27/opinion/coronavirus-trump-evangelicals.html). In listening to the president's science denials, some of his supporters even went as far as believing that only God can save them from the "plague," not face-masks, the government, or "elitist scientists." In a knowledge retreat reminiscent of leeches and bloodletting the president made outrageous claims of killing the virus by injecting poisonous disinfectants into the body, the use of "powerful lights" and UV rays, and unproven drugs, all at the shock and disbelief of medical experts, including his own medical team.

The suppression of science is not just limited to the former president's fumbling of Covid-19. Immediately after Trump was elected, the White House deleted its climate change websites, defunded science programs, gutted and reassigned federal agency science staff, and rolled back decades of environmental progress while shutting out the press that questioned Trump's nefarious motives (other than Fox News and other conservative and inflammatory talk shows). These are all warnings of how deeply entrenched the retreat of science in American politics has become. We risk democracy and informed decision making if we do not speak out.

And speak out we shall even in the darkest times of science denial. The repair of scientific integrity needs to march forward now by supporting EPA scientists and workers in the United States that have been mistreated by the previous Trump administration. Specifically, we need to enact scientific integrity principles such as those proposed by EPA's 8000 unionized employees represented by the American Federation of Government Employees (http://progressivereform.org/cpr-blog/epa-staff-clap-back-trump-workers-bill-rights/includes). The workers bill of rights includes 10 provisions as follows:

1. The right to *scientific integrity* in EPA work.
2. The right to *enforce environmental laws* without political interference.

3. The right to a fully-funded EPA budget and *full staffing levels.*
4. The right to an *end of lockouts* caused by US government shutdowns.
5. The right to work on control of greenhouse gasses, to discuss *solutions to climate change,* and to conduct climate change research.
6. The right to *whistleblower protections.*
7. The right to work-life balance that fosters productivity and *sustainability.*
8. The right to a *fair contract* that is collectively-bargained.
9. A right to a *hate-free and safe workplace.*
10. A right to protect human health and the environment, to protect environmental justice communities, and to work *without fear of reprisal.*

The theme throughout the book is just that—speaking truth to power—using science to inform decision making. It is divided into three main sections with 16 total chapters authored by scientists and advocates in the trenches of environmental and climate policy. To open the book, Section 1, *Scientists as Advocates,* includes chapters by Dominick A. DellaSala (conservation scientist on the nuts and bolts of advocacy, and politics of spotted owl conservation), Derek Lee et al. (wildlife biologist on when scientists are attacked), Franz Baumann (former Assistant Secretary-General of the United Nations on sounding the climate alarm), and Jeremy Kerr (university professor on science-based advocacy in Canada). Section 2, *Science and Policy as an Imperfect Marriage,* includes chapters by Noah Greenwald (Endangered Species Act expert on politics of the ESA), Michael Halpern (Union of Concerned Scientists on scientists as whistleblowers), Robert Hughes et al. (Courtesy/Associate Professor on why and how to advocate), and Bill Bradbury (former Oregon State Secretary on Al Gore climate reality leadership). We conclude with Section 3, *The Politics of Science in Decision Making,* with writings by Rich McIntrye (veteran campaign consultant on why scientists need to leave the ivory tower); Joel Clement (former science advisor to the Department of Interior and whistleblower), Randi Spivak and Jennifer Mamola (activists tips on effective lobbying); Kara Ayn Whittaker and Peter Goldman (forest legal advocates on shifting the burden of legal proof); Angus Duncan (former Oregon Global Warming Commission chair on net zero emissions); David Johns [veteran activist on why scientists need to learn from other successful movements (like the civil rights movement) in taking to the streets], and Augusta Wilson (Attorney at the Climate Science Legal Defense Fund). The closing chapter by DellaSala is a summary of the current state of the planet and call to action for *Speaking Truth to Power* in response to Carl Sagan's inspiring, "*Who Speaks for the Earth*?"

Personal dedication: This book is my personal shout out for advancement of women, LGBTQ, and minorities in science (especially environmental sciences); an expanded and well-funded science and ecology curriculum from K-12 to universities, including the STEM (science, technology, engineering, and mathematics) program; and an unprecedented infusion of funds for research and communications that connect planetary and human health. It is vital that we demonstrate to society that protecting biodiversity is in everyone's best interests for a safe and just world. I encourage scientific societies and universities to greatly expand training, resources, and professional awards for scientists willing to speak out for the planet and to partner with organizations schooled in communications and outreach (e.g., science-based environmental, health, climate, and social-change organizations). Scientific journals also need to adjust impact factors beyond how many times an article is cited. Impact factors need to be tailored to whether an article has policy and conservation relevance, lest we continue speaking only to the choir. Hats off to the Union of Concerned Scientists and organizations like the Public Employees for Environmental Responsibility that work with government scientists (whistleblowers) to ensure that they receive proper legal protections from retaliation when speaking out.

The book's concept was shaped during my 8-year tenure as former president of the Society for Conservation Biology, North America section, during conference debates over the proper role of scientists as advocates for conservation. The book is also a personal tribute to all scientists who have modeled advocacy while maintaining scientific credentials. My personal inspirational list of scientists that have advocated by example includes such luminaries as Albert Einstein, Aldo Leopold, Neil deGrasse Tyson, E.O. Wilson, Rachel Carson, David Suzuki, and Jane Goodall. We now stand on their shoulders as advocates for the next generation of scientists. This book is also dedicated to Carl Sagan, whose prescient writings and unique communication skills are the main reasons for why I chose a career in science and advocacy. He inspired me to act.

Finally, it is my hope that this book will be a resource for young professionals venturing into the world of science and advocacy and their mentors to apprentice them. Veteran scientists need to act more like caring parents in directing their pupils rather than bullying them into submission, which is seemingly on the rise in academic circles (https://www.nature.com/articles/d41586-018-07532-5). And like society in general, there is no excuse for bullying or character assignations as chronicled in some of the chapters of this book (see Lee et al. for an example). Young professionals need support, not condemnation.

Special thanks to Monica Bond and Alice Grant (Elsevier) for helpful editing throughout and Dan Galpern for help with the foreword concept. And last but not least, I dedicate this book to my two daughters—Ariela Fay DellaSala and Janelle Neill—my grandkids—Stella, Michael, and Nathan—and the iridescent salamanders of East Branch that jump started my personal journey of science and advocacy.

Hopefully, you will find this book inspiring enough to take action, howl for the wild things you care about, and be part of the growing movement of scientists, social justice advocates, young climate advocates, Indigenous Peoples, senior scientists, cultural creatives, and all those willing to speak up. To answer the calling, "*Who Speaks for the Earth*?" You do!

Dominick A. DellaSala
Wild Heritage, A Project of Earth Island Institute, Berkeley, CA, United States

I

Scientists as Advocates: Advocacy Should not be a Four-Letter Word

1

The nuts and bolts of science-based advocacy

Dominick A. DellaSala
Wild Heritage, A Project of Earth Island Institute, Berkeley, CA, United States

Cartoon Courtesy of Michael Halpern, Union of Concerned Scientists

A revolution in scientific thinking

Sir Francis Bacon was a 17th century English scholar and statesman (Lord Chancellor of England, Attorney General of England) whose life spanned the transition from the Renaissance to early "Modern Era" (https://plato.stanford.edu/entries/francis-bacon/). He was a unique individual, because, as a politician, Sir Francis is credited with birthing the scientific method. Herein referred to as the "Baconian" method, his groundbreaking scientific methodology was premised on knowledge derived from empirical evidence. Baconian science set the stage for a revolution in thinking captured best in Bacon's seminal treatise, *Novum Organum* (New Method) of 1620.

Conservation Science and Advocacy for a Planet in Peril.
DOI: https://doi.org/10.1016/B978-0-12-812988-3.00002-8

Sir Francis shifted critical thinking from Aristotle's deductive reasoning rooted in premises that lead to logical conclusions, to the scientific method of inductive reasoning based on falsification of ideas (hypotheses) through meticulous observations and generalizable conclusions (https://www.youtube.com/watch?v=WAdpPABoTzE). The more consistent the observations, the more robust the conclusions, which was a breakthrough in critical thinking. This way of thinking ushered in microscopes, telescopes, and chemistry flasks to unlock the inner workings of subatomic particles, advancement of medicines, and space probes that ventured out into the far reaches of the solar system (much of the early motivation was religiously based). As the centuries progressed, humanity's ecological footprint exploded to planetary proportions that became obvious to the trained eye. Scientists like Aldo Leopold (herein referred to as the "Leopoldian" method) began speaking out about our insatiable ecological footprint. In short, we became a planetary change-agent on par with major geological and climatic forcing events.

The science-policy engagement pendulum

Today, the way scientists approach policy can be illustrated along a simple philosophical continuum: on the one end, the just-the-facts Baconians choose to stay out of the sausage making of public policy, leaving it to decision makers; on the other, the take-action and ethically principled Leopoldians on a mission of influence (Fig. 1.1). Where one lies on that spectrum is a balance of personal values, comfort zones, and professional risk-taking (Table 1.1).

In this chapter, I argue that the scientific engagement pendulum needs to swing dramatically (and dare I say "radically") toward science-based advocacy (Leopold) given the precarious planetary state. I provide some basic nuts and bolts and real-world examples of speaking truth to power while maintaining scientific credibility. Notably, in some scientific circles the word "advocacy" is treated as if it were a

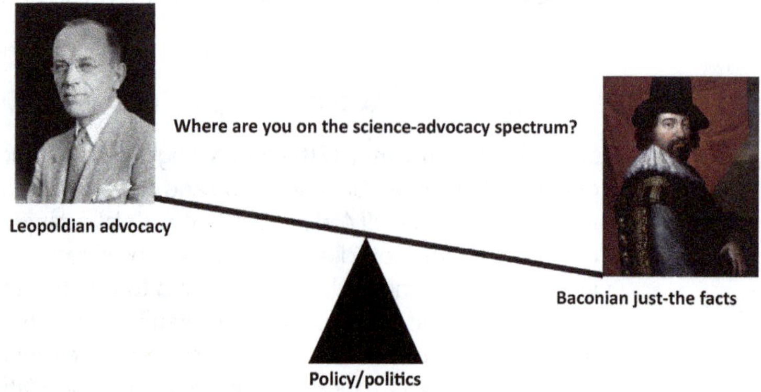
Where are you on the science-advocacy spectrum?
Leopoldian advocacy
Baconian just-the facts
Policy/politics
"Politics is a pendulum whose swings between anarchy and tyranny are fueled by perennially rejuvenated illusions" (Einstein cited in Calaprice, 2000)

Figure 1.1 A continuum of views on scientific engagement in policy, Baconian versus Leopoldian. *Photo courtesy of the Aldo Leopold Foundation. Francis Bacon image courtesy of Flickr Commons.*

Table 1.1 Some general views on science and societal engagement.

Baconian	Leopoldian	Physician
Scientists should not take positions	Land ethic guides advocacy	Hippocratic Oath to do no harm
More data before we say anything with certainty	Even if we don't have all the answers yet, the consequences to Nature override inaction	Precautionary principle—the burden of proof is on the proponent of an action to demonstrate no harm will occur
Scientists should be seen, not heard on policy	Moral imperative to save Nature—all the "cogs-in-the wheel" are important and sacred	Informed innovation

four-letter word. But there is no shame in being seen as a science-based advocate as some of the world's most brilliant scientists have taken up just causes beyond their literary works, including such luminaries as:

- Albert Einstein (see Calaprice, 2000 for multiple causes Einstein took up)
- Carl Sagan (Sagan and Turco, 1993, nuclear winter, https://www.smithsonianmag.com/science-nature/when-carl-sagan-warned-world-about-nuclear-winter-180967198/)
- Rachel Carson (Carlson, 1962, chemical contaminants, https://www.rachelcarson.org/SilentSpring.aspx)
- Jane Goodall (primate conservation, https://www.janegoodall.org.hk/)
- Neil deGrasse Tyson (climate and politics, https://www.youtube.com/watch?v = Jm_YoL9ykC4)
- Stephen Hawking (humanity facing extinction, https://www.bbc.com/news/science-environment-43408961)
- Theo Colborn (Colborn et al., 1996, endocrine disrupters and chemical industries, https://endocrinedisruption.org/about-tedx/theo-colborn-ph.d.-president/)
- James Hansen (Hansen, 2009, science and climate policy, https://www.latimes.com/archives/la-xpm-2009-dec-27-la-ca-james-hansen27-2009dec27-story.html)
- Stephen Schneider (Schneider, 2009, science and climate policy, https://whistleblower.org/politicization-of-climate-science/global-warming-denial-machine/stephen-schneider-climate-denier-gate-a-case-of-science-as-a-contact-sport/)
- Edward O. Wilson (Wilson, 2016, extinction crisis, https://www.half-earthproject.org/)
- Tyrone Hayes (atrazine science and the chemical industry, https://www.newyorker.com/magazine/2014/02/10/a-valuable-reputation)

These brave scientists took a stand in translating their work to socially and environmentally just causes even under intense criticism (sometimes from colleagues), career reprisals (see Chapter 2, When Scientists Are Attacked: Strategies for Dissident Scientists and Whistleblowers), personal threats, canceled funding, and outright nasty efforts to discredit and marginalize.

Peer review and evidence-based science as the gold standard

As scientists, the benchmark of credibility is set in stone by the arbiter (chief editor) and jury (colleagues) during peer review. This forces scientists to be impeccable, objective, precise, and relentless in the search for the truth. But what should we do when the facts dictate an emerging biodiversity [Intergovernmental Panel on Biodiversity and Ecosystem Services (IPBES), 2019] and climate crisis (Ripple et al., 2019; also see Alliance of World Scientists; https://scientistswarning.forestry.oregonstate.edu/)? How can we respond as informed members of society and planetary citizens? If we take action, will this jeopardize objectivity?

I argue that with the scientific method as our calling card, scientists can proudly step out of the Ivory Tower into the real world where decision makers operate. At times, this may even involve taking to the streets in defense of scientific integrity (see Chapter 14, The Politics of Conservation—Taking the Biodiversity Crisis to the Streets) and in response to the climate emergency (see NASA researcher James Hansen's brave arrest at a coal-mining protest: https://blogs.scientificamerican.com/news-blog/nasa-climate-researcher-hansen-arre-2009-06-25/).

What's at risk?

The antithesis of critical thinking—denial and self-interests—is prominently on display whenever we turn on the TV, browse the internet, or read the newspapers about politicians ignoring science. The most audacious example was in the daily missives of former President Donald Trump that included his shocking denial of the seriousness of the coronavirus pandemic and climate change, to name but a few! His fact twisting and suppression of science started early in his administration as White House officials deleted climate change materials from government websites (e.g., see https://www.nbcnews.com/news/us-news/two-government-websites-climate-change-survive-trump-era-n891806; https://www.independent.co.uk/news/world/americas/us-politics/trump-climate-change-government-websites-global-warming-a9020461.html), proclaiming climate change a "hoax" and China's ploy to wreck the US economy. As we will see from the examples in this chapter, the suppression of science and truth telling is not limited to any one political administration. On the flip side, ignoring misinformation tactics of government officials and others is a form of silent complicity.

Taking the planetary Hippocratic Oath to avoid silent complicity

The United Nations adopted 10 principles in 2000 known as the "global compact," which was derived from the Universal Declaration of Human Rights, the International Labour Organization's Declaration on Fundamental Principles and

Rights at Work, the Rio Declaration on Environment and Development, and the United Nations Convention Against Corruption (https://www.unglobalcompact.org/what-is-gc/mission/principles). Notably, Principle 2 calls out complicity as follows:

> Complicity means being implicated in a human rights abuse that another company, government, individual or other group is causing. The risk of complicity in a human rights abuse may be particularly high in areas with weak governance and/or where human rights abuse is widespread. However, the risk of complicity exists in every sector and every country.

While this UN principle aptly applies to human rights violations, the complicity principle also can apply to environmental and ecojustice rights (see DellaSala, 2020 for discussion of ecojustice vs environmental justice).

A planetary thought experiment

The silent complicity principle is illustrated here through a simple planetary thought experiment distinguishing Baconian from Leopoldian scientists. It goes something like this.

Humans are not causing massive and dangerous changes to the planet's life support systems (null hypothesis) as the observed changes are part of the natural cycles. To put it in Trump's terms, climate change is a hoax. We then present irrefutable evidence (high degree of certainty) to contrarians that the planet is indeed experiencing a rapidly changing climate, we can measure the speed (velocity) of climate change and the extent of related ecosystem degradation (Fig. 1.2).

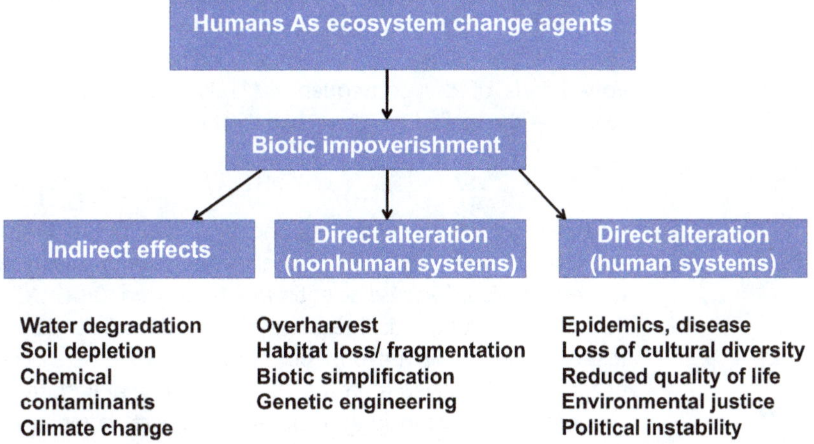

Figure 1.2 Humans as ecosystem change agents as documented by evidence-based conservation science.

Further, we show that planetary boundaries (thresholds) responsible for life on Earth are being irreparably crossed (Barnosky et al. 2012) as nearly all ecosystem services are in disrepair (Millennium Ecosystem Assessment, 2005). Deforestation and forest degradation continue apace at the expense of primary forests and intact landscapes (Mackey et al., 2015). Coral reefs and mangroves, biological "hotspots" and nursery grounds of the world's fisheries, are at imminent collapse (Millennium Ecosystem Assessment, 2005). We are on a collision course with what little remains of the natural world (Ibisch et al., 2016) as our actions push more than a million species to the brink of extinction [Intergovernmental Panel on Biodiversity and Ecosystem Services (IPBES), 2019]. We have, by some accounts, 10 years to change this alarming trajectory (Ripple et al., 2019). More than a thought experiment, this is the reality that we face in our lifetime with severe consequences to future generations. So, do we remain complicit in a Baconian sense or take action in a Leopoldian sense for the sake of the planet and future generations that have no say in the matter?

In a simplistic sense, Baconians would turn the planetary crisis data over to decision makers. Leopoldians would immediately advocate for action consistent with the facts and scale of the problem while recognizing the consequences of inactions or, even worse, bad choices.

To personalize Leopoldian versus Baconian world-view differences of the planetary crisis, let's say you went to a trusted doctor for your blood work because you were feeling ill. The results came back with calamitous news—you have a cancerous brain tumor that will metastasize if not properly treated with the best available medical science immediately. No physician in their right mind would walk you and your medical chart down the hall to speak with a medically untrained counselor—that would violate the physician's Hippocratic Oath. Why then should we expect anything less from a conservation biologist or climate change scientist reading the planetary signs that we have but a decade before it gets a lot worse? Like the medical doctor, is it irresponsible to ignore making recommendations on behalf of the patient by instead simply handing off the results, while wringing your hands of the consequences? Taking the Planetary Hippocratic Oath, we would be compelled to act on behalf of our patient, the planet.

Applying the planetary Hippocratic Oath in the real world

I have used the planetary Hippocratic Oath many times in my science-based advocacy in policy and public speaking settings as it works to counter criticisms of those that would accuse scientists of having an "agenda" (see Chapter 2, When Scientists Are Attacked: Strategies for Dissident Scientists and Whistleblowers). A real-world example (beyond our thought experiment) came up during a 2017 hearing that I testified at in the US House Natural Resources Committee regarding proposed increases in logging levels on federal lands in response to wildfires. At the time, my testimony was based on a summation of peer-reviewed literature along with my published research on how extreme fire weather (associated with climate change) and certain forms of logging were actually the prime drivers of fire increases (Bradley et al., 2016). With the CSPAN (live streaming of congressional hearings) cameras rolling, I was asked by a climate-denying, pro-logging congressman, about the efficacy of the US Endangered Species Act (ESA). The question had nothing to do with my testimony but instead was meant to intimidate and discredit me as congressional hearings are often more about theatrics than truth telling (also see Chapter 5, Blowing the Whistle on Political Interference: The Northern Spotted Owl).

The irate congressman used a hypothetical example that went something like this: if 97 out of 100 species died on your watch, would you agree that the ESA was an abysmal failure and waste of taxpayer dollars?

Like a dedicated surgeon treating patients that will live or die on the operating table, I immediately quoted the Hippocratic Oath of doing my best to get my patient off intensive care and into full recovery, which is the ethical and science-based imperative of the ESA that Congress itself approved in 1973. If 97 species died, I would mourn their loss, but I would also take solace from the three species that I saved from the brink of extinction. Rather than accept my answer to his off-topic question, I was shouted at and labeled a "fake scientist" and a paid lobbyist (see http://forestlegacies.org/programs/fire-ecology/geos-scientist-testifies-in-congress-on-climate-change-and-forest-fires/). Interestingly, my 5-minute testimony erupted into a 45-minute congressional debate covered by CSPAN and *Greenwire* as committee members challenged each other about beliefs in science. Despite the personal attacks, I felt vindicated by how the reactions to my testimony completely derailed the pro-logging agenda by having congressional members defend their views on science and, more to the point, exposed their denial of climate change. I learned later from congressional staff that my approach to a Planetary Hippocratic Oath changed the entire dynamic of the hearing putting congressional members in the spotlight on science instead of me (Leopold would be proud!).

Don't shoot the messenger

Like the rest of society, scientific debate can be quite polarizing at times as we challenge each other to sharpen our understanding of the truth. However, even scientists see what they want as there is always an inherent bias in how we perceive the world—after all, we are human, we all have an agenda of some sort. We are not some unemotional, fictitious artificial intelligence TV character like *Data*, of Star Trek, the Next Generation! Thus it's best to state our inherent values/biases up front when we step into the policy arena and get it over with so our critics can at least get the accusations right. For instance, medical doctors care about their patient's health (Hippocratic Oath) and are willing to speak on their behalf. As a scientist, I care deeply about the planet and the future of humanity and my grandkids (Planetary Hippocratic Oath). That's my agenda—oppose me at your own peril! Using my Brooklyn accent, and a dose of humor to keep it sane, my message frame goes something like this: I care about saving life on Earth and I am also an objective scientist that relies on evidence, "if you gotta problem with that—*fuggedaboutit*!" (tongue-in-cheek humor is part of my personality, what can I say!).

Seriously, those that do not like this narrative (message frame) have to defend the polar opposite frame as to why they do not care about the planet or future generations. In political campaigns, message framing separates the winners from the losers (see Lakoff, 2004).

> To put it bluntly, getting the message right is as important as getting the science right.

With data in hand, there are sound reasons to speak truth to power. But what happens when all sides claim to have the truth on their side?

Dueling science

Given my expertise in translating science for the policy arena, I was approached by a legislative aide once who was concerned that each side had its bevy of experts that would come marching into her office with a parade of facts, convinced that the other side was dead wrong and agenda driven. She would often get confused about what to do about dueling science and was motivated to act even in the absence of clear-cut evidence because her boss (a prominent US Senator) wanted to respond to his constituents in real time. Her words "whose science should I believe?" and "we cannot wait for you guys to sort all of this out" reverberated in my mind long after our meeting. Politically speaking, she is right.

This is because in science we methodically sort out our differences through painstaking and time-consuming scientific research that operates a bit like a pistol duel between archrivals—each side takes careful steps toward formulating positions (hypotheses) that counter the other. In this duel, there are always levels of uncertainty. Scientists are really good at explaining them, but short in answering the proverbial question, "what's in it for me?" (see below). My simple answer to the legislative aide was this: I am presenting you with evidence-based science rooted in the gold standard of peer-review and here's why it would be in the Senator's best interest to act responsibly with the best science—or at least if you are not, then please explain to your constituents why not.

Science versus anecdotes

Meanwhile, let's step back to the 2017 congressional hearing on forest fires mentioned above. At the hearing, a witness for the timber industry spoke about how thinning trees in a lodgepole pine (*Pinus contorta*) forest in eastern Oregon was necessary to stop intense forest fires. He presented two photos on the meeting room viewing screen—one where many of the trees were removed by mechanical thinning (logging) which, as claimed, when the fire passed through the forest it left most of the remaining trees alive (low-intensity burn). The other—an unthinned site—experienced the highest fire intensity as most of the trees were fire-killed. Despite not considering confounding factors like wind direction, aspect and slope position, time of day, humidity levels, and having absolutely no replicates or statistical design, he presented these two photos as "proof" of how thinning lowered fire intensity. In short, congressional members were faced with dueling science—to thin or not to thin—and whose science to believe. In reality, were they really faced with this choice?

In my rebuttal, I asked to have both slides put back up on the screen. I then proceeded to attack the idea, because this was not in fact science but anecdotal conjecture (the science death ray!) as the industry's "experiment" did not take place in a controlled setting nor follow any scientific methodology (hypothesis, experimental design, statistical analysis, etc.). As I explained to the committee members, lodgepole pine is actually uniquely adapted to intense forest fires. In many parts of the West, this resilient pine has seeds tightly entombed in "serotinous" cones that need blow-torch heat to energize the forest renewal process. Thinning in this forest, if it even lowered fire intensity, was actually inconsistent with the ecology of the species that needs to burn hot. I stated that peer-reviewed science and statistically based evidence is the way to sort fact from fiction, not anecdotal observations. I then cited a published dataset of >1500 forest fires gathered over four decades of fire monitoring across >9 million hectares of burned forests in 11 western states (Bradley et al., 2016). That dataset showed with statistical certainty that the forests with the most logging actually burned in the highest intensities presumably because of tight spacing of fire-prone small trees planted in dense rows after logging (plantations) and flammable slash left behind by loggers. More logging would only make the problem that congressional members wanted to solve much worse. I believe Congress learned an important lesson in peer review over hearsay and at the very least, and to their credit, were very curious about the ecology of serotinous cones, the off-topic conversation, and my highlight of the hearing (Leopold would be proud!).

Applying the precautionary principle

Like the scientific method, the precautionary principle has its roots in ancient history through—you guessed it—the Hippocratic Oath, later popularized in

1970s Germany ("Vorsorge" or foresight http://www.ejolt.org/2015/02/precautionary-principle/). Hippocrates himself noted "as to disease, make a habit of two things—to help, or at least, to do no harm" (cited in Hayes, 2005). The precautionary principle was introduced into policy at the Rio Earth Summit of 1992 as a warning to humanity to slow down the biodiversity crisis. The principle has since formed the underlying basis of environmental laws from Europe to Canada and has been inculcated in UN policy (e.g., the UN Global Compact of 2000 cited above). It has come to be defined as "if an action or policy has a suspected risk of causing harm to the public, or to the environment, in the absence of scientific consensus (that the action or policy is not harmful), the burden of proof that is not harmful falls on those taking the action" (Rio Declaration, 1992).

Kriebel et al. (2001) note that the precautionary principle has four fundamental components:

1. Take preventive action in the face of uncertainty.
2. Shift the burden of proof to the proponents of an activity.
3. Explore a wide range of alternatives instead of harmful actions.
4. Increase public participation in decision making.

To put into common language, the precautionary principle is a sort of preventive medicine for the environment that underscores:

- a look before you leap (or log) approach (i.e., assess the likely consequences before you act);
- take action with no regrets (i.e., be safe rather than sorry);
- place the burden of proof of doing no harm on the proponent of the action; and
- informed consent (similar to how doctors intervene with medicines) is better than risky, uninformed actions.

The UN also defines the precautionary approach in the Global Compact as follows (https://www.unglobalcompact.org/what-is-gc/mission/principles/principle-7):

> "The precautionary approach, Principle 15 of the 1992 Rio states that where there are threats of serious or irreversible damage, lack of full scientific certainty shall not be used as a reason for postponing cost-effective measures to prevent environmental degradation.
>
> Precaution involves the systematic application of risk assessment, risk management and risk communication. When there is reasonable suspicion of harm, decision makers need to apply precaution and consider the degree of uncertainty that appears from scientific evaluation.
>
> (*Continued*)

(Continued)

Deciding on the "acceptable" level of risk involves not only scientific-technological evaluation and economic cost-benefit analysis, but also political considerations such as acceptability to the public. From a public policy view, precaution is applied as long as scientific information is incomplete or inconclusive and the associated risk is still considered too high to be imposed on society. The level of risk considered typically relates to standards of environment, health and safety."

To take this a step further, when human activities lead to unacceptable and ethically deplorable behavior (e.g., extinction), actions need to be taken to avoid that harm and reach an acceptable level of risk. But how do we know what is "acceptable?"

US Endangered Species Act's precautionary principle

The ESA is a prime example of the precautionary principle that for decades has been under attack by decision makers (see Chapter 6, Overcoming the Politics of Endangered Species Listings). The Act's precautionary approach actually predates widespread application of the precautionary principle and generally includes many aspects of it. For instance, precaution is baked into ESA regulations through the incidental "take" provision prohibiting actions deemed to "harass, harm, pursue, hunt, shoot, wound, kill, trap, capture, or collect or to attempt to engage in any such conduct." Some have interpreted this literally to mean any action can cause harm and therefore the standard is essentially meaningless and is a deterrent to "active management" (Mealey et al., 2005). This incorrect assumption is seldom met in practice, as in reality the US Fish & Wildlife Service, an ESA regulatory agency, allows substantial incidental take provided it does not jeopardize the viability of the species. Unfortunately, the jeopardy standard for prohibiting risky actions has been weakened over time and there is now a very high bar to jeopardy such that hundreds of endangered animals can be killed by a single project using an "incidental take permit" without a jeopardy decision rendered. In fact, as we shall see in Chapter 5, Blowing the Whistle on Political Interference: The Northern Spotted Owl, opposition to the precautionary principle by the US Fish & Wildlife Service itself has led to actions that can cause irreparable harm based on the false assumption that such harm is short-lived. This is risky business for a species circling the extinction drain and the misconstrued assumption that the precautionary principle blocks all actions was never the intent of the precautionary principle. In the case of endangered species, there are irreversible consequences as the high cost of making a mistake is extinction. While decision makers certainly need to balance levels of uncertainty with risks of inaction, actions should be chosen that do no harm (Planetary Hippocratic Oath) and have the best outcomes for recovery regardless of the economic impacts (that is a fundamental premise of the ESA).

Thus to ensure proper application of the precautionary principle, especially in ESA decisions where there is little margin for error, decision makers should use the burden-of-proof standard (see Chapter 12, Shifting the Burden of Proof to Minimize Impacts During the Science-Policy Process) to evaluate a project/ action by requiring that:

- scientific uncertainties (i.e., lack of scientific consensus) are clearly factored into decisions so actions result in the lowest possible risks or, even better, no risks at all to the species;
- scenarios (or models) of harm are scientifically sound and peer reviewed (evidence-based);
- transparency is met by using best available and independently published science and public involvement;
- risks of inaction outweigh action, but actions must not cause irreparable harm;
- a well-funded monitoring plan is in place to provide a check on "active management;" and
- the benefit of doubt is given to the species at risk (i.e., always err of the side of conservation).

Should any of these burden-of-proof factors not be met, then the project proponent must substantially modify or omit the action.

What is the best available science?

Anyone familiar with federal lands or endangered species management in the USA understands very well the meaning of "best available scientific information" (BASI). It is rooted in agency directives, particularly the so-called forest planning rule of 2012 enacted by the Obama Administration. But what does BASI really mean in practice and how do we know if it is being used? Should decision makers use only peer-reviewed literature, consensus science, risk-management science, gray literature, traditional knowledge, local experience, or something else?

Esch et al. (2018) argue that when there is conflicting science or disagreement about best management practices, decision makers should use the highest standards of accuracy, reliability, and relevancy (sounds like independent peer review to me!). They also suggest a broad range of accuracy and reliability can be used by expanding potential sources of information beyond peer review and into the gray literature including traditional, local, expert, and institutional ecological knowledge. However, this last part can create dueling perspectives as what if a traditional rancher who knows the land intimately claims that through generations of cattle grazing in riparian areas, the range/habitat for fish and wildlife has improved? How then can a decision maker balance on-site knowledge of the generational rancher against the weight of scientific evidence?

According to Ryan et al. (2018), the definition of BASI depends on three primary criteria: accuracy, reliability, and relevance. In addition, the Data Quality Act (PL 106–554) can be referenced for guidance on evaluating the quality of available information. Available is defined as information that currently exists in a form useful for the planning process without further data collection, modification, or validation. Thus, in the case of the rancher, the BASI standard and, if field evidence shows impacts, can be used to falsify the ranching viewpoint (hypothesis) that conditions are improving because of grazing if in fact they are not.

To put this into an illustration, I modified Ryan et al. (2018) on assessing best available science according to relevance, accuracy, and reliance criteria (Fig. 1.3, Type I and II errors are discussed below). The relevance criterion regarding ecological context is especially germane in conservation. For instance, a local area or watershed might be the only remaining intact forest in the entire region and therefore that area's importance is elevated (or weighted) in assessing management options using best available science.

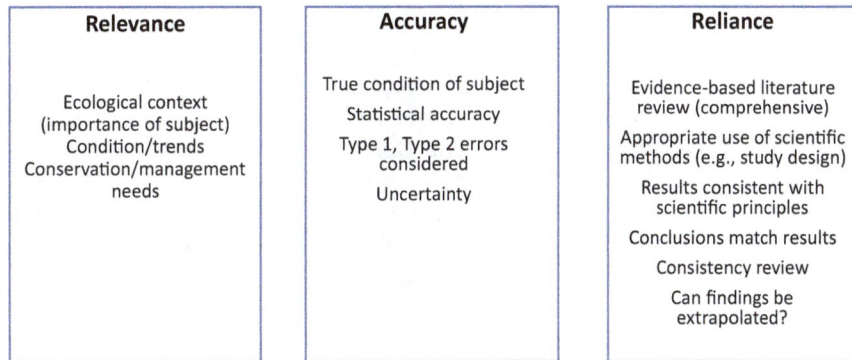

Relevance	Accuracy	Reliance
Ecological context (importance of subject) Condition/trends Conservation/management needs	True condition of subject Statistical accuracy Type 1, Type 2 errors considered Uncertainty	Evidence-based literature review (comprehensive) Appropriate use of scientific methods (e.g., study design) Results consistent with scientific principles Conclusions match results Consistency review Can findings be extrapolated?

Figure 1.3 Assessing best available science using criteria related to relevance, accuracy, and reliance. *Modified from Ryan et al. (2018).*

Agency discretion is the death ray of best available science

Absent clear direction, management decisions may not comport with best available science particularly when there is pressure on responsible officials to produce outcomes inconsistent with conservation science. During the Bill Clinton administration, the Committee of Scientists (1999), for instance, recommended that "*planning must be based on science and other knowledge of the world, including the use of scientifically based strategies for sustainability*" (emphasis added). Systematic evidence or science consistency reviews along these lines would aid responsible officials in planning for incompatibilities in management that pit natural resource extraction against fish and wildlife conservation, water quality, climate change mitigation, landscape connectivity, and ecosystem protection. They would also aid in identifying uncertainties in planning by documenting which decisions were based on evidence (with degrees of certainty) versus those where well-designed experiments are needed to sort out the differences (the dueling science example above). To better avoid these problems, land managers should engage scientists in decisions both before (planning) and after (monitoring) a decision is rendered. Absent engagement with experts, management decisions are more likely to be controversial and challenged in the courts. In this case, the decision bar should be set to the highest rung by including a requirement for *conforming* to best available science (with degrees of uncertainty, see below) similar to that inherent in the ESA and relevant statutes.

The ESA is one of several federal statutes, including the Marine Mammal Protection Act, 16 U.S.C. §§1361–1431 (2000) and the Magnuson-Stevens Fishery Conservation & Management Act, 16 U.S.C.§§ 1801–1883 (2000) that *requires* agencies to use the best scientific data available when making decisions (emphasis added).

Establishing a new credibility standard: best available independent science

The use of best available science is no guarantee that personal biases or funding commitments will not influence the questions asked or the outcomes (Okagaki and Dean, 2016). Scientists are human, they have personal biases, they compete for funding, and funding sources can influence perceptions by leading the witness or scientist in this case. An example is in wildfire research where government funding is used to promote "fuels" management with the intent of reducing fire intensity even if it is not ecologically appropriate or effective in a changing climate (and is challenged by others). Scientists bid on government funding to deliver on preconceived outcomes often specified in government contracting (e.g., fuels management is going to happen, tell us how it reduces fire intensity so we can report to Congress vs is it really the best choice ecologically and what are the tradeoffs from the treatments?). Government funding is in turn secured through congressional appropriations based on deliverables measured in acres treated to receive more appropriations (i.e., a self-perpetuating cycle of research that is then used to get more appropriations to do more of the same research and so on). Independent research that documents impacts of overly aggressive treatments—or questions the underlining assumptions of the need for the treatment—is often met with swift condemnation (see Chapter 2, When Scientists Are Attacked: Strategies for Dissident Scientists and Whistleblowers), is crushed by journal editors sending the paper out for peer review to scientists having pro-treatment biases, or is completely ignored by decision makers cherry picking the science to suit their own personal biases and politics (see Chapter 5, Blowing the Whistle on Political Interference: The Northern Spotted Owl).

To avoid this conflict of interest, I suggest adopting a higher standard of *best available independent science* (BAIS). This can be checked at the publication stage by journals requiring a statement of no conflict of interest and identification of funding sources. For instance, the statement can say up front that the research was funded by independent sources having no financial stake in the outcome. Importantly, many journals now require a conflict of interest and funding disclosure statement (Resnik et al., 2017). Government agencies should adopt the same standard in environmental impact statements and natural resource policies to hit the higher bar of independence. In addition, researchers can notify the journal editor of reviewers that may have a conflict of interest that clouds their ability to judge the paper objectively.

Degrees of uncertainty: the Intergovernmental Panel on Climate Change

Meeting the BAIS standard is often difficult to implement in practice as it takes time to conduct evidence-based reviews. However, the Intergovernmental Panel on Climate Change (IPCC, 2019) has come up with an exemplary way of dealing with this problem through explicitly stating the levels of uncertainty in the science being used. For instance, Mastrandrea et al. (2011) in the IPCC fifth climate assessment (AR5) relies on two metrics of uncertainty: (1) confidence in the validity of a finding, based on the type, amount, quality, and consistency of evidence (e.g., mechanistic understanding, theory, data, models, expert judgment) and degree of agreement where confidence is expressed qualitatively; and (2) quantified measures of uncertainty in a finding that is expressed probabilistically using statistical analysis of observations or model results, and, to a lesser extent, expert judgment (Table 1.2).

Table 1.2 Evidence and agreement statements in relation to levels of confidence.

High agreement limited evidence	High agreement medium evidence	High agreement robust evidence
Medium agreement limited evidence	Medium agreement medium evidence	Medium agreement robust evidence
Low agreement limited evidence	Low agreement medium evidence	Low agreement robust evidence

Confidence increases towards top-right corner and evidence is most robust when there are multiple, independent lines of high-quality information.

Source: Adapted from Mastrandrea et al. (2011).

Type I versus Type II errors

Lastly, scientists and decision makers will need to balance Type I versus Type II errors in weighing the consequences of action/inaction (see Field et al., 2004). In hypothesis testing a Type I error occurs when the null hypothesis (no difference) is falsely rejected when in fact it is actually true in the real world. A Type II error is when one fails to reject the null hypothesis when in fact it is false in the real world (a difference exists). Type II errors are generally considered more serious in conservation as in the case of endangered species or our thought experiment about human impacts to the planet. That is—if we fail to reject the null hypothesis (there is no detectable large die-off of species caused by humans or climate change is not serious or human caused) when in fact it is false—a Type II error (we are indeed the cause) is most consequential (Field et al., 2004). In cases where the stakes are high, erring on the side of conservation by using the precautionary principle and application of the BAIS nondiscretionary standard are of utmost importance. Without getting too far into the weeds, scientists mindful of the trade-offs can set the significance level in hypothesis testing *a priori*, that is, the probability (alpha, α) of rejecting the null hypothesis when it is in fact true (for instance, α is commonly set at 0.05 in statistical analysis—general translation—there is but a 5% chance of reaching the wrong conclusion/decision or turning this around—you will be right 95 out of 100 times). Field et al. (2004) discuss ways to minimize trade-offs between Type I and II errors by adjusting the level of significance in conservation experiments *a priori* to better avoid consequential Type II errors and achieve the optimal decision-making space.

How we communicate matters

Ok, let's be honest here. Outside preaching to the science choir (our peers), most scientists suck at public speaking to "normal" people. No offense, but it is just that we are not trained to simplify complex phenomena in Nature, and we speak in a lawyer-like jargon the general public and decisions makers just do not get. Most scientists are not schooled in the art of the snappy sound bite. We are like deer in the headlights when confronted with news cameras and only seconds to inject a sound bite. We would fail miserably in giving an elevator speech to a busy Senator who lives in an attention deficit disorder world between stops. I once had a handshake and a New York minute with President Bill Clinton to convince him why conservationists wanted some 24 million hectares of relatively pristine roadless areas protected from chain saws, bulldozers, and the like.

My brief encounter went something like this—Mr. President, thank you for your leadership on our nation's public lands, we need you to finish the job for outdoor enthusiasts, sport fishers and hunters, and future generations of Americans by enacting this historical policy that will leave a living legacy to Americans (15 seconds and he was out of there with a nod and a handshake!).

During that brief encounter, I never uttered a word of evidence-based science supporting the protection of intact areas (although I had it in my back pocket!) and if I did, the presidential encounter would have never happened in the first place.

Simply put, we scientists are out of our comfort zone when forced to communicate the way the "what's in for me" crowd think in real time.

But not to worry, there are tools for talking "normal" to people so that the complex constructs of scientific analyses can be simplified without losing scientific meaning.

One tool is to turn everything you learned about communicating as a scientist upside down. Scientists are taught a very structured and disciplinary way of communicating (Fig. 1.4). We start with a problem, develop/test hypothesis(s), state our assumptions, collect data, run statistical analyses, check for errors, examine the results, and then make conclusions (thank you Sir Francis Bacon!). I honor all scientists that communicate this way while I challenge you to think completely different when it comes to the general public and media. Fig. 1.4 is an example of how to shift your communication style from preaching to the choir to a more real-world lexicon aimed at politicos.

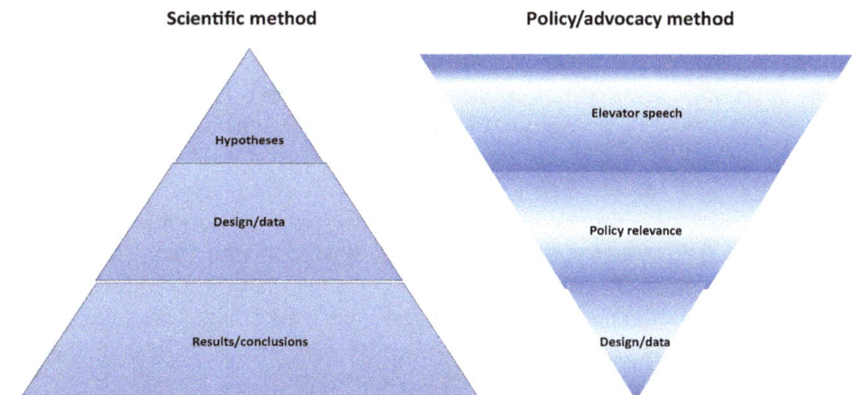

Figure 1.4 Communication styles used in science and policy.

The elevator speech

The fundamental difference between the science-policy communication versus pure science communication style (as in science conference presentations) is that you start with your elevator speech (think sound bites and "what's in it for me") by imagining that you have but two floors on a crowded elevator to convince a busy Senator why he/she should pay attention to your findings, or a news camera is stuck in your face and you have seconds of response time before the camera is turned off. This time with the communication tools herein, you approach the camera/Senator with confidence and conviction instead of bewilderment. Armed with your elevator speech, you enter the "what's in it for me" world with simplified but true to the science talking points (see Chapter 11, Essays From the Trenches of Science-Based Activism), otherwise, your long-winded proclivities about scientific methodology will fall on deaf ears, I guarantee it. Finally, have your experimental design, results, and hard evidence in your briefcase (or backpack) for the technically savvy legislative aide or contrarian congressperson so you can flip back over to the standard science communication style on the ready. That's it—it's simple in theory but hard to do in practice because we were trained on the left side of the diagram and not the right. The same holds true for congressional testimony. Over the years as an expert witness, I have reviewed some pretty arcane testimony seemingly written in Sanskrit of absolutely no interest to decision makers—it's best to cut to the chase and use the supplemental science tutorial if and when needed.

Don't be such a scientist

Olson (2009) has an excellent book that all students of science should have as their personal "Bible" called *Don't Be Such a Scientist: Talking Substance in an Age of Style*. There are five fundamental points he makes regarding how best to engage the public, media, and decision makers:

1. *Don't be so cerebral*—be ready to cut to the chase in simple, practical explanations (see below).
2. *Don't be so literal minded*—come up with clever, memorable ways to showcase results that go beyond the numbers and stats (e.g., use simple illustrations, cartoon graphics).
3. *Don't be such a poor storyteller*—learn how to tell memorable stories that convey important messages as the public will mostly remember the big picture (the story, also see Chapter 10, Out of the Ivory Tower: Campaign-Based Science Messaging for the Public) rather than the individual lines in the story (the data).
4. *Don't be so unlikeable*—this is an obvious one. No one likes an egotist: if you are one, you will most likely be criticized as having self-interest, even worse—a narcissist, or an elitist.
5. *Be the voice of science*—or in Carl Sagan's words "Who Speaks for the Earth?"

Readers may also be interested in the book by Hayes and Grossman (2006) called *A Scientist Guide to Talking With the Media*. Read it, cover to cover!

The Four Agreements

In preparing for public speaking and presentations to decision makers, I have found Ruiz's (1997) wonderfully inspiring book, *The Four Agreements*, as my personal guide. The agreements are rooted in the ancient spiritual teachings of the Toltec—Toltec itself means people of knowledge. The Toltec were one of the great Mesoamerican cultures of the 10th through mid-12th centuries that resided in Tula, Hidalgo, Mexico until they were sadly wiped out by invaders such as the Aztecs. We all have a little Toltec in us if we open our hearts to conservation and not just our minds to science. The Four Agreements teach us to live our lives with integrity as scientists, global citizens, parents, and cultural creatives by following these simple principles (I suggest you copy and paste these on your refrigerator door as affirmations!):

1. *Be impeccable with your word*—speak with integrity; say only what you mean; avoid using the *Word* to speak against yourself or to gossip about others; and use the power of your *Word* in the direction of truth and love (note—the capitalization of "*Word*" as our words can have a spiritual grounding if they are out of kindness and love or, conversely, can come from our "shadow" side if of ill intent).
2. *Don't take anything personally*—nothing others do is because of you; what others say and do is a projection of their own reality, their own dream; when you are immune to the options and actions of others, you won't be the victim of needless suffering.
3. *Don't make assumptions*—find the courage to ask questions and express what you really want; communicate with others as clearly as you can to avoid misunderstandings, sadness, and drama; and with just this one agreement, you can completely transform yourself (note—for scientists, this is modified as—*state your assumptions*!).
4. *Always do your best*—your best is going to change from moment to moment, it will be different when you are healthy as opposed to sick; and under any circumstance, simply do your best, and you will avoid self-judgment, self-abuse, and regret.

Because humor is part of what keeps me from taking things personally, and when all else fails, remember this quote from Bill Murray in the *Ghost Busters* classic movie:

"Back off Man, I'm a Scientist!" (https://www.youtube.com/watch?v=ASQvZr6Zz4o)

Here's an application example of the Four Agreements. During a congressional hearing in the House Natural Resources Committee on the ESA, I was one of several scientist whistleblowers that exposed widespread corruption and political interference in the ESA by officials in the George W. Bush Administration. I delivered a 5-minute testimony (backed by extensive written documentation) on how political appointees of the Bush administration were instructing scientists on the endangered species recovery team of the Northern Spotted Owl, on which I served, to ignore the owl's older forest habitat dependencies (see Chapter 5, Blowing the Whistle on Political Interference: The Northern Spotted Owl). Following my testimony, I was threatened by science contrarians on the Committee with contempt of Congress if anything in my testimony could be challenged by administrative officials—some of whom were deeply involved in covering up the science. While being exposed to intimidation tactics on the witness stand, I dwelled on the Toltec agreements 1, 2, and 4 to get me through the barrage of personal attacks so that my testimony and that of other scientists at the hearing could see the light-of-day in much needed scientific integrity reforms. Fortunately, when the White House changed hands from Bush to Obama, President Barack Obama issued a directive to restore scientific integrity to ESA decisions citing the testimony of scientists that spoke truth to power at the hearing that day (Leopold would be proud!).

Speaking Truth to Power: Who Speaks for the Earth?

Those provocative words ("Who Speaks for the Earth") of Carl Sagan in Cosmos: A Personal Voyage (see updated video—https://www.youtube.com/watch?v = 3QgIQzDqXmo) have inspired a generation of scientists to respond in-kind to the urgency of avoiding imminent ecological and societal collapse. And given the enormity of the problem and humanity's poor historical and contemporary track record of avoiding collapse (Diamond, 2005), it is understandable to feel helpless and overwhelmed. But again, I turn to Sagan—this time his most prescient words in the movie *Contact*—this clip from the movie is so inspiring:

> "You're an interesting species. An interesting mix. You're capable of such beautiful dreams, and such horrible nightmares. You feel so lost, so cut off, so alone, only you're not. See, in all our searching, the only thing we've found that makes the emptiness bearable, is each other." (https://www.youtube.com/watch?v = 5OTEygS02JY).

Indeed, we/you are not alone. Scientists all over the world have each other to respond in unison to the planetary crisis by taking the advocacy plunge at rallies in support of scientific integrity, joining protest movements in support of climate and ecojustice reforms, testifying in Congress with other scientists, reviewing policies, blowing the whistle on corruption, and joining science-policy committees of professional societies and conservation groups (see Chapter 7, Scientific Integrity and Advocacy: Keeping the Government Honest). We do this without personal gain even though it's clearly deserving given the risks involved in speaking out.

A Nobel Earth Prize

There are no Nobel prizes and few prestigious awards for scientists willing to speak for the Earth and future generations. Meritorious accomplishments are acknowledged not by the need to call immediate attention to the planetary crisis at hand but by the citation rate of a journal. I argue that scientists that have taken up a just cause ought to have recognition on par with prestigious academic and humanitarian awards. That is, it's time for a *Nobel Earth Prize.*

Impact to society—especially proposing effective solutions that work for Nature and society—needs to become as fundamental to performance evaluations as a journal's impact factor so as to encourage more scientists to venture out of the Ivory Tower and into the real-world of policy. Using the methods initially put in place by Sir Francis Bacon, a revolution of scientists willing to speak up would make Aldo Leopold proud! The toolkit of scientists now—more than ever—needs to include effective communications and engagement in social media and outreach to decision makers that go beyond microscopes, telescopes, and chemistry flasks.

> This may seem like a contradiction but it is reality: Science alone will not solve the planetary crisis, but we cannot solve it without more scientists willing to speak out about how best to do so.

In closing, here is my elevator speech to the readers of this book. There has never been a more urgent time to speak for the planet and future generations. Retreating into a false comfort zone of silent complicity is part of the problem, not the solution. This is your moment to shine as a contemporary Toltec scientific thinker by helping humanity transform from our record of horrible nightmares to a future of beautiful dreams. The planet awaits your innovative, creative, compassionate, and science-based problem solving.

> So, what will you do to Speak for the Earth?

References

Barnosky, A.D., et al., 2012. Approaching a state shift in Earth's biosphere. Nature 486, 52–58. Available from: https://doi.org/10.1038/nature11018.

Bradley, C.M., Hanson, C.T., DellaSala, D.A., 2016. Does increased forest protection correspond to higher fire severity in frequent-fire forests of the western United States? Ecosphere 7, 1–13.

Calaprice, A., 2000. The Expanded Quotable Einstein. Princeton University Press, Princeton, NJ, and Oxford.

Carlson, R., 1962. Silent Spring. Houghton Mifflin, Boston, MA.

Colborn, T., Dumanoski, D., Myers, J.P., 1996. Our Stolen Future: Are We Threatening Our Fertility, Intelligence, and Survival? Dutton, New York.

Committee of Scientists, 1999. Sustaining the People's Land. Recommendations for Stewardship of the National Forests and Grasslands Into the Next Century. USDA, Washington, DC.

DellaSala, D.A., 2020. Has anthropocentrism replaced ecocentrism in conservation? In: Kopina, H., Washington, H. (Eds.), Conservation: Integrating Social and Ecological Justice. Springer Nature, Switzerland AG, pp. 91–104. Available from: https://doi.org/10.1007/978-3-030-13905-6.

Diamond, J., 2005. Collapse: How Societies Choose to Fail or Succeed. Viking Press, New York, NY.

Esch, B.E., Waltz, A.E.M., Wasserman, T.M., Kalies, E.L., 2018. Using best available science information: determining best and available. J. For. 116, 473–480.

Field, S.A., Tyre, A.J., Jonzen, N., Rhodes, J.R., Possingham, H.P., 2004. Minimizing the cost of environmental management decisions by optimizing statistical thresholds. Ecol. Lett. 7, 669–675.

Hansen, J., 2009. Storms of my grandchildren. The Truth About the Coming Climate Catastrophe and Our Last Chance to Save Humanity. Bloomsbury US, New York, NY.

Hayes, A.W., 2005. The precautionary principle. Arh. Hig. Rada Toksikol. 56, 161–166.

Hayes, R., Grossman, D., 2006. A Scientists Guide to Talking With the Media. Rutgers University Press, New Brunswick, NJ, London.

Ibisch, P.L., et al., 2016. A global map of roadless areas and their conservation status. Science 354, 1423–1427.

Intergovernmental Panel on Biodiversity and Ecosystem Services (IPBES), 2019. Summary for policymakers of the global assessment report on biodiversity and ecosystem services of the Intergovernmental Science-Policy Platform on Biodiversity and Ecosystem Services. <https://ipbes.net/news/Media-Release-Global-Assessment>.

Intergovernmental Panel on Climate Change (IPCC), 2019. Refinement to the 2006 IPCC Guidelines for national greenhouse gas inventories. <https://www.ipcc.ch/report/2019-refinement-to-the-2006-ipcc-guidelines-for-national-greenhouse-gas-inventories/>.

Kriebel, D., Tickner, J., Epstein, P., Lemons, J., Levins, R., Loechler, E.L., et al., 2001. The precautionary principle in environmental science. Environ. Health Perspect. 9, 871–876.

Lakoff, G., 2004. Don't Think of an Elephant: Know Your Values and Frame the Debate. Chelsea Green Publishing Company, White River Junction, VT.

Mackey, B., DellaSala, D.A., et al., 2015. Policy options for the world's primary forests in multilateral environmental agreements. Conserv. Lett. 8, 139–147. Available from: https://doi.org/10.1111/conl.12120.

Mastrandrea, M.D., Mach, K.J., Plattner, G.-K., Edenhofer, O., Stocker, T.F., Field, C.B., et al., 2011. The IPCC AR5 guidance note on consistent treatment of uncertainties: a common approach across the working groups. Clim. Change 108, 67. Available from: https://doi.org/10.1007/s10584-011-0178-6.

Mealey, S.P., Thomas, J.W., Salwasser, H.J., Stewart, R.E., Balint, P.J., Adams, P.W., 2005. Precaution in the American Endangered Species Act as a Precursor to Environmental Decline: The Case of the Northwest Forest Plan. Taylor & Francis.

Millennium Ecosystem Assessment, 2005. Ecosystems and Human Well-Being: Synthesis. Island Press, Washington, DC.

Okagaki, L.H., Dean, R.A., 2016. The influence of funding sources on the scientific method. Mol. Plant. Pathol. 17, 652–653.

Olson, R., 2009. Don't Be Such a Scientist: Talking Substance in an Age of Style. Island Press, Washington, DC.

Resnik, D.B., Konecny, B., Kissling, G.E., 2017. Conflict of interest and funding disclosure policies of environmental, occupational, and public health journals. J. Occup. Environ. Med. Jan. 59 (1), 28–33.

Rio Declaration, 1992. Report of the United Nations conferences on environment and development, Rio de Janeiro, 3–14 June 1992, Annex I, Rio Declaration on Environment and Development. <http://www.un.org/documents/ga/conf151/aconf15126-1annex1.htm>.

Ripple, W.J.C., Wolf, T.M., Newscome, P., Barnard, W.R., Moomaw, and 11,258 scientists from 153 countries, 2019. World scientists' warning of climate emergency. Bioscience 70, 8–12.

Ruiz, D.M., 1997. The Four Agreements: A Practical Guide to Personal Freedom. Amber-Allen Publishers, San Rafael, CA.

Ryan, C.M., Cerveny, L.K., Robinson, T.L., Blahna, D.J., 2018. Implementing the 2012 Forest Planning Rule: best available scientific information in forestry planning assessments. For. Sci. 64, 159–169.

Sagan, C., Turco, R.P., 1993. Nuclear winter in the post-cold war era. J. Peace Res. 30, 369–373.

Schneider, S.H., 2009. Science as a Contact Sport. National Geographic, Washington, DC.

Wilson, E.O., 2016. Half Earth. Liveright & W.W. Norton, New York, NY.

When scientists are attacked: strategies for dissident scientists and whistleblowers

Derek E. Lee[1], Monica L. Bond[2] and Chad Hanson[3]

[1]Department of Biology, Pennsylvania State University, University Park, Pennsylvania, United States [2]Department of Evolutionary Biology and Environmental Studies, University of Zurich, Zurich, Switzerland [3]John Muir Project of the Earth Island Institute, Berkeley, CA, United States

Contrarians and dissidents

It may surprise you to learn that Albert Einstein and Carl Sagan were subjected to years of vehement criticism and smear campaigns against them in response to the theories of relativity and nuclear winter (Oreskes and Conway, 2010; Wazeck, 2013). Science is often portrayed as the work of objective, rational people who dispassionately examine evidence to reach conclusions (also see, Chapter 1, The Nuts and Bolts of Science-Based Advocacy). In fact, science is a practical method to gain knowledge—one that acknowledges all scientists are humans with inescapable subjective biases, agendas, social obligations, political beliefs, and prejudices, but the scientific method works to minimize subjectivity and helps us discover objective truth by using empirical data to dismiss false claims (Grinnell, 2009). Like democracy, science requires free speech and free press to function properly so the marketplace of ideas can consider any new, even shocking idea, and then empirically judge its veracity (Bambauer, 2017). In practice, the marketplace of ideas is severely skewed in favor of entrenched power structures and status quo ideologies (Ingber, 1984), but dissidents can change the world.

Einstein's theory of special relativity caused an uproar, and the scientific community mobilized to publish a collection of essays titled "One Hundred Authors Against Einstein." Einstein supposedly retorted: "If I were wrong, one would have been enough," because if his theory were wrong, any one person could collect and publish data that would disprove all or part of the theory. That so many people weighed in against Einstein's new theory shows the power of

Conservation Science and Advocacy for a Planet in Peril.
DOI: https://doi.org/10.1016/B978-0-12-812988-3.00009-0

public relations in the marketplace of ideas. In Einstein's case his theory prevailed against the public relations campaign that was waged against it. In the case of climate change, where near-total scientific consensus has existed for decades, a public relations campaign by the fossil fuels industry successfully misinformed a huge portion of the public and stopped policies to reduce greenhouse gases at the time such efforts would have been most effective. We are providing information in this chapter to help readers understand the institutional, political, and psychological barriers to honest open scientific discussion, what to expect when a scientist becomes a contrarian or dissident, and how to successfully advance and promote your dissident ideas to facilitate conservation in the coming decades.

The genesis and evolution of scientific understanding on a topic often follow a standard life cycle (Fleck, 1935; Kuhn, 1970). Early observations and their interpretations on a topic form a body of evidence and theory that become the foundation of a worldview or dominant paradigm regarding the topic. The dominant paradigm shapes the discourse, research, and policy related to the topic, and is guarded by gatekeepers in the thought collective of people from academia, industry, and government that have a vested interest in the paradigm. Inevitably, anomalies accumulate showing failures or omissions in the paradigm, initiating one or more of three possible responses: the existing paradigm can be corrected; a revolution can occur that replaces the old paradigm with a new one (Kuhn, 1970); or the gatekeepers might ignore or suppress the anomalies (Martin, 1999a).

There is a long history of intellectual gatekeeping of ideas in science going back at least to Galileo. Whenever a power structure is threatened by novel ideas, we can expect the powerful to attempt to suppress those ideas. The modern synergies among government, industry, academia, and agency scientists comprise the current power structure in science. The dominant players are governments and large corporations that provide most of the funding for science, the community of professional scientists themselves, and the scientific elites who control funding decisions. Martin (1998) suggests that it is useful to think of the scientific community, and any associated thought collective defending a dominant paradigm, in terms of interests such as money, power, status, privilege, or other advantages (Barnes, 1977). Because interests are powerful shapers of people's world views and self-identities, interests exert strong pressure on the direction of research and shape the responses by gatekeepers to those who challenge a dominant paradigm.

Scientists who challenge a dominant paradigm generally fall into two categories that differ by degree: contrarians or dissidents (Delborne, 2008).

Contrarian scientists practice agonistic engagement, which are conventional behaviors within the scientific community (e.g., publishing data or critical commentaries on others' papers). Contrarian disagreements are common when there is a lack of consensus within the relevant scientific community. Contrarian science goes against dominant scientific paradigms by challenging accepted theories, introducing new methods or theories, or exposing inconsistencies in assumptions. Contrarians are potentially disruptive, but their work is usually conducted and deployed with at least some hope of convincing a mainstream scientific community of a new fact or approach (Delborne, 2008). Some contrarian scientists known as whistleblowers discover a fact that threatens a powerful industry such as genetically modified foods or pesticides (Ewen and Pusztai, 1999; Hayes et al., 2002), and efforts by the industry to discredit and suppress the whistleblower's findings may turn whistleblowers into dissidents.

Dissident scientists are characterized by scientific dissent, which refers to views that run contrary to widely accepted scientific theories, methods, or assumptions. Dissents challenge the knowledge claims of a dominant paradigm and call for some degree of reform in the relationships among science, politics, industry, and the public. Dissident behavior often provokes sanction by the scientific community, but credible dissident science combined with effective activism represents a powerful strategy to influence scientists, the public, government, and industry (Frickel, 2004), and to advance knowledge against countervailing political and economic forces.

Dissent and controversy have long been recognized to play crucial roles in the production of scientific advances (Kuhn, 1970). Dissenting views can correct false assumptions and ensure consideration of a wider range of theories, models, and explanations (Popper, 1963; Feyerabend, 1975; Longino, 2002; Kitcher, 2011). Dissident science explicitly acknowledges the politics within and around scientific controversy and advocates for new relationships among scientists, the public, interest groups, and academic institutions to reform the mechanisms involved in knowledge production. Dissident science represents practices that merge intellectual struggle with social and political action, incorporating a variety of strategies (Delborne, 2008). However, because contrarian and dissident scientists challenge the dominant paradigm, both can expect gatekeepers to practice impedance, suppression, co-opting, or defamation.

Scientists have an ethical obligation to stand up for their data and the greater truth embodied in empirical evidence. If one's science uncovers evidence of harm to human or nonhuman life due to actions, nonactions, products, or byproducts of an industry, the discovery must be disseminated to the perpetrators, the public, policymakers, and regulatory authorities so that the harm can be stopped or remedied (i.e., the Hippocratic Planetary Oath of DellaSala, Chapter 1, The Nuts and Bolts of Science-Based Advocacy). Whistleblowing or activist scientists are required to broadcast their claims such that the harm can be identified and ameliorated.

Brian Martin is a theoretical physicist turned historian of science who has spent decades studying scientific suppression and whistleblowing (Martin, 1981, 1999a,b; Campanario and Martin, 2004). He advises dissenters to understand the systems of power in which they operate and to carefully consider their tactical options (also see Chapter 7, Scientific Integrity and Advocacy: Keeping the Government Honest). Dissenting scientists can expect gatekeeping behavior that ignores or suppresses their work from journal editors, peer reviewers, conference organizers, professional associations, and academic and government bureaucracies.

Martin also delineated strategies for dissident scientists including publish and publicize your dissenting data and interpretations wherever possible, expose attempts at suppression, and build a social movement (Martin, 1998). Delborne (2008) suggested the academic-industrial complex hinders the production and dissemination of dissenting science through traditional outlets such as scientific journals, so creating space for dissenting science is also an issue of intellectual and academic freedom. Transparency and public participation are also required in order to challenge or overcome entrenched paradigms within our institutions.

Contradictory or dissenting science will generally be rejected and ignored at first, but an effective public relations campaign and proactive popularization of the new idea can build a constituency that helps hasten, if not the adoption of the new idea, at least its consideration. Groups of scientists have constructed new political voices by organizing public interest science organizations (e.g., Union of Concerned Scientists: see Chapter 7, Scientific Integrity and Advocacy: Keeping the Government Honest) outside of traditional professional and academic societies (Moore, 1996; Frickel, 2004). Scientists, like other citizens, have political preferences and values that guide their actions, and bringing like-minded scientists together increases their political influence and reduces the risk of marginalization.

When gatekeeping scientists attack a challenge to the status quo made by contrarian or dissenting science, targets of criticism may include methodology, interpretation, application of theories or models, the credibility of the contrarian scientists, the appropriateness of the research question, the forum of the publication, or the policy or management implications of the findings. Gatekeepers acting as peer reviewers, journal editors, conference organizers, professional organization executives, or bureaucrats will employ diverse means to exclude the contrarian claim or dissident scientist from the zone of scientific legitimacy (see Chapter 1, The Nuts and Bolts of Science-Based Advocacy, regarding best available science biases). If the dissenting idea shows any promise of gaining popularity or threatens an industry, the gatekeepers will attempt to smear and discredit the proponents of that idea in ad hominem attacks. Contrarian, whistleblowing, or dissenting scientists can expect to have their life, career, and reputation assaulted by the gatekeepers of a paradigm. Marginalized scientists may, however, be able to expose unjust or repressive tactics used by more powerful forces.

The thought collective that defends a scientific paradigm can be purely intellectual such as geologists who refused to accept plate tectonics (Oreskes, 2001) or paleontologists who refused to believe dinosaurs could be anything other than cold-blooded (Desmond, 1975). Intellectual collectives such as scientific societies can be very dangerous to open-minded advocates of new ideas. The "old boys club" of senior famous scientists whose identities and prestige are enmeshed with paradigmatic ideas can use scientific professional societies and their influence with employers and journals to crush careers and block publications while protecting and promoting ossified or corrupt concepts.

Thought collectives can also be linked to powerful industries such as tobacco, forestry, agriculture, petrochemicals, or weapons. In these cases, the dominant scientific paradigm may be either supported or opposed by the industry. In the cases of the tobacco and fossil fuels industries, the scientific community was largely united around the ideas that smoking causes cancer and burning fossil fuels causes climate change. In these cases, the industries funded a small group of scientific skeptics or useful stooges who inflated the uncertainty about these ideas, manufactured the illusion of doubt, and provided cover for the industries' damage to human health and welfare (Oreskes and Conway, 2010).

Challenging the dominant forestry paradigm: a case study

In the case of forestry, the dominant paradigm is defended by a powerful integrated complex composed of the timber industry, governments that subsidize and promote the industry while simultaneously being tasked with regulating it and protecting human and environmental health, universities that receive funding from the industry and government, and academic and government agency scientists employed or funded by the industry or by land management agencies that are financially involved in timber commodity production. Other industries that use or profit from timber (i.e., construction and transportation) are also indirectly involved in defending the dominant forestry paradigm.

For more than 100 years, the dominant paradigm of forestry has been built on the assumption that cutting and removal of trees is necessary. The reasons proclaimed by forestry proponents have changed regularly (e.g., remove decadence, clear land for agriculture, forest health, slow succession, speed succession, fire risk reduction, save endangered species, restore the forest, increase grazing, improve fish habitat), but the action underlying every rationale has always been to cut the trees and sell them. An enormous body of science is produced every year in forestry detailing how to maximally grow and cut profitable trees while maintaining the minimum populations of endangered species or noncommercial tree species required by law. The US Forest Service has a long history of suppressing the science that contradicts its preconceived management actions (Schiff, 1962; Huggard and Gómez, 2001), and even today few papers are published questioning the fundamental assumptions of forestry or forest management.

Our work to protect the native plants and animals that require severely burned forests in the western USA brought us into conflict with the gatekeepers of the forestry paradigm. Our research showing that spotted owls (*Strix occidentalis*: Fig. 2.1A and B) can benefit from forest fires (Bond et al., 2009; Lee et al., 2012; Lee and Bond, 2015; Bond et al., 2016; Lee, 2018) demolished the primary reason given by the US Forest Service to justify lucrative postfire logging and logging in the last remaining stands of old-growth forest—fire risk reduction to save spotted owls. We also publicly criticized the faulty work of academic scientists who considered themselves the only legitimate experts on the topic, and who are funded by the US Forest Service. The gatekeepers responded by vilifying us within

(A)

(B)

Figure 2.1
(A) California spotted owl inhabiting forests burned by the 2002 McNally Fire, Sequoia National Forest. (B) Monica Bond measures trees in a California spotted owl nest stand that remained active despite having been burned by the McNally Fire. *(A) Photo by Brett Hartl; (B) Photo by Derek Lee.*

the forest science and management communities, smearing our reputations and credibility publicly, attempting to damage our careers and livelihoods, and publishing criticisms of our work while blocking our rebuttals. Our work has saved many trees from the forester's chainsaws and the field of forest ecology is forever changed by our new ideas, but the fight has been asymmetrical and personal instead of a purely scientific disagreement.

One of the tactics used against us by the gatekeepers of the forestry paradigm is to claim that we were "agenda-driven." This term is an attempt to vilify us for something every human does, because all human beings have agendas and motivations of some sort that drive their actions. Our agenda is to ensure the best available science is applied toward land management decisions that influence spotted owl habitat. The agenda of our opponents, who are funded by the US Forest Service, is to support logging spotted owl habitat under the pretense that it might reduce fire severity. In our advocacy, we had urged the government to utilize new science that challenged the dominant paradigm. Our opponents responded by personally attacking us and attempting to discredit our science by questioning our motives. Calling us "agenda-driven" and claiming our advocacy for our data was unprofessional or "outside scientific norms" are classic gatekeeper tactics. In reality, who better than the scientists themselves to present their scientific findings to management agencies in a comment letter or expert declaration? It is a scientist's obligation and duty to stand up for their data. Moreover, providing input on land management projects is a critical public service offered by scientific experts to the government, free-of-charge. Anyone who claims that a scientist who advises the legal process of managing public lands is acting "against norms," is trying to silence the dissident and is out of touch with the hundreds of scientists that routinely do this as exemplified by the chapters in this book (e.g., see Chapter 7, Scientific Integrity and Advocacy: Keeping the Government Honest and Chapter 8, Why Advocate and How?). This is a chilling phenomenon that has no place in an open marketplace of ideas that is not distorted by corruption or unacknowledged biases.

Before you dissent

You might become a dissident scientist accidentally by simply following your own curiosity and moral compass. We believed that we were merely contrarians engaged in a healthy scientific debate about spotted owls and forest fire until our adversaries published an ad hominem smear article defaming us, and their widespread vilification of us and our work within the industry and US Forest Service became known to us. If you have the luxury of knowing you will become a dissident for your ideas, take time to learn about what will likely transpire, and make preparations.

Carefully examine the situation and devise a plan to: (1) publish and publicize your dissenting data and interpretations wherever possible, (2) expose attempts at suppression, and (3) build a social movement to promote your idea. Study the methods of effective social actions and public relations campaigns. Contact experienced activists and public relations experts and obtain their advice or services. Find like-minded scientists and activists and build a community for mutual support.

Ask yourself: who is your opposition and what is their network of support? Who are your current supporters? Who are potential supporters? What actions can be taken to win greater support or to undercut opposition? What will be the financial costs? How much time will it take? What are the various options? What happens when roadblocks and significant adversity are encountered? What are the options down the track? Does this action contribute to long-term goals? (Martin, 1997).

The pamphlet Courage Without Martyrdom: A Survival Guide for Whistleblowers, published by the Government Accountability Project and the Project on Government Procurement, offers the following advice to would-be whistleblowers and dissidents (also see Chapter 7, Scientific Integrity and Advocacy: Keeping the Government Honest).

- Before taking action, see if there is some way to achieve your goal by working within the system.
- Try to find out if there are other people, especially coworkers, who share your concerns.
- Before taking any action that may lead to an attack on you, consult with family and close friends. You need their support.
- Keep a detailed record of events. When something important happens, write up a statement including witnesses, if possible.
- Make copies of as many of the important documents that you can. Your case may depend on them.
- Find allies among honest supporters including politicians, journalists, and community organizations.
- Develop a plan for taking the initiative; do not just respond to actions by the other side.
- Obtain advice about taking legal action.
- Do not overstate your case.

From interviews with whistleblowers (Martin, 1997):

- Do not trust the system. "The system" here refers to the organizational hierarchy of the workplace and the external agencies for pursuing complaints.
- Be prepared for any conceivable attacks. Many whistleblowers learned a bitter lesson: that when they spoke out, there was almost no limit to what might be done to shut them up.

- Many whistleblowers expected to be listened to openly and treated fairly. Instead, they found they had few allies and were attacked in unexpected ways. This was summed up as: "don't be naive."
- "Document everything." Many whistleblowers wished that they had collected more records and held off speaking out until their documentation was greater.

Other useful hints mentioned by one or more whistleblowers:

- Always be tactful and polite.
- Do not despair in adversity.
- Never give up.
- Timing is crucial.
- Do not blame yourself.
- Publicity is valuable.
- Do what you believe is right in the pursuit of truth, even when the path is difficult and strewn with challenges.

Getting your ideas out

Publish and publicize your ideas whenever and wherever possible. Many of the contests in the marketplace of ideas were won by exposure and repetition. Publishing your idea in a scientific journal or book is not the end of the battle, it is the beginning, but even that can be a challenge. Dissenting scientists should expect more rejections than average and should prepare for an extended cycle of resubmissions. Scientific publishing is a well-known minefield for outsiders and the process is often used to censor challenging ideas (Horrobin, 1990; Alvesson and Sandberg, 2014; Siler et al., 2015).

Perseverance and careful adherence to formatting and other author guidelines for each journal should eventually result in a publication. Rejections can also be refuted via a response to the editor point by point, and it is worth challenging rejections that are weak or unsubstantiated. Contrarian and dissident scientists may have to seek publication outside the high-impact journals, but the open science movement and digital publishing make it easier than ever to publish novel ideas (Bartling and Friesike, 2014; McKiernan et al., 2016; Tennant et al., 2017). Postpublication peer review sites such as PubPeerPublons can also provide an outlet for dissident science when it directly challenges a published article that supports a dominant paradigm. As Justice Hugo Black observed in Associated Press v. United States, the First Amendment "rests on the assumption that the widest possible dissemination of information from diverse and antagonistic sources is essential to the welfare of the public, that a free press is a condition of a free society."

A public relations campaign must accompany any new idea if you hope to gain any traction in the marketplace of ideas. Prepare a number of press releases (https://www.wikihow.com/Write-a-Press-Release) and diverse communications for social media, blogs, op-ed pieces, etc. Explainer pieces, attention-getting imagery, and short videos should all be included in your public relations arsenal. Cultivate different audiences and constituencies to cast a wide net for proponents to join to your idea. If you are not already a part of a larger community devoted to bringing about the changes your dissenting ideas lead to, join an appropriate group that is sympathetic to your cause. Storytelling and narrative offer clues on how to improve science communication (Dahlstrom, 2014, also see Chapter 10, Out of the Ivory Tower: Campaign-Based Science Messaging for the Public). Stories about science breakthroughs that resonate broadly with the public do so because they stir our imagination and elicit emotion.

In addition to communicating your science, always include a call to action so that receivers can be mobilized to make the paradigmatic social or institutional changes your idea initiates. Create many different calls to action that would appeal to different groups or people with diverse motivations. Weigold's (2001) review of science communication and Montgomery's (2017) guide to communicating science will help orient newcomers to this field. Understanding personality-based marketing will help you craft more effective messaging that appeals to specific personality types (Hajnik, 2014; Moss, 2017).

After the first blow lands

If your idea gains some traction within the marketplace of ideas, expect gatekeepers of the status quo or defenders of any industry threatened by your idea to attempt to smear or discredit you. As Martin (1997) stated, "When dissenters first come under attack, often they have a strong impulse to seek redress through 'proper channels.'" This includes appeal procedures, grievance procedures, writing letters to top management, seeking support from professional bodies, ombudspersons, official tribunals, or the courts. Most people believe that the formal structures in organizations and society can provide justice. Many dissidents and whistleblowers speak out precisely because they believe that if they speak the truth, people will listen and take action. They are shocked when the response is to attack them instead. The belief that someone somewhere is looking out for injustices and can correct them is a dangerous illusion. Sometimes official channels do work. Sometimes it is wise to use them as part of a wider campaign, but it is important to realize their disadvantages, and not to expect any solutions from them. With a campaign, formal channels may not be necessary. Politicians and top administrators can always intervene if the urgency is great enough. A noisy campaign is more likely to trigger the involvement of powerful actors than a case following standard bureaucratic protocol.

Understand that powerful actors opposing your new idea will not hesitate to try and defame you to protect their vested interests. Scientific publishers, whether private or part of a professional association, have only disincentives to publishing comments critical of any paper they already published and considered to be accurate and "true." Thus they too can act as gatekeepers to ideas that challenge the dominant paradigm, simply out of a desire to maintain the appearance that their previous work was of the highest quality and error-free.

Continuing the fight

Some gatekeepers will be protecting their financial interests when they attempt to suppress a new idea. One effective strategy for dealing with powerful scientists tied to powerful industries and/or agencies is public exposure of these links. Such exposure has occurred in controversies over issues like nuclear power, tobacco and cancer, food additives, and pesticides. When the public was made aware of conflicts of interests in the roles of scientists, the ability of scientific experts to legitimize policies and practices of government and industry was greatly reduced (Martin, 1981).

Other gatekeepers will be fighting to preserve their self-identity which has become entangled with the paradigm they are defending. The way our brains are wired into our self-identity makes it exceptionally difficult for dissonant facts to enter our consciousness (Lord et al., 1979; Tesser, 2000; Kahneman, 2011). Presenting facts that dispute a person's innate beliefs may even cause them to adhere more strongly to their misinformed opinion (Nyhan and Reifler, 2010; Hart and Nisbet, 2012). The arguments during the COVID-19 pandemic about whether masks are essential to reduce coronavirus transmission and spread versus it is a hoax and an infringement on personal rights speaks to the depth and controversy of cognitive dissonance. US government climate change denial is also exemplary of the corruption that can taint decision-makers when they are financially supported by fossil fuel companies.

Lewandowsky et al. (2012) offer seven principles for overcoming cognitive biases:

1. Create a narrative that your idea completes.
2. Counter the misinformation from your opposition.
3. Emphasize the facts you wish to communicate rather than the myths from the other side.
4. Provide an explicit warning before mentioning a myth, to ensure that people are cognitively on guard and less likely to be influenced by the misinformation.
5. Ensure that your material is simple and brief. Use clear language and graphs where appropriate. If the myth is simpler and more compelling than your debunking, it may be cognitively more attractive, and you will risk an overkill backfire effect.
6. Consider whether your content may be threatening to the worldview and values of your audience. If so, you risk a worldview backfire effect, which is

strongest among those with firmly held beliefs. The most receptive people will be those who are not strongly fixed in their views.

7. If you must present evidence that is threatening to the audience's worldview, present your content in a worldview-affirming manner (e.g., by focusing on opportunities and potential benefits rather than risks and threats) and/or by encouraging self-affirmation.

Speaking Truth to Power

There are positive signs of progress in making science more transparent and open, which in turn could help dissolve the power that the gatekeepers and thought collectives wield to suppress contrarian and dissenting scientists that have legitimate and provocative research views to contribute. Many universities and funding entities are calling upon researchers to adhere to open science policies, which include making raw data publicly available. Some science journals are reforming the peer-review process by using either the double-blind method where peer reviewers do not know the authors of a submitted manuscript or a completely public and transparent process where the authors and the peer reviewers are identified and all versions, reviews, and responses are published to provide the historical evolution of the ideas therein (e.g., PeerJ https://peerj.com/). There are also postpublication peer review websites where the ongoing discussions about published studies can be seen, even when the publishers do not provide this service (e.g., Pubpeer https://pubpeer.com/). In a digital publishing age, there can be no excuse for limiting discourse by subjecting comments and replies to a gatekeeping peer-review procedure. These are steps in the right direction. Conservation science in general would greatly benefit from more openness, transparency, and fairness in the scientific publishing process and help to change false and entrenched paradigms and embrace scientific advancement. All science and scientists benefit when dominant paradigms are upheld or taken apart—that is the very nature of science, the search for truth beyond reasonable doubt.

References

Alvesson, M., Sandberg, J., 2014. Habitat and habitus: boxed-in versus box-breaking research. Organ. Stud. 35, 967–987.

Bambauer, J.R., 2017. The empirical first amendment. Ohio State Law J. 78, 947.

Barnes, B., 1977. Interests and the Growth of Knowledge. Routledge and Kegan Paul, London.

Bartling, S., Friesike, S., 2014. Opening Science: The Evolving Guide on How the Internet is Changing Research, Collaboration and Scholarly Publishing. Springer Open.

Bond, M.L., Lee, D.E., Siegel, R.B., Ward, J.P., 2009. Habitat use and selection by California spotted owls in a post-fire landscape. J. Wildl. Manag. 73, 1116–1124.

Bond, M.L., Bradley, C., Lee, D.E., 2016. Foraging habitat selection by California spotted owls after fire. J. Wildl. Manag. 80, 1290–1300.

Campanario, J.M., Martin, B., 2004. Challenging dominant physics paradigms. J. Sci. Explor. 18, 421–438.

Dahlstrom, M.F., 2014. Using narratives and storytelling to communicate science with nonexpert audiences. Proc. Natl Acad. Sci. U. S.A. 111, 13614–13620.

Delborne, J.A., 2008. Transgenes and transgressions: scientific dissent as heterogeneous practice. Soc. Stud. Sci. 38, 509–541.

Desmond, A.J., 1975. The Hot-Blooded Dinosaurs: A Revolution in Palaeontology. Blond & Briggs, London.

Ewen, S.W.B., Pusztai, A., 1999. Effects of diets containing genetically modified potatoes expressing Galanthus Nivalis Lectin on rat small intestine. Lancet 354, 1353–1354.

Feyerabend, P., 1975. Against Method: Outline of an Anarchistic Theory of Knowledge. Humanities Press, Atlantic Highlands, NJ.

Fleck, L., 1979. Genesis and Development of a Scientific Fact. University Chicago Press, Chicago, IL.

Frickel, S., 2004. Just science? Organizing scientist activism in the US environmental justice movement. Sci. Cult. 13, 449–469.

Grinnell, F., 2009. Everyday Practice of Science: Where Intuition and Passion Meet Objectivity and Logic. Oxford University Press, Oxford.

Hajnik, Z., 2014. Big Five Personality Traits in Marketing: A Literature Review. University of Vienna.

Hart, P.S., Nisbet, E.C., 2012. Boomerang effects in science communication: how motivated reasoning and identity cues amplify opinion polarization about climate mitigation policies. Commun. Res. 39, 701–723.

Hayes, T., Haston, K., Tsui, M., Hoang, A., Haeffele, C., Vonk, A., 2002. Herbicides: feminization of male frogs in the wild. Nature 419, 895–896.

Horrobin, D.F., 1990. The philosophical basis of peer review and the suppression of innovation. J. Am. Med. Assoc. 263, 1438.

Huggard, C.J., Gómez, A.R., 2001. Forests Under Fire: A Century of Ecosystem Mismanagement in the Southwest. University of Arizona Press, Tucson.

Ingber, S., 1984. The marketplace of ideas: a legitimizing myth. Duke Law J. 1984 (1), 1–91. Available from: https://doi.org/10.2307/1372344.

Kahneman, D., 2011. Thinking, Fast and Slow. Macmillan, New York.

Kitcher, P., 2011. Science in a Democratic Society. Prometheus Books, Amherst, NY.

Kuhn, T.S., 1970. The Structure of Scientific Revolutions, 2nd ed. University Chicago Press, Chicago, IL.

Lee, D.E., 2018. Spotted owls and forest fire: a systematic review and meta-analysis of the evidence. Ecosphere 9, e02354. Available from: https://doi.org/10.1002/ecs2.2354.

Lee, D.E., Bond, M.L., 2015. Occupancy of California spotted owl sites following a large fire in the Sierra Nevada, California. The Condor 117, 228–236.

Lee, D.E., Bond, M.L., Siegel, R.B., 2012. Dynamics of California spotted owl breeding-season site occupancy in burned forests. The Condor 114, 792–802.

Lewandowsky, S., Ecker, U.K., Seifert, C.M., Schwarz, N., Cook, J., 2012. Misinformation and its correction: continued influence and successful debiasing. Psychol. Sci. Public Interest 13, 106–131.

Longino, H.E., 2002. The Fate of Knowledge. Princeton University Press, Princeton, NJ.

Lord, C.G., Ross, L., Lepper, M.R., 1979. Biased assimilation and attitude polarization: the effects of prior theories on subsequently considered evidence. J. Pers. Soc. Psychol. 37, 2098–2109.

Martin, B., 1981. The scientific straightjacket: the power structure of science and the suppression of environmental scholarship. Ecologist 11, 33–43.

Martin, B., 1997. Suppression Stories. Fund for Intellectual Dissent, Wollongong.

Martin, B., 1998. Strategies for dissenting scientists. J. Sci. Explor. 12, 605–616.

Martin, B., 1999a. Suppression of dissent in science. Res. Soc. Probl. Public Policy 1, 105–135.

Martin, B., 1999b. The Whistleblower's Handbook: How to Be an Effective Resister. Envirobook, Sydney, Australia.

McKiernan, E.C., Bourne, P.E., Brown, C.T., Buck, S., Kenall, A., Lin, J., et al., 2016. Point of view: how open science helps researchers succeed. Elife 5, e16800.

Montgomery, S.L., 2017. The Chicago Guide to Communicating Science. University of Chicago Press.

Moore, K., 1996. Organizing integrity: American science and the creation of public interest organizations, 1955–1975. Am. J. Sociol. 101 (6), 1592–1627.

Moss, G., 2017. Personality, Design and Marketing: Matching Design to Customer Personal Preferences. Routledge, New York.

Nyhan, B., Reifler, J., 2010. When corrections fail: the persistence of political misperceptions. Political Behav. 32, 303–330.

Oreskes, N., 2001. Plate Tectonics: An Insider's History of the Modern Theory of the Earth. Westview Press, Oxford.

Oreskes, N., Conway, E.M., 2010. Merchants of Doubt: How a Handful of Scientists Obscured the Truth on Issues From Tobacco Smoke to Global Warming. Bloomsbury Press, New York.

Popper, K., 1963. Conjectures and Refutations: The Growth of Scientific Knowledge. Routledge, London.

Schiff, A.L., 1962. Fire and Water: Scientific Heresy in the Forest Service. Harvard University Press, Cambridge.

Siler, K., Lee, K., Bero, L., 2015. Measuring the effectiveness of scientific gatekeeping. Proc. Natl. Acad. Sci. U. S. A. 112, 360–365.

Tennant, J.P., Dugan, J.M., Graziotin, D., et al., 2017. A multi-disciplinary perspective on emergent and future innovations in peer review [version 3; peer review: 2 approved]. F1000Research 6, 1151.

Tesser, A., 2000. On the confluence of self-esteem maintenance mechanisms. Pers. Soc. Psychol. Rev. 4, 290–299.

Wazeck, M., 2013. Marginalization processes in science: the controversy about the theory of relativity in the 1920s. Soc. Stud. Sci. 43, 163–190.

Weigold, M.F., 2001. Communicating science: a review of the literature. Sci. Commun. 23, 164–193.

Further Reading

Watson, D.L., 1938. Scientists are Human. Watts, London.

Sounding the climate alarm—scientists and politics

Franz Baumann
New York University, Graduate School of Arts and Science, Program in International Relations, New York, NY, United States

"Little would be wanting to the happiness of life, if every man could conform to the right as soon as he was shown it"

(Samuel Johnson)

The context

The corona pandemic was no black swan. It was neither unforeseen nor unexpected in its trajectory (Osterholm and Olshake, 2020) but it revealed the fragility of everything: polities, economies, societies, and existences. Life is exposed as a high-wire act, even in some of the world's richest countries. Its brittleness and unsustainability have come into sharp relief. In the United States, an unprecedented number of people line up at foodbanks, millions of workers have lost their jobs and health insurance, hundreds of thousands of small businesses have been driven into bankruptcy, industries are contracting, airplanes have been grounded, and cruise ships mothballed. As Herbert Stein, a chairman of the White House Council of Economic Advisers under Presidents Nixon and Ford, legendarily observed: "If something cannot go on forever, it will stop." The fetish of endless growth in a finite world is revealed as such. Inequality within and among countries has been exposed as unjust and unsustainable. Courtesy of the corona pandemic, the history of the past decades will be rewritten. The proverbial impossible has become the inevitable overnight. The Catch-22 that has haunted politics for too long, namely, the unattainability of reforms needed for sustainability and the inadequacy of those that are attainable, may conceivably be overcome.

Conservation Science and Advocacy for a Planet in Peril.
DOI: https://doi.org/10.1016/B978-0-12-812988-3.00007-7

Humanity has been on a collision course with nature for decades. Also, for decades, scientists have sounded the alarm about the Greenhouse Effect, Climate Change, and Global Heating—the term has evolved. Instead of slowing down or tacking away from inevitable catastrophe, we have been picking up speed. For half a century or more, countless scientists have explained the reason for this relentless expansion and sounded the alarm. In 2020, epidemiologists are finally being heard, though not yet climatologists. Because being listened to is not the same as being heard.

This chapter will trace the ever more granular evidence that the climate is overheating the planet, that humans are causing it and that the consequences will be catastrophic. It will then track the environmental policies of the activist 1960s, the neoliberal backpedaling, the economic expansion of the past decades and, finally, some remarkable advocacy initiatives of scientists to influence public policy.

The science

What did we know and when did we know it? The basics of anthropogenic global warming have been understood since the 19th century, as can easily be established by googling for instance Joseph Fourier, Claude Pouillet, John Tyndall, Svante Arrhenius, or Eunice Foote (American Institute of Physics, 2020; Arrhenius, 1896; Pierrehumbert, 2004; Schwartz, 2020; Tyndall, 1859/1860). The essence of their discovery is that greenhouse gas molecules in the atmosphere—mainly nitrous oxide (N_2O), water vapor (H_2O), carbon dioxide (CO_2), and methane (CH_4)—absorb and trap the sun's heat that hits the Earth and is reflected back into space (NASA, undated). The more greenhouse gases there are in the atmosphere, the less heat escapes from Earth into outer space and the higher the Earth's median temperature. This is taught to school kids (NASA, Visit to an Ocean Planet, undated). The temperature rises as foreseen by the models, perhaps a bit faster (Brysse et al., 2013; Cheng et al., 2019). Ever more sophisticated models enhance their predictive precision (Voosen, 2020).

The gaseous composition of the atmosphere is measured in parts per million. Nitrogen (78%) and Oxygen (21%) account for 99% (NASA, Earth Factsheet, undated). Carbon dioxide, an invisible, odorless trace gas—called so because it is in the 0.03%–0.04% range—was about 280 parts per million (ppm) in 1750. In the 200 or so years until 1959, it rose by 12% to 316 ppm, while in the sixty or so years since then, it rose by 76% to 416 ppm, as per the famous Keeling Curve (Scripps Institution of Oceanography, undated). It is now higher than at any point during the past 800,000 years, perhaps three million years (Willeit et al., 2019) when temperatures "were 2°C–3°C (3.6°–5.4°F) higher than during the preindustrial era and the sea level was 15–25 m (50–80 feet) higher than today" (Lindsay, 2020). Annual increments of 2–3 ppm, as they have been for the past decades (UK Met Office, 2020), or 71 ppm during the 20th century, are historically unprecedented. They were "larger during the Industrial Era than during any comparable period of at least the past 16,000 years" (Joos and Spahni, 2008). At current emission levels, and in the absence of negative emission technologies deployed at scale, it will be around 10 years before we blow past the 450 ppm goalpost that corresponds to 1.5°C warming, the assumptions being that a carbon budget of 740 billion tons of carbon dioxide remains for a 33% probability of limiting global warming to 1.5°C, of 480 billion tons for a 50% probability or of 320 billion tons for a 66% probability (Rogelj et al., 2019).[1]

It would be simplistic to consider greenhouse gases bad tout court. On the contrary, they are enablers of life on Earth and controllers of the climate that, for the past 10,000 or so years, have made human civilization possible. Without greenhouse gases, the average temperature would be −18°C (−0.4°F), instead of the current +15°C (+59°F). But there can be too much of a good thing. Methane, black carbon, and fluorinated gases are short-lived because they linger less in the atmosphere than carbon dioxide.

Methane, like carbon dioxide, is an invisible, odorless trace gas. It is present in the atmosphere at just under two parts per million molecules, this is a 400th of carbon dioxide (Global Monitoring Laboratory, undated). Methane traps nearly 90 times as much heat as carbon dioxide but dissipates in about a decade. Still, it is responsible for approximately a quarter of global warming since 1750. Numbers are rising steeply, mainly on account of increasing meat consumption and oil industry offshoots, including fracking (Jackson et al., 2020; The World Bank, I, 2020).

Carbon dioxide, the biggest contributor to Global Heating, is an "infinite-lifetime gas," so called because, as a cumulative pollutant, it stays in the atmosphere for centuries (Allen et al., 2016). This is highly relevant in the context of international negotiations on climate mitigation. Since today's warming climate is both a stock issue (the cumulative buildup over time) and a flow issue (current emissions), the burden of reductions cannot in fairness be the same for all countries. The Industrial Revolution, which was highly energy-intensive, laid the basis for the current wealth in North America, Europe, and Japan. The world's developed countries account for three-quarters of historic emissions (1850–2002), developing countries for one quarter (Baumert et al., 2005; Saussay, 2019).

Until the Industrial Revolution, primary energy sources supporting human life were plants for food, either directly or through meat, and biomass for warmth and light. They were climate-neutral. Animal and human muscles, water, wind as well as the burning of wood and animal dung, had practically no effect on the composition of the atmosphere, "because it only slightly speeded up the natural decay processes that continually recycle carbon from the biosphere to

---

[1] *The Guardian Carbon Countdown Clock estimates 650Gt CO_2 remaining to limit warming to 2°C. https://www.theguardian.com/environment/datablog/2017/jan/19/carbon-countdown-clock-how-much-of-the-worlds-carbon-budget-have-we-spent.*

the atmosphere" (President's Science Advisory Committee, 1965). The use of coal ruptured the shackles of direct photosynthesis and real-time use of biomass as humanity's main energy source. They allowed humans to reach into the Earth's store of millions of years of plant matter converted by solar energy into coal, oil, and gas (United States Energy Information Administration, 2019).

Coal fueled the Industrial Revolution. It converted heat into mechanical power (such as in steam engines) and movement (such as in railways and ships).[2]

Oil was first struck in 1859 in Pennsylvania.[3] It supplanted coal in importance after World War II when road and air transport mushroomed, thus changing the world. Oil was the largest single item in the dollar budget of most Western European countries. It was also a high priority in the Marshall Plan, as it helped to preserve market share for U.S. companies (Painter, 2009).

Coal, oil, and gas are organic materials, the result of millions of years of stored photosynthesis—plants having converted solar energy into biomass—which returned "to the atmosphere and oceans the concentrated organic carbon stored in sedimentary rocks over hundreds of millions of years" (Revelle and Suess, 1957). At the end of the 19th century, fossil fuels began to power electrification; during World War I, they were also used as raw materials for explosives and fertilizers. In the 1920s chemical conversion turned coal into gasoline, industrial grease, synthetic rubber, and even margarine.

Excavating and burning fossil fuels—the sun's energy stored for eons—in ever greater quantity laid the basis for the upsurge of productivity, production, and consumption of the past decades. To illustrate: one gallon of gasoline (3.79 L) consists of one hundred tons of ancient plants and contains about 35 kW/h of energy.[4] A healthy, well-fed and motivated manual laborer generates an output of about 1 kW/day, which means that one gallon of gasoline is the equivalent of nearly 2 months of labor.

---

[2] *"Between 1839 and 1850 some 6000 miles of railways were opened in Britain." Eric J. Hobsbawm,* Industry and Empire: From 1750 to the Present Day *(London: Weidenfeld & Nicolson, 1968; Penguin, 1990), p. 110.*
"In 1860, British tonnage had been a little larger than American, 6 times as large as the French, 8 times as large as the German; in 1890, it was over twice as large as the American tonnage, 10 times as large as the French, and still roughly 8 times as large as the German." (ibid., *p. 179).*
"British shipyards in 1870 built 343,000 tons of vessels for British owners, and in 1913 almost a million tons." (ibid., *p. 207).*

[3] *The world's first oil well began to operate in Titusville, Pennsylvania, in August 1859; cf. Samuel T. Pees, "Drake's Well,"* Oil History, *undated. http://www.petroleumhistory.org/OilHistory/pages/drake/drakewell.html.*

[4] *U.S. Energy Information Administration, Units & Calculators explained: British Thermal Units (BTU). https://www.eia.gov/energyexplained/units-and-calculators/british-thermal-units.php.*
BTU/hr to Watt conversion. https://www.rapidtables.com/convert/power/BTU_to_Watt.html.
Watt/hr to kcal. http://convert-to.com/conversion/power/convert-w-to-kcal-per-hr.html#:~:text=1%20watt%20(W)%20%3D%200.86,per%20hour%20(kcal%2Fh).

No doubt, scientific and technological advances were additional catalysts of the great acceleration after World War II, but it was the burning of fossil fuels in ever greater quantities that catapulted humanity into the modern age of mass production and consumption. A few generations of humanity are burning the fossil fuels that were generated over several hundred million years, a stunning excess that is problematic not because it exhausts finite resources, but because it destroys the natural life-support systems on which animals, plants, and humans depend.

Annual emissions in 1945 were 4.3 billion tons carbon dioxide, more than double that in 1960 (9.4 billion tons) and eight times as much now (36.6 billion tons). The emissions of the past decade exceed those of all human history up to the mid-1960s (Global Carbon Atlas, undated; Our World in Data, undated). Not only do emissions continue to rise but also the rate of emissions. Present-day global temperatures are the warmest they have been in at least 12,000 years, and possibly far longer (Kaufman et al., 2020). 2016 was the hottest year on record, 2019 the second hottest, and chances are that 2020 will come out on top (National Oceanic and Atmospheric Administration (NOAA), 2020a). 2015–19 were the 5 warmest years, 2010–19 the warmest decade, and since the 1980s, each successive decade has been warmer than any preceding decade going back to 1850 (World Meteorological Organization, 2020). The Arctic, believed to be warming twice as fast as the global average, was in flames during the summer of 2020 (World Meteorological Organization, II, 2020). Emissions, the carbon dioxide and methane content in the atmosphere, temperature rise, land degradation (especially forest losses), ocean warming, and sea-level rise are reaching dangerous levels. The curve is not being flattened. It is still going up. At present rates of emissions, there will be more carbon dioxide in Earth's atmosphere by 2025 than at any time in at least the last 3.3 million years (Vega et al., 2020).

Wallace S. Broecker coined the term "Global Warming" in 1975 (Broecker, 1975). A decade later, he warned about what subsequently became known as tipping points, namely, "the possibility that the main responses of the system to our provocation of the atmosphere will come in jumps, whose timing and magnitude are unpredictable" (Broecker, 1987). In 1996 the Intergovernmental Panel on Climate Change (IPCC) introduced the term tipping points, levels of "change in system properties beyond which a system reorganizes, often abruptly, and does not return to the initial state even if the drivers of the change are abated" (IPCC, I, 2019). In other words, tipping points are thresholds that, once crossed, push a system into an entirely new state, irreversibly so. There are several identified tipping points, among them the shutdown of the Atlantic Meridional Overturning Circulation, the disintegration of the West Antarctic ice sheet, the dieback of the Amazon rainforest, the shift of the West African monsoon, the permafrost loss, the coral reef die-off, the Indian monsoon shift, the Greenland ice sheet disintegration, and the boreal forest shift (McSweeney, 2020). Risks are mounting that one or more tipping points will be crossed and that we are "approaching a global cascade of tipping points that [lead] to a new, less habitable, 'hothouse' climate state. . . . We are in a state of planetary emergency: both the risk and urgency of the situation are acute" (Lenton et al., 2019).

As if the climate catastrophe was not terrifying enough, there are the related but distinct crises of collapsing biodiversity, dying oceans, and degraded land. No fewer than a million animal and plant species are at risk of extinction around the globe. Biodiversity is declining faster than at any time in human history (Intergovernmental Science-Policy Platform on Biodiversity and Ecosystem Services (IPBES), 2019). This is not a matter of polar bears disappearing but of endangered species acting as canaries in the coalmine (Molnár et al., 2020). If they vanish, entire ecosystems, on which human life depends, are endangered (Trisos et al., 2020). The same goes for warming oceans that are already killing off marine ecosystems, raising sea levels, and making hurricanes more destructive (Intergovernmental Panel on Climate Change (IPCC), II, 2019). Current land-use practices endanger food security and exacerbate warming by transforming natural carbon sinks into potent sources of greenhouse gas emissions, such as forest degradation and deforestation (IPCC, III, 2019). These calamities—Global Heating, biodiversity loss, dying oceans, land degradation, disappearing rain forests—are not natural phenomena like earthquakes or erupting volcanoes, but the consequences of purposeful human pressures on Earth's life-support systems: a stable climate, fertile soils, intact forests and marine systems, ample unsoiled water, and a protective ozone layer.

The politics

In the momentous year 1965, the causal relationship between burning fossil fuels and warming the Earth's climate moved into the political domain. Civil unrest rocked the United States while the first U.S. combat troops arrived in Vietnam. At the same time, public policy activism unfolded on a scale unseen since Roosevelt's New Deal three decades earlier.

President Lyndon B. Johnson declared in February that "this generation has altered the composition of the atmosphere on a global scale through radioactive materials and a steady increase in carbon dioxide from the burning of fossil fuels" (Johnson, 1965). In late July, Johnson signed into law the bill that led to Medicare and Medicaid, 1 week later the *Voting Rights Act* that empowered African-Americans and in early October the *Immigration and Nationality Act* that abolished quotas [since 1924, 70% of immigration slots had been reserved for Northern Europeans; Social Security Act Amendments (1965), Voting Rights Act (1965), Immigration and Nationality Act, October (1965)]. In 1966 over the fierce opposition of car manufacturers, the National Traffic and Motor

Vehicle Safety Act of 1966 was promulgated, and in April 1968, the Civil Rights Act of 1968 that prohibited any discrimination in respect of housing. In November 1965, the White House published the President's Science Advisory Committee's report *Restoring the Quality of our Environment* with an introduction by President Johnson. It warned of momentous changes—melting of Antarctic ice caps, sea-level rise, ocean acidification—pervasive in nature and disregarding political boundaries (President's Science Advisory Committee, 1965). Its analyses and forecasts have borne out well, yet political action has lagged behind, even though there were some advances.

The first Earth Day on April 22nd, 1970 brought millions into the streets. The Republican President Richard Nixon signed the *National Environmental Policy Act* (Kershner, 2011), established the National Oceanic and Atmospheric Administration (NOAA), 1970, and the Environmental Protection Agency (EPA), 1970. Bipartisan, science-based, and hugely successful policies were implemented, such as The Clean Air Act, 1970, The Clean Water Act, 1972, and The Endangered Species Act, 1973.

Environmental policy was a technical matter, not an ideological issue. Science informed public policy and prevailed over economics and special interest politics. It played an important role in evaluating environmental impacts, setting air pollution standards, and deciding on the protection of species. The U.S. Senate's discussion on the dangerous warming effect of carbon dioxide emissions on April 3rd, 1980 was covered by Walter Cronkite on the CBS evening news.[5] Forty years on, the analysis, narrative, and graphs are still on target.

On June 1st, 1988, the joint statement of U.S. President Ronald Reagan and Soviet Union President Mikhail S. Gorbachev, after their summit in Moscow, pledged to expand "cooperation with respect to global climate and environmental change, including ... environmental protection, such as protection and conservation of stratospheric ozone and a possible global warming trend" (Joint Statement, 1988). A few weeks later, NASA scientist James Hansen testified before the U.S. Senate Committee on Energy and Natural Resources that the greenhouse effect was measurable, causing extreme weather events (Hansen, 1988). Before the end of the month, in Toronto, the *International Conference of the Changing Atmosphere: Implications for Global Security* resolved to reduce emissions. The conference's strongly worded statement identified human pressures on nature as "an unintended, uncontrolled, globally pervasive experiment, whose ultimate consequences would be second only to those of a global nuclear war" (International Conference of the Changing Atmosphere: Implications for Global Security, 1988). Within months, the United Nations

---

[5]*Walter Cronkite on Climate Change, CBS, Thursday, April 3rd 1980. https://www.youtube.com/watch?v=MU9s0XyEctI.*

General Assembly noted with concern in its first resolution on climate change, "that the emerging evidence indicates that continued growth in atmospheric concentrations of "greenhouse" gases could produce global warming with an eventual rise in sea levels, the effects of which could be disastrous for mankind if timely steps are not taken at all levels" (United Nations General Assembly Resolution, 1988). The IPCC was established and *Time Magazine* declared Earth *Planet of the Year* with a compelling cover photo.[6] Yet, more carbon dioxide has been emitted since 1988 than in all of prior human history.

The IPPCC, together with former U.S. Vice President Al Gore, won the Nobel Peace Prize in 2007. Its first assessment report in 1990 anticipated that warming of 0.3°C (0.6°F) per decade would occur in the 21st century, and 1°C (1.8°F) by 2025. The sea level would be up 30 cm by 2030 (Houghton et al., 1990). These predictions will be exceeded. In 1992 the United Nations Earth Summit at Rio de Janeiro agreed on the *Framework Convention on Climate Change*, and the "stabilization of greenhouse gas concentrations in the atmosphere at a level that would prevent dangerous anthropogenic interference with the climate system" (United Nations Framework Convention on Climate Change (UNFCCC), 1992). The Kyoto Protocol, concluded in 1997 and entered into force in 2005, operationalized that target (Kyoto Protocol, 1997/2005). It was stillborn because it required only industrialized countries to reduce carbon dioxide emissions, yet not emerging giants such as China, India, or the Republic of Korea. However, this might have been intellectually plausible and historically fair since, between 1850 and 2002, the world's developed countries account for three-quarters of historic emissions and developing countries for only one quarter (Baumert et al., 2005; Saussay, 2019).

Scientific insights or ecological necessities do not change debates on their merits, only if they are amplified politically. The United States signed, yet never ratified, the Kyoto Protocol. Canada did ratify it in 2002 but withdrew in 2011 (United Nations Treaty Collection). The Kyoto Protocol proved a nonstarter and quietly expired in 2012. As the causes of a warming atmosphere—fossil fuel combustion—became clearer, the climate debate evolved from an esoteric meteorological subject to a highly charged political one, ever more so as the costs of prevention, mitigation and adaptation measures came into focus.

China's, India's, and South Korea's stunning economic growth also colossally increased their carbon emissions. Between 1990 and 2018, China's gross domestic product grew from $361 billion to $13,608 billion, India's from $320 billion to $2720 billion, and the Republic of Korea's from $485 billion to $1620 billion (World Bank – Data GDP in current US$). China's carbon dioxide emissions quadrupled from 2420 million to 10,065 million tons, as did India's from

[6]Time Magazine, *"Planet of the Year," January 2nd 1989. http://content.time.com/time/covers/0,16641,19890102,00.html.*

617 million to 2654 million tons, while Korea's more than doubled from 247 million to 659 million tons (Global Carbon Atlas, undated). In 1990 U.S. emissions (5100 million tons) were more than twice China's. Now China's are nearly double those of the United States (5416 million tons). In 1990 the EU emitted 4471 million tons or 50% more than China; now only about one third, namely, 3445 million tons.

These dynamics, and the unwillingness of most players, especially China and the United States, hampered international agreement and led to the near-collapse of the Copenhagen conference in 2009 (COP15). Surprisingly, six years later at COP21 in Paris was a splendid success diplomatically. Instead of assigning emission reduction targets to some countries only—industrialized market economies—the Paris Agreement applies to all countries and held "the increase in the global average temperature to well below 2°C (3.6°F) above preindustrial levels and to pursue efforts to limit the temperature increase to 1.5°C (2.7°F) above preindustrial levels" (Paris Agreement, 2015) All countries are required to submit voluntary emissions pledges, called nationally determined contributions (NDCs). These NDCs are to be ever more ambitious, reported for the first time when ratifying the agreement, subsequently in 2020 (now in 2021, since the Glasgow COP26 has been postponed) and every 5 years thereafter. Importantly, a Green Climate Fund is to support developing countries' mitigation and adaptation efforts. All of this was unanimously agreed.[7]

Yet agreeing is not the same as doing. In the world of diplomacy the task is to negotiate, promulgate targets, and to arrange words inoffensively in the expectation that they buy time while sufficient momentum develops to overcome underlying problems and inherent contradictions. In the case of the Paris Agreement, the hope was unsurprisingly short-lived because it, like Kyoto, the Agenda 2030 (United Nations General Assembly, 2015), the Global Migration Pact (United Nations General Assembly, 2018), and all other multilateral agreements suffer from the triple defect of being uncoordinated, nonbinding and nonenforceable. To be sure, this does not mean that they are inoperable, only that they have limited utility. Quite like New Year's resolutions. After marshaling the international community to reach agreement on scientifically validated targets, the actual achievement of negotiated agreements requires different mechanisms. Multilateral fora have important convening, standard-setting, monitoring, and validating functions, which is not to be sneezed at. But also, not to be overrated.

The United Nations Environment Program publishes two gap reports annually. The *Emissions Gap Report* focuses on countries' projected carbon dioxide

[7]As of August 27th 2020, 189 of the 197 parties to the United Nations Framework Convention on Climate Change have ratified the Paris Agreement. United Nations Treaty Collection. https://treaties.un.org/Pages/ViewDetails.aspx?src = TREATY&mtdsg_no = XXVII-7-d&chapter = 27&clang = _en.

emissions and tracks the difference between submitted NDCs and the emission levels required to keep the mean global temperature rise in 2100 to below 2°C or 1.5°C above preindustrial levels. The 2019 report evidenced that, to meet the targets of the Paris Agreement, emissions must drop 7.6% every year from 2020 to 2030 for the 1.5°C goal and 2.7% yearly for the 2°C goal. To achieve *the well below 2°C goal,* "countries must increase their NDC ambitions threefold and more than fivefold to achieve the 1.5°C goal. ... Evidently, greater cuts will be required the longer that action is delayed" (United Nations Environment Program (UNEP), 2019). The Corona Pandemic is expected to reduce global carbon dioxide emissions in 2020 by about 8% (2.6 billion tons), which means to the level of 10 years ago. This "reduction would be the largest ever, ... twice as large as the combined total of all previous reductions since the end of World War II" (International Energy Agency, 2020). No cause for celebration, though, since, achieving the 1.5°C goal requires cuts of this magnitude every year for the present decade and beyond.

The *Production Gap Report* presents an even more troubling picture, since it does not rely on voluntary NDCs but on public planning figures of projected fossil power production. These reports show that the planned fossil fuel production of countries by 2030 will lead to the emission of 39 billion tons of carbon dioxide, 53%, more than would be consistent with a 2°C pathway, and 21 billion tons of carbon dioxide, 120%, more than would be consistent with a 1.5°C pathway (SEI et al., 2019). Still, exploration continues apace. Since the 2015 Paris agreement, 35 major global banks have collectively poured $2.7 trillion into fossil fuels, with financing increasing every year (Rainforest Action Network, 2020). The biggest single investment ever in Africa was announced in July 2020 by the French energy giant *Total.* The $20 billion project to extract and export gas from Mozambique is financially backed by the United States, Japan, Britain, Italy, Netherlands, South Africa, and Vietnam (Total, 2020). This is a stunningly out of synch move that speaks volumes about these countries' decarbonization commitment, and that of the World Bank. It supported the venture with a $50 million technical assistance grant (The World Bank, II, 2020).

Considering such staggering numbers, there is the hope, or the expectation, for a deus ex machina: Carbon Capture & Sequestration, the removal of carbon dioxide from the atmosphere. However, at current low prices of carbon dioxide, and current high prices for direct air capture, it is too expensive and too modest in impact (European Academies Science Advisory Council, 2018). Of the more than 1000 possible scenarios considered by the IPCC, only 116 (~12%) limit warming to below 2°C, and of these a hundred and eight involve negative emissions, all with presently still unavailable technology (Kolbert, 2017). The IPCC warns that carbon dioxide removal "deployed at scale is unproven, and reliance on such technology is a major risk in the ability to limit warming to 1.5°C" (IPCC, 2018). A strategy built around a "get-out-of-jail-free-card" is unsound (Jeffery et al., 2020).

The World Meteorological Organization computes a one in five chance of global temperatures rising by 1.5°C (2.7°F) before 2025 (World Meteorological Organization, I, July 2020), a threshold beyond which millions of homes become uninhabitable and hundreds of millions of lives are at stake (IPCC, 2018). The gap is widening between the sustainable and the actual because decarbonizing the world's economy is terribly urgent, yet wickedly difficult. Ratcheting down global carbon emissions to net zero by 2050 is a gargantuan challenge without parallel in human history. The scale and upfront price tag are unprecedented, as is the requirement to redesign existing production, trade, and consumption patterns. The task is exceedingly hard, because the costs—financial, jobs, lifestyles—are to be borne now, and because the losers are identifiable, organized, noisy, and politically connected, while the beneficiaries are diffuse and many are not yet even born.

Fossil fuels are key drivers in the economic development of poor countries, especially oil-producing ones. In industrialized countries, fossil fuels power mass production, mass consumption, housing, and mobility. Neoliberal policies—pioneered by Margaret Thatcher in the United Kingdom and President Reagan in the United States—pursued in the 1980s and thereafter in many Western countries, encouraged short-termism, weakened regulations, reinforced profit-maximizing, damaged social cohesion, offshored well-paying blue-collar jobs, accelerated economic as well as political polarization, and raised the fortunes of the financial economy. At the same time, it sapped multilateral arrangements and mutually beneficial cooperation. The modus operandi of this winner-takes-all globalized economy, namely, strip-mining productive businesses in the interest of quick returns to shareholders—aptly called "fishing with dynamite" (University of Virginia, 2020)—sacrificed resilience for efficiency, inequality, and environmental destruction.

America First, Brazil First, Britain First, Russia First, or whichever country first cannot contain the Corona pandemic. It is certainly cannot prevent the climate catastrophe. Public leadership, science-based policies, international cooperation and compromise, mission-driven national and global investment in public goods are anathema to an ideology that glorifies selfishness and presumes that markets are the optimal arbiters of policy, infinite economic growth, the natural order of things, and balanced budgets as a goal in itself. Only free markets, as the mantra goes, can allocate resources efficiently, reward individuals based on what they deserve, and pick winners as well as losers dispassionately in accordance with their contribution to the economy—nota bene not to society. The "magic of the marketplace" was to be unleashed by deregulation, tax cuts, privatization of state functions, and low government debt. Instead, unregulated markets led to monopolies that concentrated political power. They also led to less competition, lower taxes for the rich, more deregulation, value extraction, and rent-seeking. And to environmental destruction.

Climate Science became politicized, with many industry-funded outfits—for instance the deceptively named *Global Climate Coalition* (Savage, 2019; Climate Investigations Center, 2019; Revkin, 2009)—masquerading as serious think tanks sowing doubt and spreading false information to discredit and delay meaningful climate policies.[8] Exxon's and Shell's internal research, knowledge, and camouflage of the dangers of Global Heating are well documented (Banerjee, 2015; Climate Files, undated; Greenpeace, undated; Franta, 2018; Song et al., 2015). Associating climate concerns with the antibusiness political left has poisoned and polarized the discussion in the United States, much like the wearing of facemasks to contain Covid-19. Unsurprisingly, it is downplayed or ignored that the U.S. intelligence and defense establishments also assess Global Heating as a most serious threat (Coats, 2018; United States Army War College, 2019).

Marginal emission reductions would have sufficed 30 or so years ago, at the time of the first IPCC report (IPCC, 1990/1992), when Republican Congresswoman Claudine Schneider, supported by 24 Republicans and 114 Democrats, introduced the *Global Warming Prevention Act of 1989* (Schneider, 1989), when British Prime Minister Margaret Thatcher warned that "we have unwittingly begun a massive experiment with the system of this planet itself" (Thatcher, 1988), when the U.S. Environmental Protection Agency issued its landmark 400 page report to Congress, *The Potential Effects Of Global Climate Change On The United States* (Environmental Protection Agency (EPA), 1989), when President George H.W. Bush established the U.S. Global Change Research Program that produces the *National Climate Assessment*, (USGCRP, 2018), or when the CDU, Germany's governing conservative party, included carbon pricing in its platform, yet never acted on it (CDU, 1995). Not any longer.

Much more decisive action is warranted but not forthcoming. "We continue to commit our planet—for centuries or longer—to more Global Heating, sea-level rise, and extreme weather events every year. If humans were to suddenly stop emitting CO_2, it would take thousands of years for our CO_2 emissions so far to be absorbed into the deep ocean and atmospheric CO_2 to return to preindustrial levels" (National Oceanic and Atmospheric Administration (NOAA), 2020b). In the last three decades, annual emissions of carbon dioxide have risen by more than 60%. "Set against the scientific logic of carbon budgets, the global community has spent virtually all of the Paris-compliant emission space and now faces a decarbonization agenda far removed from any economic equilibrium" (Anderson et al., 2020). *Laissez-faire* policies will not right the listing ship. Instead, governments must assume a lasting role in directing, stabilizing, and remodeling economic life. And they must do it soon. Further delay narrows options and jeopardizes the possibility to organize the inevitable transformation in an orderly way. But will governments do what

[8] *In 2016 19 Senators introduced the* Web of Denial *Resolution. https://www.desmogblog.com/sites/beta.desmogblog.com/files/Web%20of%20Denial%20Resolution.pdf.*

logic, national interest and global survival demand? Unlikely the U.S. government, the most extreme case of dysfunctionality among big countries, leaving aside kleptocratic (corrupt leaders that use political power to appropriate wealth) Russia, dictatorial China, and commandeered Brazil.

Politics is a power contest over goals and means, competencies and resources, whether in families, communities, countries, or indeed the world. In this power contest, unevenly distributed resources prejudice outcomes. An old insight, even a trite one, is that wealth constitutes power, shapes debates, and determines outcomes (Smith, 1880/1776). Regarding Global Heating, powerful interests set the agenda, determine the pace, scope and speed of change, the weight of science, and whether and how victims are compensated. Think Puerto Rico versus Miami.

Until the 1980s, the top marginal tax rate on individual incomes in the United States was 70%. Today, it is 37%. Corporate profits were taxed at 50%, today at 21% (Hungerford, 2013). Chief Executives in 1965 earned 20 times more than a typical worker, in 1989 it was 58 times more. Today's ratio is 278-to-1 (Mishel and Wolfe, 2019). It is stunning, yet not surprising that:

- The top 1% of U.S. households own over 30% of total wealth, the bottom half about 1% (Batty et al., 2019).
- Nearly 80% of U.S. workers live paycheck to paycheck, that three in four workers are in debt and that more than half of minimum wage workers have to hold down more than one job to make ends meet (Career Builder, 2017).
- Perniciously, the gap in life expectancy in the United States between the richest 1% and the poorest 1% is about 15 years for men and 10 years for women (Chetty et al., 2016).
- In affluent suburbs in the United States, emissions can be 15 times higher than in nearby neighborhoods (Goldstein et al., 2020).
- The world's wealthiest 10% emit as much carbon dioxide as the 90% (Motesharrei et al., 2016; Hubacek et al., 2017).
- The average carbon footprint of the richest 1% globally is more than 170 times that of the poorest 10% (Oxfam, 2015).
- In Germany, six of ten people never or rarely fly; only two of ten twice or more per year (Infratest, 2019).
- In Britain, over half the population never flies, whereas 10% take four or more flights per year (UK Department of Transportation, 2014). The numbers and patterns in other countries are similar or even more extreme.

The defense of privileged behavior is a major obstacle for decarbonization policies within and among countries.

In the United States, special interest groups have successfully lobbied for policies that benefit them directly, while dispersing the costs across everyone else.

Beginning with President Reagan, special interests—the rich, the banks, the fossil fuel industry, the airlines, car manufacturers—captured many policy areas that benefitted only them, not the country as a whole. They managed to make off-limits energy taxes or government regulations, even though their unsustainable business models are patently damaging to society and nature. The hijacked state has little capacity left, yet must urgently rebuild it because the task at hand—rapid and thorough decarbonization of the global economy—is the greatest challenge humanity has ever faced. Put more technically, containing the Earth's warming at 1.5°C is an unprecedented global public policy responsibility.

No U.S. government, even one not in cahoots with big business and big finance, is well-positioned to deal with systemic challenges, such as income inequality or Global Heating. Its modus operandi is incrementalism, and its time horizon is short. Both are inadequate for the problem at hand. Normally, whether in personal life or in politics, time does heal wounds and offers the space for solutions to emerge. In Global Heating, time is against us. We are running out of it. Normal conflicts can be resolved through dialog and compromise. Not so Global Heating. Nature does not negotiate. Nature also does not compromise. Deep changes and the imposition of immediate sacrifices for uncertain prospective benefits, while warranted by the magnitude and urgency of the problem, are not winning political propositions, no matter that the future cost of today's inaction will be steep. Baked into the political system is the short-sightedness that, aptly, has been called the "tragedy of the horizon" (Carney, 2015). Today's decision-makers will not have to live with the outcomes of their actions—and inactions. With mitigation costs due now, yet benefits accruing in the future, Global Heating is politically framed mainly in cost terms—what to spend, not even what to invest, now—rather than as a matter of environmental preservation and risk management. Risk management in two respects, namely, mitigation, which means avoiding the unmanageable, and adaptation, which means managing the unavoidable (Stern, 2015). And, of course, in today's integrated world, both aspects must be tackled nationally as well as internationally.

The economics

After the Second World War, mass consumption became a reality in much of the Western world and an aspiration everywhere else. The extraordinary growth of productivity, production, and consumption has no parallel in the human experience:

- The world's population increased from 3 billion in 1960 to 7.7 billion (World Bank – Data Population).
- Life expectancy globally rose from 52 years in 1960 to 73 years (World Bank – Data Life Expectancy at Birth).

- The world's gross domestic product increased from $1.4 trillion in 1960 ($450 per capita of the then living 3 billion people) to more than $88 trillion (over $11,000 per capita of today's 7.7 billion people, albeit very unevenly distributed) (World Bank – Data GDP in current US$, World Bank – Data GDP per capita) (in current dollars).
- The number of motor vehicles globally quintupled from 250 million in 1970 to 1.3 billion in 2015. Before the Corona pandemic, it was expected to double again by 2050 (International Organization of Motor Vehicle Manufacturers (OICA), undated).
- The number of airline passengers increased more than 12-fold from 310 million in 1970 to over 4.5 billion in 2019 (World Bank – Data Air transport, passengers carried; IATA, 2020).
- International tourist arrivals more than doubled from 563 million in 1996 to 1.4 billion in 2018 (United Nations World Tourism Organization, 2018; World Bank – Data International Tourism, Number of Arrivals). The Corona pandemic might have put a temporary stop to this expansion, yet the industry is primed to return to business-as-usual and to resume an expansionary course as soon as possible (International Civil Aviation Organization (ICAO), 2020).
- Global production of plastic, an oil derivative, grew from around 1.5 million tons in 1950 to 381 million tons in 2018 (Borrelle et al., 2017; Our World in Data Global Plastics Production, 1950–2015, undated; PlasticsEurope, 2019), much of which single-use. Between 4 million and 12 million tons are discarded annually into oceans (Jambeck et al., 2015).
- Ruminants (mainly cattle, sheep, and goats) have increased by over 50% since 1960, responding to the growing demand for meat and dairy products (Ripple et al., 2013).[9]

The phenomenal growth since World War II—of people, health, life expectancy, wealth, production, and consumption—is due to human ingenuity and work, but also the exploitation of nature, especially the burning of fossil fuels. The flipside of an extraordinary economic success story is becoming ever more evident. An inconvenient truth, since the vigor of the world economy as it is structured today depends on ever-growing resource extraction, industrial production, global trading, and consumption of goods as well as services (Wiedmann et al., 2020).

The beneficiaries of a high-energy lifestyle, consumers no less than producers, are happy to stay the course, as are those who aspire to join the better off, namely, the disadvantaged in developed and even more in developing countries. Governments are pro-growth for many reasons, not least because distributional conflicts are easier managed in an expanding than in a shrinking

---

[9] *See also Ripple's* Global Livestock Methane Counter. *https://wolfkind.neocities.org/methane/counter.html.*

economy. Investments in the fossil fuels industry are in danger of becoming stranded assets by decarbonization. For a 50% chance of remaining within the 1.5°C warming budget, the burning of coal, oil, and gas would have to cease in about 10 years, assuming current production levels. Proved reserves, again at current production levels, are for coal: 132 years; gas: 51 years; and oil: 50 years (Coffin and Grant, 2019). Banks are realizing that their investments in fossil fuels are underperforming. Wells Fargo, one of America's largest financial institutions, reported that 47% of its past-due corporate loans were from the oil, gas, and pipeline industry in the second quarter of 2020, even though that industry makes up only 3% of its outstanding commercial loans (Weinstein, 2020). The writing is on the wall.

Also, in danger of being stranded are the economies of Saudi Arabia, Nigeria, Venezuela, Russia, and other countries that depend on the extraction of oil and gas. Important sectors and employers in OECD countries, such as car manufacturers, airlines, mechanized agriculture, meat factories, etc. need to undergo colossal transformations to mitigate Global Heating and to adapt to it. Transitioning to a cleaner, net-zero emissions economy and strengthen resilience to the impacts of climate change requires targeted incentives to drive private investment away from extraction and toward conservation. Taxes, subsidies, regulations, and other public policies must be aligned to reduce environmentally damaging and to foster sustainable activities across entire economies.

Given the worldwide extent of Global Heating, repositioning policies must be national as well as international. Their success depends on a mix of regulations and targeted incentives as well as disincentives, on leveraging technology and on choreographing the cooperation of key countries. William Nordhaus proposes a *Climate Club* to reduce free-riding, induce cooperation, and penalize nonparticipants (Nordhaus, 2020).

A necessary, though in itself insufficient, policy would be rigorously applying the user-pays principle to remedy the unconscionable market failure that fossil fuels are not properly priced. Coal, oil, and gas have been and still are sold below cost, since the damage caused by carbon dioxide emissions is externalized, that is, paid not by those who cause the carbon combustion but by others who neither benefit nor have a say in the matter. The atmosphere was and still is used as a free good in economic terms or a dumping ground ecologically speaking. The social cost per ton of carbon dioxide is assessed by the German Environmental Protection Agency (Umweltbundesamt) at between €180 (≥$210) and €730 (≥$860), depending on the applicable timeframe and discount rate (Matthey and Bünger, 2019). While several carbon pricing regimes exist, the average global price is only about $2 (≤€1.7) per ton of carbon dioxide (The World Bank, III, 2020). On the other hand, environmentally damaging and economically distorting fossil fuel subsidies of over $5 trillion (or 6.5% of GDP) are still handed out globally. "Efficient fossil fuel pricing in 2015 would have lowered global carbon emissions by 28% and fossil fuel air pollution deaths by 46%, and increased government revenue by 3.8% of GDP" (Coady et al., 2019). Corona relief payments for fossil fuels by G20 countries exceed by far those to renewable energy sources, thus missing an historic opportunity to realign incentives (Energy Policy Tracker, 2020). Consequently, the clumsily obsolete practice of burning stored solar energy in the form of fossil fuels continues needlessly.

Proper carbon pricing will catalyze the conversion of the sun's energy. Capturing it directly will, in a manner of historical irony, reclaim the preindustrial technique. At inordinately higher levels of efficiency than burning biomass, wind turbines and photovoltaic cells convert solar energy by order of magnitude better than the preindustrial way of capturing it. With the sun providing more energy every hour than humanity consumes in a year, the need and scope for innovation are immense (Sivaram, 2018).

In the United States, no significant climate legislation has been adopted in over 10 years. President Barack Obama's *American Recovery and Reinvestment Act* of February 2009 responded to the financial meltdown of 2008, aiming to reinvigorate the economy and save millions from losing their jobs and homes (H.R.1, 2009). Changing the trajectory of renewables in America, $90 billion of the Act's $800 billion were earmarked for clean energy generation, electric vehicles, transit, and training for green jobs. As a result, in the decade after the Recovery Act passed, wind power capacity more than tripled, and solar capacity increased by an astonishing factor of 94 (United States Energy Information Administration, 2020). There are at least three lessons here:

- Government incentives (and disincentives) work.
- Shifting from fossil fuels to renewables makes environmental and economic sense, since it yields substantial additional benefits that exceed policy costs: cleaner air, improved health, and fewer premature deaths (Lelieveld et al., 2019).
- Emissions do not go down as much as they could, unless they are specifically targeted, as was intended by the 2009 Waxman-Markey Bill, overwhelmingly approved by the House of Representatives, but never brought to the Senate for discussion or a vote (H.R.2454, 2009).

There is a prejudice, not a serious argument, regarding a contradiction between ecology and economy. The latter cannot possibly thrive against the former, as there will be neither jobs nor dividends on a dead planet. Since infinite growth collides with the physics of a finite world, it is not only the climate change deniers who are anti-science. So, too, are the technology enthusiasts and renewable energy optimists who entertain the fantasy that it will be possible for ten billion people to live in the style of the American or European middle class. At least, they are putting up arguments, which are better than denying the problem (The Global Commission on the Economy and Climate, 2018). But not much, since they assert—which is different from providing evidence—that an orderly transition to a post-carbon economy is possible or realistic with the existing incrementalist approach. Scientists will have to hold many feet to the fire.

The scientists

In the United States, science plays an ambivalent role. On the one hand, some of the best universities are in the country. There are nearly 1400 universities around the world. Nineteen of the top 30, 15 of the top 20, or 7 of the top 10 are American (Times Higher Education, 2020). U.S. nationals have won by far more Nobel prizes in the sciences than others. Of 616 Nobel Prizes in Physics, Chemistry, and Medicine, 269 were won by U.S. citizens, which is more than the next three top winners—Britain, Germany, and France—combined (Nobel Prize Organization, undated; The Economist, 2018). On the other hand, the anti-science sentiment in the United States is formidable, unparalleled elsewhere. It is a sign of malaise, not a fun fact, that there are twice as many Republican Senators called John[10] as there are Ph.D. holders in the current Senate.[11] To be sure, the point is not the surfeit of Johns but the dearth of scientists in key political positions.

Section 49–1922 of the Tennessee Code, prohibiting the teaching of evolution, was repealed as late as 1967. Struck down in 1975 was the Tennessee state law requiring textbooks to include biblical creationism; in 1982 and 1987, laws in Arkansas as well as Louisiana mandating balancing evolution with creationism. A Pennsylvania school district's policy compelling the presentation of intelligent design survived until 2005 (Branch and Reid, 2017). Even today, nearly half of Republicans believe in evolution (Pew Research Center, 2014). Not one of the dozen or so Republican presidential candidates in the 2016 campaign acknowledged that human evolution is a fact (Corneliussen, 2015). Likewise, none of these Republican presidential candidates spoke clearly about anthropogenic climate change. Instead, most denied it outright; others allowed that they were skeptics. The Republican leadership, thus reflects the over 60% of Republicans, who think protecting the environment should not be a priority for the President and Congress, or the only 21% who consider addressing climate change a top priority (Pew Research Center, 2014). It rounds out the picture that the oil and gas industry make 87% of its political donations to the Republican Party.

Since populism and anti-intellectualism lend themselves to the rejection of scientific expertise, one suspects considerable overlap between anti-evolutionists,

---

[10] *John Barrasso (R) Wyoming; John Boozman (R) Arkansas; John Cornyn (R) Texas; John Hoeven (R); North Dakota; John Kennedy (R) Louisiana; John Thune (R) South Dakota.*
[11] *Tammy Duckworth (D, Illinois; Human Services), Ben Sasse (R, Nebraska; History), and Kyrsten Sinema (D, Arizona; Justice Studies). Congressional Research Service; Membership of the 116th Congress (2019–20). https://fas.org/sgp/crs/misc/R45583.pdf.*

climate change deniers, and those who refuse to wear facemasks during the corona pandemic. With people caught in the trap of "confirmation bias," (e.g., Chapter 1, The Nuts and Bolts of Science-Based Advocacy; Chapter 2, When Scientists Are Attacked: Strategies For Dissident Scientists and Whistleblowers; Chapter 8, Why advocate and how?), the tendency to embrace information that supports one's beliefs and to reject information that contradicts them, scientists will not get their message across with better data alone. They must link up with—and provide ammunition to—those who will act (e.g., Chapter 7, Scientific Integrity and Advocacy: Keeping the Government Honest, essays in Chapter 11, Essays from the Trenches of Science-Based Activism, Chapter 14, The Politics of Conservation—Taking the Biodiversity Crisis to the Streets). Battling vested interests, carefully nurtured anti-intellectualism, polarization, and alternative facts, the fight for science fight is uphill, not the least since it has also been financially gutted. The federal government's research and development spending as a share of GDP has dropped to a quarter of what it was at its height in the 1960s (Pethokoukis, 2020). That is the landscape.

Science is above all a method of inquiry, the posing of questions, and the rigorous testing of hypotheses against data (Chapter 1, The Nuts and Bolts of Science-Based Advocacy; Chapter 2, When Scientists Are Attacked: Strategies For Dissident Scientists and Whistleblowers). With new and better data, more reliable conclusions are obtained. Scientists by disposition hope for a broad consensus in favor of measures to save the planet. A related hope is that data will speak for themselves and that fact-based rational discourse will result in reasonable public policy (as in Baconian science, see Chapter 1, The Nuts and Bolts of Science-Based Advocacy). Advocacy—expert voices mobilizing public pressure to influence public policy—has diminishing returns and, thus is best reserved for existential threats.

Albert Einstein wrote in 1947: "We scientists believe that a clear and widespread understanding of the facts and implications of the atomic discoveries is indispensable to a reasonable public stand on questions of international politics" (Einstein, 1947). With nuclear annihilation the most immediate danger for human survival, Einstein and others mobilized to halt the march to destruction (The Russell-Einstein Manifesto, 1955). The Pugwash Conference was established in 1957; it won the Nobel Peace Prize in 1995 but has sadly faded. In 1953 the United States and then the Soviet Union began testing thermonuclear weapons. The *Bulletin of Atomic Scientists* set its Doomsday Clock at 2 minutes to midnight. Today, after many gyrations, it is back at less than 2 minutes, with Global Heating having been added to the existential threats facing humanity. Actions "of many world leaders continue to increase global risk, at a time when the opposite is urgently needed" (Bulletin of the Atomic Scientists, 2020). Nuclear war, as the maximum impact and irrationally high probability risk, has in recent years been joined by Global Heating, a decidedly probable and certainly devastating threat (World Economic Forum, 2020).

In connection with the Rio Earth Summit in 1992, some 1700 of the world's leading scientists, including the majority of Nobel laureates in the sciences, issued an urgent *Warning to Humanity*: "Human beings and the natural world are on a collision course. Human activities inflict harsh and often irreversible damage on the environment and on critical resources" (Union of Concerned Scientists, 1992). More than two decades later, the National Science Academies of Brazil, Britain, Canada, China, India, France, Germany, Italy, Japan, Russia, and the United States urged world leaders to reduce the causes of climate change and to prepare for its consequences (Joint Science Academies' Statement, 2005). There were several such warnings from the world's science academies before and since.[12]

Ripple et al. (2017) picked up the thread in 2017, issuing the *World Scientists' Warning to Humanity: A Second Notice* that was signed by over 15,000 scientists from 184 countries around the world. A more detailed sequel was issued in 2020, endorsed by more than 11,000 scientists from around the world. It stated unequivocally, based on myriad indicators, that Earth is facing a climate emergency (Ripple et al., 2020).

It could be hoped that data would tell the story that the implications are understood and that politicians in positions of authority would act to mitigate inevitable damage. But that is not how it works. Facts, insights, projections are—without power—nothing more than data points, concepts, models, and musings, floating down from ivory towers without much impact. As the saying goes, some people change when they see the light, others when they feel the heat. Science provides light, not pressure. The formidable abolitionist Frederick Douglass noted: "The whole history of progress shows that all concessions yet made have been born of earnest struggle. Power concedes nothing without a demand. It never did and it never will" (Douglass, 1857). Assembling an unimpeachable inventory of data and getting thousands of scientists to rally behind it is a signal achievement. But in a world run by interests, rather than by ideas, it does not effect change.

A dramatic turn of events was Greta Thunberg's address to the United Nations Climate Change Conference (COP24) in Katowice on December 12th, 2018. The then 15-year-old girl who, a few months earlier had begun skipping school

[12]*Scientists for Science-based Policy,* Statement to restore science-based Policy in Government by Concerned Members of the U.S. National Academy of Sciences, *April 2018 (updated June 2020). https://scientistsforsciencebasedpolicy.org/.*
Joint Statements on Climate Change from National Science Academies around the World, January 27th 2017. https://www.skepticalscience.com/joint-statements-on-climate-change-from-nas-around-world.html.

each Friday in a solitary act of civil disobedience to shame the world into addressing climate change, declared: "You are not mature enough to tell it like it is. Even that burden, you leave to us children" (Thunberg, 2018). Four months later, the *Fridays for Future* movement burst onto the global scene. Hundreds of thousands of young people around the world began to organize, march and exert political pressure. Ideas come alive through icons: Mahatma Gandhi, Martin Luther King, Willy Brandt, Nelson Mandela, and Mikhail Gorbachev. Like them, Greta Thunberg represents an idea whose time has come. Her authenticity and integrity catalyzed action. What a gift to the world!

Thunberg's call resonated: "Listen to the science, and act on what you hear!" In Germany, Austria, and Switzerland, within days, a group of academics formed *Scientists for Future* (S4F) to support the students. S4F put out a statement that was endorsed by 26,800 scientists (Hagedorn et al., 2019).

There were tangible results both in Europe and in the United States, with green platforms unveiled that would have been unthinkable 2 years ago (European Commission, 2019; Biden-Sanders Unity Task Force, 2020). Building a clean energy economy that is oriented around the future is a good way to put people to work and to address the climate crisis that cannot wait any longer. Good beginnings, surely, yet far still from the goal of climate neutrality. We are in the opening skirmishes of a marathon battle over goals, means, timelines, and resources. The outcome is uncertain.

Speaking truth to power

With rare unanimity, scientists warn that possibly human life, certainly human civilization, is imperiled, overwhelmed by an economic machine in overdrive. The devastating climate crisis—not looming but ongoing and accelerating—emphatically is not the result of bad science, lacking data, absent information, or inadequate communication. It is the result of greed, hubris, indifference, and policy failure.

Perhaps, there is too much information and not enough interpretation or integration. Tens of thousands of scientists work on climate issues, putting out ever more detailed—and ever more alarming—analyses. From a policy perspective, the quantity of studies, their wide range, and their conservatism—erring on the side of detachment—might actually impede action in a world of short attention spans, accelerating news cycles and vested interests. An overwhelmed public tired of bad news unwittingly creates openings for slow walkers, opportunists, deniers, obstructionists, and profiteers, who sit back, waiting for the uproar du jour to pass, like so many before. Yet, nature's clock is ticking, and the living conditions of our children and grandchildren are imperiled.

Politicians frame issues in terms of what is doable, not necessity. Scientists, on the authority of their work, must call them out. With the Global Heating endgame in sight, scientists need to leave no doubt that incremental greening will not do the trick, that disruptive change is unavoidable and that options are receding. Very soon, nature will have taken over and removed possibilities for human influence altogether. In the short time remaining before the climate tips irreversibly, scientists, but also physicians and educators, must deploy their moral authority, innate credibility, and the professional responsibility to prevent avoidable harm (the Planetary Hippocratic Oath, Chapter 1, The Nuts and Bolts of Science-Based Advocacy). They must communicate effectively the dangers facing human civilization and critically evaluate proposed political solutions.

Barry Commoner's four laws of ecology are as valid today as when they were formulated 50 years ago (Butler, 2012). But they are ever so much more urgent:

1. Everything is connected to everything else (there is one ecosphere for all living organisms and what affects one, affects all).
2. Everything must go somewhere (there is no "waste" in nature and there is no "away" to which things can be thrown).
3. Nature knows best (any major human-made change in a natural system is likely to be detrimental to that system).
4. There is no such thing as a free lunch (exploitation of nature will inevitably involve the conversion of resources from useful to useless forms).

References

Allen, M.R., Fuglestvedt, J.S., Shine, K.P., Reisinger, A., Pierrehumbert, R.T., Forster, P.M., 2016. New use of global warming potentials to compare cumulative and short-lived climate pollutants. Nat. Clim. Change 6, 773–776. Available from: http://10.1038/nclimate2998.

American Institute of Physics, 2020. The discovery of global warming, The carbon dioxide greenhouse effect, January 2020. https://history.aip.org/history/climate/co2.htm.

Anderson, K., Broderick, J.F., Stoddard, I., 2020. A factor of two: how the mitigation plans of 'climate progressive' nations fall far short of Paris-compliant pathways. Clim. Policy 20 (10). Available from: https://www.tandfonline.com/doi/full/10.1080/14693062.2020.1728209.

Arrhenius, S., 1896. On the influence of carbonic acid in the air upon the temperature of the ground. London, Edinburgh, Dublin Philos. Mag. J. Sci., Series 5 41, 237–276. Available from: https://www.rsc.org/images/Arrhenius1896_tcm18-173546.pdf.

Banerjee, N., 2015. More exxon documents show how much it knew about climate 35 years ago: documents reveal Exxon's early CO_2 position, its global warming forecast from the 1980s, and its involvement with the issue at the highest echelons, *Inside* Climate News, December 1st 2015. https://insideclimatenews.org/news/01122015/documents-exxons-early-co2-position-senior-executives-engage-and-warming-forecast.

Batty, M., Briggs, J., Pence, K., Smith, P., Volz, A., 2019. The distributional financial accounts, FEDS Notes. Washington: Board of Governors of the Federal Reserve System, August 30th 2019. https://www.federalreserve.gov/econres/notes/feds-notes/the-distributional-financial-accounts-20190830.htm.

Baumert, K.A., Herzog, T., Pershing, J., 2005. Navigating the numbers: greenhouse gas data and international climate policy. World Resources Institute, table 4, p. 113. http://pdf.wri.org/navigating_numbers.pdf.

Biden-Sanders Unity Task Force Recommendations: *Combating the climate crisis and pursuing environmental Justice*. July 8th 2020. https://joebiden.com/wp-content/uploads/2020/07/UNITY-TASK-FORCE-RECOMMENDATIONS.pdf.

Borrelle, S.B., Rochman, C.M., Liboiron, M., Bond, A.L., Lusher, A., Bradshaw, H., et al., 2017. Why we need an international agreement on marine plastic pollution. Proc. Natl Acad. Sci. U. S. A. 114 (38), 9994–9997. Available from: http://www.pnas.org/content/114/38/9994.

Branch, G., Reid, A., 50 Years ago: repeal of Tennessee's 'Monkey Law', Scientific American, May 10th 2017. https://blogs.scientificamerican.com/observations/50-years-ago-repeal-of-tennessees-monkey-law/.

Broecker, W.S., 1975. Climatic change: are we on the brink of a pronounced global warming? Science 189 (4201), 460–463. Available from: https://doi.org/10.1126/science.189.4201.460.

Broecker, W.S., 1987. Unpleasant surprises in the greenhouse? Nature 328, 123–126. Available from: https://www.nature.com/articles/328123a0.

Brysse, K., Oreskes, N., O'Reilly, J., Oppenheimer, M., 2013. Climate change prediction: erring on the side of least drama? Glob. Environ. Change 23 (1), 327–337. Available from: https://doi.org/10.1016/j.gloenvcha.2012.10.008.

Bulletin of the Atomic Scientists, *It is 100 seconds to midnight. 2020 Doomsday Clock Statement Science and Security Board Bulletin of the Atomic Scientists*. January 23rd 2020. https://thebulletin.org/wp-content/uploads/2020/01/2020-Doomsday-Clock-statement.pdf.

Butler, S., Barry commoner: scientist, activist, radical ecologist, Green Left, issue 941, October 4th 2012. https://www.greenleft.org.au/content/barry-commoner-scientist-activist-radical-ecologist.

Career Builder Press release, August 24th 2017. http://press.careerbuilder.com/2017-08-24-Living-Paycheck-to-Paycheck-is-a-Way-of-Life-for-Majority-of-U-S-Workers-According-to-New-CareerBuilder-Survey.

Carney, M., Breaking the tragedy of the horizon: climate change and financial stability. Speech at Lloyd's of London, September 29th 2015. https://www.bankofengland.co.uk/speech/2015/breaking-the-tragedy-of-the-horizon-climate-change-and-financial-stability.

CDU. 7. Parteitag der CDU Deutschlands, *Sicher in die Zukunft*, October 1995; Decision C 83 Steuerreform – Einführung einer CO_2-Energiesteuer, p. 289. https://www.kas.de/c/document_library/get_file?uuid = 4d88ca6e-5b15-9747-e10e-ccddb9981815&groupId = 252038.

Cheng, L., Abraham, J., Hausfather, Z., Trenberth, K.E., 2019. How fast are the oceans warming? Science 363 (6423), 128–129. Available from: https://doi.org/10.1126/science.aav7619.

Chetty, R., Stepner, M., Abraham, S., Lin, S., Scuderi, B., Turner, N., et al., 2016. The association between income and life expectancy in the united states, 2001–2014. J. Am. Med. Assoc. 315 (16), 1750–1766. Available from: https://doi.org/10.1001/jama.2016.4226.

Climate Files, 1988. Shell confidential report The Greenhouse Effect. http://www.climatefiles.com/shell/1988-shell-report-greenhouse/.

Climate Investigations Center, Global climate coalition: climate denial legacy follows corporations. April 25th 2019. https://climateinvestigations.org/global-climate-coalition-industry-climate-denial/.

Coady, D., Parry, I., Le, N.-P., Shang, B., Global fossil fuel subsidies remain large: an update based on country-level estimates. *IMF Working Paper*. May 2nd 2019. https://www.imf.org/en/Publications/WP/Issues/2019/05/02/Global-Fossil-Fuel-Subsidies-Remain-Large-An-Update-Based-on-Country-Level-Estimates-46509.

Coats, D.R., Worldwide threat assessment of the US Intelligence Community. February 13th 2018. https://www.dni.gov/files/documents/Newsroom/Testimonies/2018-ATA-Unclassified-SSCI.pdf.

Coffin, M., Grant, A., Balancing the budget: why deflating the carbon bubble requires oil and gas companies to shrink. Carbon Tracker, November 1st 2019. https://carbontracker.org/reports/balancing-the-budget/.

Corneliussen, S., Do all possible Republican presidential candidates really deny evolution?, *Physics Today*, February 20th 2015. https://physicstoday.scitation.org/do/10.1063/pt.5.8101/full/.

Douglass, F., Two speeches: one on the Dred Scott decision, delivered in New York on the occasion of the anniversary of the American Abolition Society, May, 1857, and one on West India emancipation, delivered at Canandaigua, August 4th 1857. *University of Rochester, Frederick Douglass Project.* https://rbscp.lib.rochester.edu/4398.

Einstein, A., Letter dated November 29th 1947. https://www.docsteach.org/documents/document/letter-from-albert-einstein-and-the-emergency-committee-of-atomic-scientists.

Energy Policy Tracker, *Public money commitments to fossil fuels, clean and other energy in recovery packages.* July 22nd 2020. https://www.energypolicytracker.org/region/g20/.

Environmental Protection Agency (EPA), Established December 2nd 1970. https://www.epa.gov/history.

Environmental Protection Agency (EPA), *The potential effects of global climate change on the United States,* December 1989. https://www.nrc.gov/docs/ML1434/ML14345A597.pdf.

European Academies Science Advisory Council, Negative emission technologies: what role in meeting Paris Agreement targets? EASAC policy report 35. February 2018. https://easac.eu/fileadmin/PDF_s/reports_statements/Negative_Carbon/EASAC_Report_on_Negative_Emission_Technologies.pdf.

European Commission, *The European Green Deal,* December 11th 2019. https://eur-lex.europa.eu/legal-content/EN/TXT/?qid = 1576150542719&uri = COM%3A2019%3A640%3AFIN.

Franta, B., Shell and Exxon's secret 1980s climate change warnings. Newly found documents from the 1980s show that fossil fuel companies privately predicted the global damage that would be caused by their products. *The* Guardian, September 19th 2018. https://www.theguardian.com/environment/climate-consensus-97-per-cent/2018/sep/19/shell-and-exxons-secret-1980s-climate-change-warnings.

Global Carbon Atlas, undated. Territorial emissions. http://www.globalcarbonatlas.org/en/CO2-emissions.

Global Monitoring Laboratory, Earth System Research Laboratories; Carbon cycle greenhouse gases, global CH_4 monthly means. https://www.esrl.noaa.gov/gmd/ccgg/trends_ch4/.

Goldstein, B., Gounaridis, D., Newell, J.P., 2020. The carbon footprint of household energy use in the United States. Proc. Natl Acad. Sci. U. S. A. 117, 19122–19130. Available from: https://doi.org/10.1073/pnas.1922205117.

Greenpeace, undated. Exxon's climate denial history: a timeline. https://www.greenpeace.org/usa/global-warming/exxon-and-the-oil-industry-knew-about-climate-change/exxons-climate-denial-history-a-timeline/.

H.R.1 – American Recovery and Reinvestment Act of 2009. https://www.congress.gov/bill/111th-congress/house-bill/1.

H.R.2454 – American Clean Energy and Security Act of 2009. https://www.congress.gov/bill/111th-congress/house-bill/2454.

Hagedorn, G., et al., 2019. The concerns of the young protesters are justified. A statement by scientists for future concerning the protests for more climate protection. GAIA 28 (2), 79–87. Available from: https://www.oekom.de/_uploads_media/files/gaia_2019_02_79_1_010542.pdf.

Hansen, J., NASA Goddard Space Institute. Testimony to the U.S. Senate Committee on Energy and Natural Resources on June 23rd 1988. https://www.sealevel.info/1988_Hansen_Senate_Testimony.html.

Houghton, J.T., Jenkins, G.J., Ephraums, J.J., 1990. Climate Change: The IPCC Scientific Assessment. Cambridge University Press, p. XI. Available from: https://www.ipcc.ch/site/assets/uploads/2018/03/ipcc_far_wg_I_full_report.pdf.

Hubacek, K., Baiocchi, G., Feng, K., Muñoz Castillo, R., Sun, L., Xue, J., 2017. Global carbon inequality. Energy, Ecol. Environ. 2, 361–369. Available from: https://doi.org/10.1007/s40974-017-0072-9.

Hungerford, T.L., Corporate tax rates and economic gro wth since 1947. Economic Policy Institute, June 4th 2013. https://www.epi.org/publication/ib364-corporate-tax-rates-and-economic-growth/.

IATA Industry Statistics. Fact sheet, June 2020. https://www.iata.org/en/iata-repository/pressroom/fact-sheets/fact-sheet-industry-statistics/.

Immigration and Nationality Act, October 3rd 1965. https://www.govinfo.gov/content/pkg/STATUTE-79/pdf/STATUTE-79-Pg911.pdf.

Infratest dimap; Flughäufigkeit, July 2019. https://www.infratest-dimap.de/umfragen-analysen/bundesweit/umfragen/aktuell/sechs-von-zehn-fliegen-selten-oder-generell-nicht/.

Intergovernmental Panel on Climate Change (IPCC), 1990/1992. The 1990/1992 assessments. https://www.ipcc.ch/report/climate-change-the-ipcc-1990-and-1992-assessments/.

Intergovernmental Panel on Climate Change (IPCC), Global warming of 1.5°C: an IPCC special report on the impacts of global warming of 1.5°C above pre-industrial levels and related global greenhouse gas emission pathways, in the context of strengthening the global response to the threat of climate change, sustainable development, and efforts to eradicate poverty, Summary for Policymakers, October 6th 2018. https://www.ipcc.ch/sr15/.

Intergovernmental Panel on Climate Change (IPCC), I, Global warming of 1.5°C: an IPCC special report on the impacts of global warming of 1.5°C above pre-industrial levels and related global greenhouse gas emission pathways, in the context of strengthening the global response to the threat of climate change, sustainable development, and efforts to eradicate poverty. Full report, 2019, p. 559. https://www.ipcc.ch/site/assets/uploads/sites/2/2019/06/SR15_Full_Report_High_Res.pdf.

Intergovernmental Panel on Climate Change (IPCC), II, Summary for policymakers. In: Pörtner, H.-O., Roberts, D.C., Masson-Delmotte, V., Zhai, P., Tignor, M., Poloczanska, E., Mintenbeck, K., Nicolai, M., Okem, A., Petzold, J., Rama, B., Weyer, N. (Eds.), IPCC Special Report on the Ocean and Cryosphere in a Changing Climate. In press. September 24th 2019. https://www.ipcc.ch/site/assets/uploads/sites/3/2019/11/03_SROCC_SPM_FINAL.pdf.

Intergovernmental Panel on Climate Change (IPCC), III, Climate change and land: an IPCC special report on climate change, desertification, land degradation, sustainable land management, food security, and greenhouse gas fluxes in terrestrial ecosystems: summary for policymakers, August 7th 2019. https://www.ipcc.ch/site/assets/uploads/2019/08/4.-SPM_Approved_Microsite_FINAL.pdf.

Intergovernmental Science-Policy Platform on Biodiversity and Ecosystem Services (IPBES), Media Release: *Nature's dangerous decline 'unprecedented;' species extinction rates 'accelerating'*, May 6th 2019. https://www.ipbes.net/news/Media-Release-Global-Assessment.

International Civil Aviation Organization (ICAO) News Release. *ICAO Council agrees to the safeguard adjustment for CORSIA in light of COVID-19 pandemic.* Montréal, June 30th 2020. https://www.icao.int/Newsroom/Pages/ICAO-Council-agrees-to-the-safeguard-adjustment-for-CORSIA-in-light-of-COVID19-pandemic.aspx.

International Conference of the Changing Atmosphere: Implications for Global Security, (June 27th–30th 1988, Toronto). https://www.academia.edu/4043227/The_Changing_Atmosphere_Implications_for_Global_Security_Conference_Statement_1988.

International Energy Agency (IEA), *Global energy review 2020*, April 2020. https://www.iea.org/reports/global-energy-review-2020.

International Organization of Motor Vehicle Manufacturers (OICA), undated. World vehicles in use (all). http://www.oica.net/wp-content/uploads//Total_in-use-All-Vehicles.pdf.

Jackson, R.B., Saunois, M., Bousquet, P., Canadell, J.G., Poulter, B., Stavert, A.R., et al., 2020. Increasing anthropogenic methane emissions arise equally from agricultural and fossil fuel sources. Environ. Res. Lett. 15 (7). Available from: https://doi.org/10.1088/1748-9326/ab9ed2.

Jambeck, J.R., Geyer, R., Wilcox, C., Siegler, T.R., Perryman, M., Andrady, A., et al., 2015. Plastic waste inputs from land into the ocean. Science 347 (6223), 768–771. Available from: http://science.sciencemag.org/content/347/6223/768?ijkey = 86112180a6c6e6eb14fd994ccfe28e0fc3edb9cd&keytype2 = tf_ipsecsha.

Jeffery, L., Höhne, N., Moisio, M., Day, T., Lawless, B., Options for supporting carbon dioxide removal, NewClimate Institute, July 28th 2020. https://newclimate.org/2020/07/28/options-for-supporting-carbon-dioxide-removal-discussion-paper/.

Johnson, L.B., *Special message to congress on conservation and restoration of natural beauty*, February 8th 1965. https://www.presidency.ucsb.edu/documents/special-message-the-congress-conservation-and-restoration-natural-beauty.

Joint Science Academies' Statement, *Global response to climate change*, June 2005. https://royalsociety.org/-/media/Royal_Society_Content/policy/publications/2005/9649.pdf.

Joint Statement following the Soviet-United States Summit Meeting in Moscow, June 1st 1988. https://www.reaganlibrary.gov/research/speeches/060188b.

Joos, F., Spahni, R., 2008. Rates of change in natural and anthropogenic radiative forcing over the past 20,000 years. Proc. Natl Acad. Sci. U. S. A. 105 (5), 1425–1430. Available from: https://doi.org/10.1073/pnas.0707386105.

Kaufman, D., McKay, N., Routson, C., Erb, M., Dätwyler, C., Sommer, P.S., et al., 2020. Holocene global mean surface temperature, a multi-method reconstruction approach. Sci. Data 7, 201. Available from: https://doi.org/10.1038/s41597-020-0530-7.

Kershner, J., NEPA, the National Environmental Policy Act, August 27th 2011, HistoryLink.org. https://www.historylink.org/File/9903.

Kolbert, E., Can carbon-dioxide removal save the world?, *The* New Yorker, November 20th 2017. https://www.newyorker.com/magazine/2017/11/20/can-carbon-dioxide-removal-save-the-world.

Kyoto Protocol to the United Nations Framework Convention on Climate Change of December 11th 1997, entered into force on February 16th 2005 (1997/2005). https://unfccc.int/sites/default/files/kpeng.pdf.

Lelieveld, J., Klingmüller, K., Pozzer, A., Burnett, R.T., Haines, A., Ramanathan, V., 2019. Effects of fossil fuel and total anthropogenic emission removal on public health and climate. Proc. Natl. Acad. Sci. U. S. A. 116 (15), 7192–7197. Available from: https://doi.org/10.1073/pnas.1819989116.

Lenton, T.M., Rockström, J., Gaffney, O., Rahmstorf, S., Richardson, K., Steffen, W., et al., 2019. Climate tipping points—too risky to bet against. The growing threat of abrupt and irreversible climate changes must compel political and economic action on emissions. Nature 575, 592–595. Available from: https://www.nature.com/articles/d41586-019-03595-0.

Lindsay, R., Climate change: atmospheric carbon dioxide, National Oceanic and Atmospheric Administration, February 20th 2020. https://www.climate.gov/news-features/understanding-climate/climate-change-atmospheric-carbon-dioxide.

Matthey, A., Bünger, B., *Methodological convention 3.0 for the assessment of environmental costs*—cost rates version 02/2019, February 2019. https://www.umweltbundesamt.de/sites/default/files/medien/1410/publikationen/2019-02-11_methodenkonvention-3-0_en_kostensaetze_korr.pdf.

McSweeney, R., Nine 'tipping points' that could be triggered by climate change, Carbon Brief, February 10th 2020. https://www.carbonbrief.org/explainer-nine-tipping-points-that-could-be-triggered-by-climate-change.

Mishel, L., Wolfe, J., CEO compensation has grown 940% since 1978. Typical worker compensation has risen only 12% during that time, Economic Policy Inst*itute*, August 14th 2019. https://www.epi.org/publication/ceo-compensation-2018/.

Molnár, P.K., Bitz, C.M., Holland, M.M., Kay, J.E., Penk, S.R., Amstrup, S.C., 2020. Fasting season length sets temporal limits for global polar bear persistence. Nat. Clim. Change 10, 732–738. Available from: https://doi.org/10.1038/s41558-020-0818-9.

Motesharrei, S., Rivas, J., Kalnay, E., Asrar, G.R., Busalacchi, A.J., Cahalan, R.F., et al., 2016. Modeling sustainability: population, inequality, consumption, and bidirectional coupling of the Earth and human systems. Natl Sci. Rev. 3 (4), 470–494. Available from: https://doi.org/10.1093/nsr/nww081.

NASA, Earth Factsheet, undated. https://nssdc.gsfc.nasa.gov/planetary/factsheet/earthfact.html.

NASA, undated. Global climate change, vital signs of the planet. The causes of climate change. https://climate.nasa.gov/causes/.

NASA, Visit to an Ocean Planet, undated. Making a greenhouse. https://sealevel.jpl.nasa.gov/files/archive/activities/ts1hiac1.pdf.

National Oceanic and Atmospheric Administration (NOAA), Established October 3rd 1970. https://www.history.noaa.gov/legacy/noaahistory_1.html.

National Oceanic and Atmospheric Administration (NOAA), National Centers for Environmental Information. *State of the climate: global climate report for March 2020*, published online April 2020a. https://www.ncdc.noaa.gov/sotc/global/202003/supplemental/page-2.

National Oceanic and Atmospheric Administration (NOAA). Rise of carbon dioxide unabated. Seasonal peak reaches 417 parts per million at Mauna Loa Observatory, NOAA Research News, June 4th 2020b. https://research.noaa.gov/article/ArtMID/587/ArticleID/2636/Rise-of-carbon-dioxide-unabated.

National Traffic and Motor Vehicle Safety Act of 1966, September 9th 1966. https://uscode.house.gov/statutes/pl/89/563.pdf.

Nobel Prize Organization, undated. *Nobel prize facts.* https://www.nobelprize.org/prizes/facts/nobel-prize-facts/.

Nordhaus, W., 2020. The climate club: how to fix a failing global effort. Foreign Aff. 99 (3), 10–17. Available from: https://www.foreignaffairs.com/articles/united-states/2020-04-10/climate-club.

Osterholm, M.T., Olshake, M., 2020. Chronicle of a pandemic foretold. Learning from the COVID-19 failure—before the next outbreak arrives. Foreign Aff. 99 (4), 10–24. Available from: https://www.foreignaffairs.com/articles/united-states/2020-05-21/coronavirus-chronicle-pandemic-foretold.

Our World in Data, undated. Annual CO_2 emissions 1751–2017. https://ourworldindata.org/grapher/annual-co2-emissions-per-country?year = latest&time = .

Our World in Data Global Plastics Production, 1950-2015, undated. https://ourworldindata.org/grapher/global-plastics-production.

Oxfam Media Briefing, Extreme carbon inequality. Why the Paris climate deal must put the poorest, lowest emitting and most vulnerable people first, December 2nd 2015. https://oi-files-d8-prod.s3.eu-west-2.amazonaws.com/s3fs-public/file_attachments/mb-extreme-carbon-inequality-021215-en.pdf?cid = aff_affwd_donate_id85386&awc = 5991_1594500806_d206a0e7eb36cdcdc0d086e3f80872cf.

Painter, D.S., 2009. The Marshall plan and oil. Cold War Hist. 9 (2), 159–175. Available from: https://doi.org/10.1080/14682740902871851.

Paris Agreement, FCCC/CP/2015/L.9, December 12th 2015, Article 2 (a), p. 22. https://unfccc.int/resource/docs/2015/cop21/eng/l09r01.pdf.

Pethokoukis, J., US Federal Research Spending is at a 60-year low, Am*erican* Enterpr*ise* Inst*itute*, May 11th 2020. https://www.aei.org/economics/us-federal-research-spending-is-at-a-60-year-low-should-we-be-concerned/.

Pew Research Center, *Republicans' views on evolution*, January 3rd 2014. https://www.pewresearch.org/fact-tank/2014/01/03/republican-views-on-evolution-tracking-how-its-changed/.

Pierrehumbert, R.T., 2004. Warming the world. Nature 432, 677. Available from: https://doi.org/10.1038/432677a.

PlasticsEurope, 2019. *Plastics—the facts 2019.* https://www.plasticseurope.org/application/files/9715/7129/9584/FINAL_web_version_Plastics_the_facts2019_14102019.pdf.

President's Science Advisory Committee, Environmental pollution panel, Restoring the quality of our environment, *The White House*, Washington, November 1965, p. 112. http://www.climatefiles.com/climate-change-evidence/presidents-report-atmospher-carbon-dioxide/#:~:text = 1965%20President%20Science%20Advisory%20Committee%20Report%20on%

20Atmospheric%20Carbon%20Dioxide&text = %E2%80%9CRestoring%20the%20Quality %20of%20Our,role%20in%20addressing%20the%20future.

Rainforest Action Network, *Banking on climate change: Fossil fuel finance report card 2020*, March 18th 2020. https://www.ran.org/wp-content/uploads/2020/03/Banking_on_Climate_Change__2020_vF.pdf.

Revelle, R., Suess, H.E., 1957. Carbon dioxide exchange between atmosphere and ocean and the question of an increase of atmospheric CO_2 during the past decades. Tellus 9 (1), 18–27, here page 19. Available from: doi/abs/10.3402/tellusa.v9i1.9075. Available from: https://www.tandfonline.com/doi/pdf/10.3402/tellusa.v9i1.9075?needAccess = true.

Revkin, A.C., Industry ignored its scientists on climate, *The N*ew York Times, April 23rd 2009. https://www.nytimes.com/2009/04/24/science/earth/24deny.html.

Ripple, W.J., Smith, P., Haberl, H., Montzka, S.A., McAlpine, C., Boucher, D.H., 2013. Ruminants, climate change and climate policy. Nat. Clim. Change 4, 2–5. Available from: https://doi.org/10.1038/nclimate2081.

Ripple, W.J., Wolf, C., Newsome, T.M., Barnard, P., Moomaw, W.R., 2020. World scientists' warning of a climate emergency. BioScience 70 (1), 8–12. Available from: https://doi.org/10.1093/biosci/biz088.

Ripple, W.J., Wolf, C., Newsome, T.M., Galetti, M., Alamgir, M., Crist, E., et al., 2017. World scientists' warning to humanity: a second notice. BioScience 67 (12), 1026–1028. Available from: https://doi.org/10.1093/biosci/bix125.

Rogelj, J., Forster, P.M., Kriegler, E., Smith, C.J., Séférian, R., 2019. Estimating and tracking the remaining carbon budget for stringent climate targets. Nature 571, 335–342. Available from: https://doi.org/10.1038/s41586-019-1368-z.

Saussay, A., 2019. Global historical emissions map. https://aureliensaussay.github.io/historica-lemissions/.

Savage, K., Global climate coalition: fighting global climate action in favor of fossil fuels' survival, *The Climate Docket*, April 25th 2019. https://www.climatedocket.com/2019/04/25/gcc-global-climate-coalition-un-fossil-fuels/.

Schneider, C., H.R. 1078 (101st): Global Warming Prevention Act of 1989: to establish national policies and support and encourage international agreements that implement energy and natural resource conservation strategies appropriate to preventing the overheating of the Earth's atmosphere, known as the "greenhouse effect." Introduced on February 22nd 1989. https://www.govtrack.us/congress/bills/101/hr1078.

Schwartz, J., Overlooked no more: Eunice Foote, climate scientist lost to history, *The N*ew York Times, April 21st 2020. https://www.nytimes.com/2020/04/21/obituaries/eunice-foote-overlooked.html.

Scripps Institution of Oceanography, undated. Full record. https://scripps.ucsd.edu/programs/keelingcurve/wp-content/plugins/sio-bluemoon/graphs/mlo_full_record.png.

SEI, IISD, ODI, Climate Analytics, CICERO, UNEP, 2019. *The production gap: the discrepancy between countries' planned fossil fuel production and global production levels consistent with limiting warming to 1.5°C or 2°C.* http://productiongap.org/wp-content/uploads/2019/11/Production-Gap-Report-2019.pdf.

Sivaram, V., Can't stop the shining solar power is the world's most promising clean energy solution, but governments must abandon outdated policies for it to succeed, Foreign Policy, March 1st 2018. http://foreignpolicy.com/2018/03/01/cant-stop-the-shining/.

Smith, A., 1880/1776. *An inquiry into the nature and causes of the wealth of nations*, 1880/1776, vol. 1, p. 31. https://books.google.com/books?id = mlsPAQAAIAAJ&printsec = frontcover&source = gbs_ge_summary_r&cad = 0#v = onepage&q&f = false.

Social Security Act Amendments, 1965. July 30th 1965. https://www.ourdocuments.gov/doc.php?flash = false&doc = 99#.

Song, L., Banerjee, N., Hasemyer, D., Exxon: the road not taken, Inside Climate News, September 22nd 2015. https://insideclimatenews.org/news/18092015/exxon-confirmed-global-warming-consensus-in-1982-with-in-house-climate-models.

Stern, N., 2015. Economic development, climate and values: making policy. Proc. R. Soc. B 282 (1812). Available from: http://rspb.royalsocietypublishing.org/content/282/1812/20150820.

Thatcher, M., Speech to the Royal Society, September 27th 1988. https://www.margaret-thatcher.org/document/107346.

The Civil Rights Act of 1968, also known as the Fair Housing Act, April 11th 1968. https://www.hud.gov/program_offices/fair_housing_equal_opp/aboutfheo/history#:~:text = The %201968%20Act%20expanded%20on,Housing%20Act%20(of%201968).

The Clean Air Act of December 31st 1970. https://www.govinfo.gov/content/pkg/STATUTE-84/pdf/STATUTE-84-Pg1676.pdf.

The Clean Water Act of October 18th 1972. https://www.boem.gov/environment/environmental-assessment/clean-water-act-cwa.

The Economist. The hierarchy of countries winning Nobels in the sciences is shifting, May 10th 2018. https://www.economist.com/graphic-detail/2018/05/10/the-hierarchy-of-countries-winning-nobels-in-the-sciences-is-shifting.

The Endangered Species Act of December 27th 1973. https://www.fws.gov/international/pdf/esa.pdf.

The Global Commission on the Economy and Climate. Unlocking the inclusive growth story of the 21st century: accelerating climate action in urgent times, August 2018. https://newclimateeconomy.report/2018/wp-content/uploads/sites/6/2018/09/NCE_2018_FULL-REPORT.pdf.

The Russell-Einstein Manifesto. July 9th 1955. http://umich.edu/~pugwash/Manifesto.html.

The World Bank—Data Air transport, passengers carried. https://data.worldbank.org/indicator/IS.AIR.PSGR.

The World Bank—Data GDP in current US$. https://data.worldbank.org/indicator/NY.GDP.MKTP.CD?locations = CN-IN-KR.

The World Bank—Data, GDP per capita (in current dollars). https://data.worldbank.org/indicator/NY.GDP.PCAP.CD.

The World Bank—Data International Tourism, Number of Arrivals. https://data.worldbank.org/indicator/ST.INT.ARVL.

The World Bank—Data Life Expectancy at Birth. https://data.worldbank.org/indicator/SP.DYN.LE00.IN.

The World Bank—Data Population. https://data.worldbank.org/indicator/SP.POP.TOTL?locations = 1W.

The World Bank, I, 2020. Global gas flaring jumps to levels last seen in 2009, Press Release, July 21st 2020. https://www.worldbank.org/en/news/press-release/2020/07/21/global-gas-flaring-jumps-to-levels-last-seen-in-2009.

The World Bank, II, 2020. Mozambique mining and gas technical assistance project (P129847): implementation status & results report, March 30th 2020. http://documents1.worldbank.org/curated/en/714271585576119813/pdf/Disclosable-Version-of-the-ISR-Mozambique-Mining-and-Gas-Technical-Assistance-Project-P129847-Sequence-No-13.pdf.

The World Bank, III, 2020. State and trends of carbon pricing 2020, May 2020, p. 8. https://openknowledge.worldbank.org/bitstream/handle/10986/33809/9781464815867.pdf?sequence = 4&isAllowed = y.

Thunberg, G., 2018. Address to the United Nations climate change conference in Katowice, December 12th 2018. https://www.youtube.com/watch?v = VFkQSGyeCWg.

Times Higher Education. World university rankings 2020. https://www.timeshighereducation.com/world-university-rankings/2020/world-ranking#!/page/0/length/25/sort_by/rank/sort_order/asc/cols/stats.

Total Press Release, Total announces the signing of mozambique lng project financing, July 17th 2020. https://www.total.com/media/news/communiques/total-announces-the-signing-of-mozambique-lng-project-financing.

Trisos, C.H., Meros, C., Pigot, A.L., 2020. The projected timing of abrupt ecological disruption from climate change. Nature 580, 496–501. Available from: https://doi.org/10.1038/s41586-020-2189-9.

Tyndall, J., 1859/1860. Note on the transmission of radiant heat through gaseous bodies. Proc. R. Soc. Lond. 10, 37–39. Available from: https://www.jstor.org/stable/111604?seq = 1#metadata_info_tab_contents.

UK Department of Transportation, Statistical Release, *Public experiences of and attitudes towards air travel: 2014*, July 23rd 2014, p. 2. https://assets.publishing.service.gov.uk/government/uploads/system/uploads/attachment_data/file/336702/experiences-of-attitudes-towards-air-travel.pdf.

UK Met Office, Mauna Loa carbon dioxide forecast for 2020. https://www.metoffice.gov.uk/research/climate/seasonal-to-decadal/long-range/forecasts/co2-forecast.

Union of Concerned Scientists, *1992 World scientists' warning to humanity*, July 16th 1992. https://www.ucsusa.org/resources/1992-world-scientists-warning-humanity.

United Nations Environment Programme (UNEP), *The emissions gap report 2019*, November 2019, p. XX. https://wedocs.unep.org/bitstream/handle/20.500.11822/30797/EGR2019.pdf?sequence = 1&isAllowed = y.

United Nations Framework Convention on Climate Change (UNFCCC), 1992. article 2. https://unfccc.int/files/essential_background/background_publications_htmlpdf/application/pdf/conveng.pdf.

United Nations General Assembly, Resolution A/RES/43/53, *Protection of global climate for present and future generations*, December 6th 1988. https://undocs.org/A/RES/43/53.

United Nations General Assembly, Resolution A/RES/70/1, *Transforming our world: the 2030 agenda for sustainable development*, September 25th 2015, Resolution. http://undocs.org/A/RES/70/1.

United Nations General Assembly, Resolution A/RES/73/195, *Global compact for safe, orderly and regular migration*, December 19th 2018. http://undocs.org/A/RES/73/195.

United Nations Treaty Collection, Chapter XXVII: 7. A Kyoto protocol to the United Nations framework convention on climate change. https://treaties.un.org/Pages/ViewDetails.aspx?src = TREATY&mtdsg_no = XXVII-7-a&chapter = 27&clang = _en.

United Nations World Tourism Organization, *All countries: inbound tourism: arrivals 1995–2016* (01.2018). https://www.e-unwto.org/doi/abs/10.5555/unwtotfb0000270019952016201801.

United States Army War College, *Implications of climate change for the U.S. Army*, May 2019. https://climateandsecurity.files.wordpress.com/2019/07/implications-of-climate-change-for-us-army_army-war-college_2019.pdf.

United States Energy Information Administration, *Natural gas explained*. December 9th 2019. https://www.eia.gov/energyexplained/natural-gas/.

United States Energy Information Administration, *Electric power monthly, Table 1.1.A. Net generation from renewable sources*: Total (all sectors), 2010–April 2020. https://www.eia.gov/electricity/monthly/epm_table_grapher.php?t = epmt_1_01_a.

University of Virginia, Institute of Business in Society, *Fishing with dynamite*, 2020. https://www.darden.virginia.edu/ibis/fishing-with-dynamite.

USGCRP, 2018. Impacts, risks, and adaptation in the United States: Fourth National Climate Assessment, Volume II. In: Reidmiller, D.R., Avery, C.W., Easterling, D.R., Kunkel, K.E., Lewis, K.L.M., Maycock, T.K., Stewart, B.C. (Eds.). Washington, DC, USA: U.S. Global Change Research Program. https://doi.org/10.7930/NCA4.2018.

Vega, Edl, Chalk, T.B., Wilson, P.A., Bysani, R.P., Foster, G.L., 2020. Atmospheric CO_2 during the Mid-Piacenzian Warm Period and the M2 glaciation. Sci. Rep. 10, 11002. Available from: https://doi.org/10.1038/s41598-020-67154-8.

Voosen, P., 2020. Earth's climate destiny finally seen more clearly. Science 369 (6502), 354–355. Available from: https://doi.org/10.1126/science.369.6502.354.

Voting Rights Act, August 6th 1965. https://www.ourdocuments.gov/doc.php?flash = false&doc = 100&page = transcript.

Weinstein, A., Wells Fargo plans $10 billion in cuts, posts first quarterly loss since 2008, *The* Charlotte Observer, July 14th 2020. https://www.charlotteobserver.com/news/business/banking/article244196997.html?mbid = &utm_source = nl&utm_brand = tny&utm_mailing = TNY_ClimateCrisis_072220&utm_campaign = aud-dev&utm_medium = email&bxid = 5be9dd3a24c17c6adf447307&cndid = 28623401&hasha = 2e5dd19b681392269185448a9674eabb&hashb = f1a425e4dad7cf486260c7a3268c6dcb12049f83&hashc = 93536bbc8bf4de1a68b7365e995e368f0f67918f6d815bdb5ae3d741ea8209d2&esrc = no_source_code&utm_term = TNY_ClimateCrisis.

Wiedmann, T., Lenzen, M., Keyßer, L.T., Steinberger, J.K., 2020. Scientists' warning on affluence. Nat. Commun. 11, 3107. Available from: https://doi.org/10.1038/s41467-020-16941-y.

Willeit, M., Ganopolski, A., Calov, R., Brovkin, V., 2019. Mid-Pleistocene transition in glacial cycles explained by declining CO_2 and regolith removal. Sci. Adv. 5 (4), eaav7337. Available from: https://doi.org/10.1126/sciadv.aav7337.

World Economic Forum, *The Global Risks Report 2020* (15th ed., January 13th 2020). http://www3.weforum.org/docs/WEF_Global_Risk_Report_2020.pdf.

World Meteorological Organization, Multi-agency report highlights increasing signs and impacts of climate change in atmosphere, land and oceans. Press Release Number: 10032020; March 10th 2020. https://public.wmo.int/en/media/press-release/multi-agency-report-highlights-increasing-signs-and-impacts-of-climate-change.

World Meteorological Organization, I, New climate predictions assess global temperatures in coming five years, Press Release, July 8th 2020. https://public.wmo.int/en/media/press-release/new-climate-predictions-assess-global-temperatures-coming-five-years.

World Meteorological Organization, II, Siberia: heat, fire and melting ice. *Latest WMO News*, July 24th 2020. https://public.wmo.int/en/media/news/siberia-heat-fire-and-melting-ice-0.

Further reading

OpenSecrets.org, undated. *Most Heavily Partisan Industries, Election Cycle 2018*. https://www.opensecrets.org/overview/partisans.php.

Pew Research Center, *As economic concerns recede, environmental protection rises on the public's policy agenda*. February 13th 2020. https://www.pewresearch.org/politics/2020/02/13/as-economic-concerns-recede-environmental-protection-rises-on-the-publics-policy-agenda/.

Science integrity and environmental decision-making in Canada: a fragile renaissance

Jeremy T. Kerr
University of Ottawa, Ottawa, ON, Canada

Scientific integrity and public policy

Global changes have precipitated a sixth mass extinction (Barnosky et al., 2011) and launched a new geological epoch—the Anthropocene—in which the flows of energy and material at a planetary scale now depend substantially on human activities (Crutzen, 2002). Whether through climate change (Kerr et al., 2015; Kerr, 2020), habitat loss (Kerr and Currie, 1995), dangerous species introduction into new environments (Vitousek et al., 1997), or various forms of pollution (Kerr, 2017), the array of environmental challenges confronting humanity now imperils human well-being and the persistence of society itself (Díaz et al., 2006). Policies to address such sophisticated challenges must be based on the best available evidence from all sources.

Scientific integrity is indispensable for the development of effective policy (Carroll et al., 2017). Scientific integrity is simply the capacity to collect, use, and communicate scientific evidence without suppression, misdirection, or other interference (Carroll et al., 2017). It is a normal and required ingredient for effective decision-making, whether for environmental or other policies. For anyone who has ever engaged with a policy-maker or even begun to think seriously about doing so, it is obvious that policies are nearly always based on more than scientific evidence, and competing values may yield differences of opinion about how to address challenges.

Simply put, scientific evidence does not dictate policy in isolation and never has. Even when the evidence is clear, it is possible to make an authentic, thoughtful choice to *not* change a policy for a variety of reasons. For instance, the costs of a proposed solution might prove impractically high, the benefits

Conservation Science and Advocacy for a Planet in Peril.
DOI: https://doi.org/10.1016/B978-0-12-812988-3.00001-6

too small, or the solution could impose unacceptable ethical compromises in other respects. Civil infrastructure dedicated to providing potable water in most areas of Canada, the United States, or Europe, for example, tolerates the presence of bacteria at low levels (Reasoner, 1990). It would be possible to reduce those levels further and make risks associated with the water supply even more remote, but the costs would be very high relative to the benefits in most instances. The scientific evidence is clear, but a reasonable person could decide—while fully acknowledging the breadth of scientific evidence—that the risk:reward ratio is sufficiently unappealing that society should not invest more in making clean drinking water still cleaner.

Scientific integrity is simply a specialized form of integrity. In general terms, integrity requires that someone be able to distinguish between honest and dishonest positions and to act accordingly (Integrity, 2020). Political leaders who undermine scientific integrity—such as those who argue that climate change is some kind of conspiracy (Wong, 2016), or who suppress routine communication of factual information as the government led by defeated Prime Minister Harper did in Canada (Palen, 2017)—are practically certain to *also* lack integrity in more fundamental ways. The story of attacks on scientific integrity over the past several decades is inextricably tied to the rise of ideological, often mercenary (and self-dealing), interests that simply cannot coexist with basic principles of integrity, let alone scientific integrity.

Assaults on scientific integrity in western democracies have become a de facto requirement for such ideological and mercenary decision-making. Simply put, when the evidence gets in the way of implementing the solution that is desired, it might be necessary to ignore the evidence, attack its reliability, and to attack those who have collected it. Deliberate efforts to poison the relationship between facts and decisions enable the creation of policies that might, in the short term, appeal to some political or funding base. Yet, policies created under such circumstances are unlikely to solve the problems they were ostensibly developed to address and are frequently nothing more than public relations screens to ensure inaction on major issues, which then grow predictably (and sometimes dangerously) worse (Weiland, 2020).

Perhaps the best examples of these issues come from the realm of climate science and the carefully orchestrated creation of public relations campaigns to create societal doubts about the reliability of accepted scientific understanding [ranging from "debates" over whether thermometers accurately record temperature (Watts, 2019), to scientifically illiterate claims that rising CO_2 concentrations do not contribute to warming temperatures (Hussain, 2012)]. Effective policy intervention to address environmental challenges, whether small or large, requires that policies target specific properties of a problem and change them so that the problem becomes smaller or less immediate. Such progress is far more challenging when political leaders express disagreements about observable reality (Kerr and Côté, 2015), such as climate change. Climate change has not slowed because climate change deniers poisoned public discourse around policy intervention, pretending that evidence was uninformative, sea levels were not rising, and the weather was not becoming more extreme. Wildlife populations imperiled by these and other threats do not spontaneously recover while such pressures continue to grow (Soroye et al., 2020) because these threats exist independently of fantasy-based rhetoric manufactured to prevent policy action: real problems are only solved by evidence-based policies.

Here, I focus on recent, concerted assaults on scientific integrity and the critical steps that were taken to restore it subsequently. I draw particularly on the Canadian experience, but the attacks on scientific integrity in Canada mirror those in the United States, Australia, the United Kingdom, and more broadly. It is no accident that assaults on scientific integrity were followed rapidly by efforts to dismantle environmental protections, which were viewed as hindrances to unfettered industrial resource extraction. Terminating environmental protections and politicizing environmental management is much more achievable if public access to scientific evidence is reduced and filtered through politically polarizing communications. In Canada, strategic attacks on scientific integrity enabled the elimination or degradation of nearly every form of federal environmental legislation, either through legislative change or by refusal to enforce legislated environmental protections at the federal level.

Ideological policy-making and public assaults on scientific integrity [and sometimes on scientists themselves (also see in Chapter 2, When Scientists Are Attacked: Strategies for Dissident Scientists and Whistleblowers; Union of Concerned Scientists, 2014)] impose strong selective pressures on researchers to evolve new skills around communication and engagement with both policymakers and the public and particularly to avoid common pitfalls and divisive approaches that undermine progress toward shared objectives. The recovery of scientific integrity at the federal level in Canada was prioritized because of the concerted efforts of then-opposition political leaders and allied scientists in advance of the federal election in 2015 and subsequently. Though evidence-based environmental decision-making and federal environmental legislation had been badly degraded, improvements to both were strongly facilitated by restoring—and then building strong safeguards to protect—scientific integrity.

Fabrications versus facts: how scientists can defend scientific integrity

> *"Those who can make you believe absurdities, can make you commit atrocities."*
>
> **-Voltaire, Miracles and Idolatry**

Science training and scientific integrity: vaccines against disinformation

An underlying motivation for repeated assaults on scientific integrity is that factual information gets in the way of decisions demanded by adherents to extreme ideologies and those with mercenary interests (Oreskes and Conway, 2011). While this axiom applies across the political spectrum, some of its most harmful impacts are evident in the elimination or erosion of environmental policies. Such attacks on integrity generally and scientific integrity specifically create extraordinary challenges for scientists (Oreskes, 2019). Yet, scientific training provides powerful foundations to counter disinformation tactics and misinformation more broadly.

Researchers have been trained to use technical expertise to gather evidence in service of discovery: the unveiling of new knowledge. This is the creative side of the scientific process. The destructive side can be equally important: by challenging existing ideas with the best possible evidence and the most effective techniques, those ideas can be refined or replaced with alternatives that provide better insights and make stronger predictions. Scientists who advance ideas with poor evidence can expect to see them demolished decisively, sometimes in very public ways. Finally, if a researcher knowingly advances their work using falsified data (George and Buyse, 2015), there will likely be career-ending consequences and criminal prosecutions are not unknown in instances when public funds were used fraudulently (Bernstein, 2015).

In other words, scientific integrity is written into the DNA of professional scientists and cannot be separated from it without fundamentally altering the practice and philosophy of science. Scientific training is an extraordinarily rigorous, lengthy process whose purpose is to help generate and test new ideas. Moreover, science has always operated on the idea of reproducibility (National Academies of Science, 2019): methods and results are described in research in sufficient detail to enable a knowledgeable practitioner to replicate the research. While this principle of reproducibility continues to evolve rapidly in this era of "big data" and sophisticated analysis, it has been present in science since its inception in the modern era. The basic philosophy of science involves profound skepticism and provisional belief that even the best-supported hypotheses could be incorrect (Popper, 1963).

Ideology can affect science and scientists, leading to the development of—and belief in—profoundly wrong hypotheses. However, science is a vaccine against disinformation for the simple reason that scientists are trained to be very good at demolishing weak (or falsified) evidence. Ironically, the scientific capacity to disprove poor ideas is the very thing that makes some policy-makers uncomfortable, contributing to organized, politicized attacks on scientific integrity around environmental issues.

Holding the line on scientific integrity

Some political leaders attack academic scientists in an effort to silence them or push them out of the public space (Phelps, 2020). Scientists must not retreat from their role in assembling and advancing the best possible evidence, whether in service of the creation and testing of new hypotheses or through their application to specific policy challenges. Such a retreat would diminish the value of research for society profoundly, potentially turning much of it into a kind of indulgence whose practical value might simply be insufficient to merit continued support. Yet, this is exactly the pressure that researchers must routinely confront as they bring evidence forward to policy-makers.

Being attacked is never pleasant, but attacks can be a sign that efforts to defend scientific integrity are having an impact. After all, there is no reason to attack someone whose work is irrelevant. During some of the most difficult years for scientific integrity in Canada, I participated in many media interviews to identify specific instances where federal government decisions had clearly undermined scientific integrity. During one such interview on CBC Radio 1 in September, 2013, I was attacked—mid-interview and on live air—by the federal Minister of Natural Resources because I publicly stated an internal policy that research scientists in his Department had been forced to follow around the use of the term "carbon." The interviewer interrupted the Minister mid-denunciation and demanded that he address my comment in specific terms, rather than simply repeating demonstrable falsehoods in his patronizing (and factually wrong) way. It was a memorable moment and another embarrassment for a government that had made attacks on scientific integrity into a signature policy. I left the studio and went to Parliament Hill to deliver my speech at the first "Stand up for Science" demonstration (see Makuch, 2013).

The point is simple: reality is not supposed to be a partisan issue (Kerr and Côté, 2015). Yet, presenting evidence and factual information when it is relevant to policy is a core scientific function that has been attacked repeatedly by those with interest ("shills") in perverting understanding of scientific evidence to poison the policy process (Oreskes and Conway, 2011). The "stay in your lane" argument that is sometimes trotted out to discourage and silence scientists from engaging with policy-makers and the public (Woo, 2015) is simply another way of arguing that science and scientists should agree to have minimal societal relevance. For academic scientists to remain silent under such circumstances is, in other words, a road to diminish science and scientists into a rump role as technical enablers of decisions made using a filtered subset of evidence that might have been cherry-picked to support particular world views. In reality, science and scientists are indispensable to the development of sound policy (Gaieck et al., 2020), and greater scientist engagement with policy-makers would be extremely valuable.

Silencing science and scientists in the policy process make it difficult for the public to distinguish between evidence and nonsense, reality and fantasy.

Some policy-makers already use every opportunity to foment conspiracy theories that serve their worldview and to reward political allies (or themselves) (Desikan et al., 2020). Increased accessibility of scientific information can be part of strategies to help society and policy-makers make decisions based on a shared understanding of reality.

The Trump years in the United States were so extreme in terms of creating unhinged conspiracy theories that it is easy to overlook that this practice began long before. Joe Oliver, then Canada's Minister of Natural Resources, published an open letter (2012) accusing "environmentalists and other radicals" of seeking to "hijack" Canada's regulatory process over the construction of massive diluted bitumen pipelines, and participating in a conspiracy with foreign powers, particularly celebrities (Staff, 2012). Today, such a communication reads like a precursor to QAnon. However, this letter and associated public communications were not a rhetorical exercise. The Conservative government explicitly grouped environmentalists, representing a large swath of civil society, with terrorists in public statements in 2012 by then-Public Safety Minister Vic Toews (McCarthy, 2012). Such conspiratorial thinking was followed up in practical ways, including with a directive to the Canadian Revenue Agency to audit registered charities that worked on environmental issues to limit their role in public discourse around the topics they were founded to discuss (Solomon and Everson, 2014).

In other words, whatever the role of scientists in informing policy, reality should not be a partisan issue, Stephen Colbert's satirical remarks ["Reality has a well-known liberal bias;" (Colbert, 2006)] notwithstanding. Concerted attacks on scientists who advocate for the use of the evidence lead to the key consequence that encouraging transparent use of factual and scientific evidence is interpreted by some as a partisan issue. For a scientist working on issues with immediate policy relevance, this creates a bizarre and untenable situation: if providing the best available scientific evidence ("facts") for a decision is redefined as politically partisan, and scientists must avoid partisan activities, then science loses any reliable relevance in the development of public policy. Why study how to conserve biological diversity or to mitigate climate change if evidence emerging from that work must be kept carefully away from policy-makers because facts are partisan? The implications for democracy of strategies to undermine scientific integrity and dismantle environmental protections are profoundly corrosive. Ideology in service of power and control in systems with few other checks and balances then becomes paramount.

Scientific integrity in Canada: a dark age

Suppression of science

In 2006, Canadians elected the first of a series of Conservative governments, led by Stephen Harper. The Harper government did unprecedented harm to scientific integrity in federal decision-making, silencing the public service, terminating science programs, and firing many scientists, particularly those involved with aspects of environmental monitoring. Many of these impacts have been exhaustively documented (Linnitt, 2013), but some of the most important and insidious changes are routinely overlooked.

Traditionally, the public service, including thousands of scientists, had a duty to the public that was at least partly independent of political leadership. Such responsibilities were spelled out in the Values and Ethics Code for the Public Sector (President of the Treasury Board, 2011). Prior to 2011 (and the election of a majority Conservative government), this code included a number of such provisions that ensured the public interest was paramount. For example, the code required that the public service "shall make decisions in the public interest." The public service was even encouraged to report instances where members believed they had been asked to act in ways that were not consistent with the public interest: "Furthermore, any public servant who believes that he or she is being asked to act in a way that is inconsistent with the values and ethics set out in Chapter 1 of this Code can report the matter in confidence and without fear of reprisal to the Senior Officer..." (Minister of Public Works and Government Services, 2003).

In 2011, in the early phases of the Conservative majority, the Values and Ethics Code was rewritten so that service to political leaders became paramount. References to the duty of making decisions in the public interest were replaced by a requirement that public servants "shall follow policy" (President of the Treasury Board, 2011). Such changes do not need to be interpreted as nefarious, and it seems unlikely that public service leadership sought to degrade scientific integrity through rewriting the Values and Ethics Code. However, this change certainly contributed to that outcome: the updated Code required loyalty to a far greater extent than service to the public, which was largely eliminated in terms of a specified requirement.

Policies in the public service undermining scientific integrity by preventing scientists from communicating freely about their research began almost immediately after the election of the Harper government in 2006. However, such policies would have been hard to sustain in the absence of the profound change in the core values and ethics governing the federal public service. In case there was any uncertainty on that point, information control was enabled specifically in the new Values and Ethics Code, particularly through mechanisms such as the revised section 3.4, which required public servants to act in such a way as to "maintain their employer's trust." This kind of policy would resurface much later in the United States, when many public servants in the United States would be transferred to "Schedule F" status and required to demonstrate loyalty to Trump (Davidson, 2020). Routine activities could then be inferred to create distrust in a political master. While, clauses of this kind can be innocuous under some kinds of political leadership, they facilitated expansive control over individual behavior under the Harper government and contributed to profound damage to scientific integrity.

The array of policies designed to shut down even the most routine forms of scientific communication to the public, or other scientists, came to be known as "muzzling" (Office of the Information Commissioner of Canada, 2018). The pattern for smothering the public communication of federal science was well documented, and many examples have been publicized. At scientific conferences, communications staff from the Ministry of Environment or from Fisheries and Oceans sometimes literally followed scientists around at meetings, observing what they said, who they said it to, and even what scientific posters they read (Learn, 2017). No word on whether they followed scientists into bathrooms. Researchers who communicated with the public faced discipline and potential dismissal (Gatehouse, 2013). In one instance, a researcher in the Department of Fisheries and Oceans notified his Department's communications group of a request to speak about Great White Sharks the next morning. He received no reply, so he participated in the interview. This researcher had previously been the "science spokesperson of the year" for the Department, so this kind of activity was routine, but he was disciplined and threatened with dismissal.

Media interest in science led by federal researchers is widespread and routine, but access to scientists who did that work was prevented even when the issue to be discussed bore limited relation to any policy that might reasonably make the elected leaders of the day nervous (Woo, 2015). *Didymosphenia geminata,* affectionately known as "rock snot" by those who study it, is an algal species that grows on rocks in streams and lakes across broad geographical regions, from New Zealand to Canada. While the deposition of nutrients, particularly phosphorus, in aquatic ecosystems is associated with algal blooms and a cascade of biotic consequences, *Didymosphenia* has the highly unusual property of blooming when soluble reactive phosphorus concentrations drop to very low levels, less than 2 parts per billion, but these blooms might become more likely as a result of climate change. The imposition of muzzling policies meant that the lead author of this study could not respond to media requests for information about his world-leading research on this topic without first completing an internal communications review. In this instance, that review involved 110 pages of email messages and 16 different communications staffers, a long delay, and an eventual denial of permission to speak about the work. Many examples of this kind have been documented. Some of these assaults on scientific integrity, including the "rock snot" example, have been highlighted in the recent TED talk by Minister of Science, Dr. Kirsty Duncan, who spoke about her work to end this "war on science" federally (Duncan, 2018).

In a particularly bizarre turn, government leaders during this period vehemently denied that muzzling was taking place. Propaganda statements of this kind from the government were normalized throughout this period, leading to uncounted instances where government communications staff and their political masters made assertions that were precisely the opposite of actual events. Then-Minister of State for Science and Technology, Gary Goodyear, issued a statement saying that "Government scientists and experts are readily available to share their research with the media and the public" (Gatehouse, 2013), even as scientists across Canada were reporting—necessarily anonymously—that their capacity to communicate had essentially been eliminated. It is unclear if anyone believed the Minister's media line despite the overwhelming evidence that it was profoundly misleading. The specific reference from Orwell's study of propaganda and its capacity to poison society is "doublethink" (Orwell, 1949): in this instance, simultaneously knowing the truth but believing the lie. To my knowledge, no one ever asked political operatives who crafted statements like this, and who helped muzzle scientists, if they also managed such feats of doublethink or if communications such as these were merely cynical. The facts contradicted the propaganda: media reporting on topics like climate change from federal sources dropped by 80% after muzzling policies were fixed in place (Gatehouse, 2013).

Attacks on scientific integrity and environmental protection

Suppressing scientific evidence is much easier if it is not collected in the first place, and attacks on scientific integrity were also associated with the termination of many monitoring programs. This most famously included eliminating the requirement that Canadians complete the long-form census, which had been providing data on the social, health, and economic status of Canadians for decades. Completion of the census had been a basic civic duty, but it was replaced by a "National Household Survey" in 2011, just as the Conservatives won their only majority mandate. The optional survey was more expensive to run than the census, costing $22–29M more. More importantly, the total expense for the National Household Survey was $652M to collect data that was so plagued by low response rates (rising to 26%) that its statistical outcomes were unreliable and could not be matched to historical trends detected using previous census data (Wherry, 2013). The Chief Statistician of Statistics Canada, Munir Sheikh, warned the Government of these problems in advance and resigned rather than disingenuously defend a system that could not fulfill its purpose: "I want to take this opportunity to comment on a technical statistical issue which has become the subject of media discussion. This relates to the question of whether a voluntary survey can become a substitute for a mandatory census. It can not. Under the circumstances, I have tendered my resignation to the Prime Minister" (Chase and Grant, 2010).

While some scientists resigned, retired, or moved to academic institutions during this period, many others were fired and environmental monitoring programs ended. By 2013 the Conservatives had cut 5332 professionals, including scientists (Nelson, 2013), from the public service or reassigned them away from environmental monitoring work. Many programs were terminated, consistent with the apparent desire to limit the provision of evidence that would interfere with the creation of ideologically driven policies. The National Round Table on the Environment and the Economy was one such casualty of this "war on science." It produced a series of reports identifying mechanisms to enable environmental protection while minimizing economic impacts (or encouraging economic growth). This platform for environmental economics advice was terminated after it produced a series of reports suggesting that pricing carbon was necessary to reduce CO_2 emissions, including a report on mechanisms to enable a transition to a low carbon economy (National Round Table on the Environment and the Economy, 2012). One Government Minister commented that the elimination of NRTEE reflected a desire to not receive such evidence-based advice (Scoffield and Ditchburn, 2012).

The most well-known example of cuts aimed to limit federal environmental research in Canada was the attempt to close the Experimental Lakes Area (ELA). This region in northwestern Ontario, Canada, is on the boreal shield and has long been managed by the federal Department of Fisheries and Oceans. Like many ecologists, I have long taught about classic experiments performed in the ELA that demonstrate the effects of nutrient additions, like phosphorus, on lake eutrophication. Experimental additions of nitrogen, phosphorus, and carbon to a lake in the ELA demonstrated decisively that phosphorus was a critical determinant of algal productivity (Schindler, 1974). Subsequent work in the ELA focused on a wide range of vital topics with urgent societal and environmental relevance, such as the effects of climate change (Schindler et al., 1996). The ELA was described as the "crown jewel" of whole ecosystem experimentation. Provincial governments in Ontario and Manitoba joined forces to maintain the ELA until its federal funding was restored after the Conservative Government was defeated in 2015.

Dismantling environmental legislation

Federal environmental legislation was repealed and weakened systematically and almost comprehensively after the election of the majority Conservative Government in 2011. In a majority Parliament, there are few limits on legislative power: Westminster Parliamentary systems are not structured to have the checks and balances that are familiar to United States citizens. The Fisheries Act was stripped of nearly every provision protecting either fish species or their habitat across Canada (Hutchings and Post, 2013). Protections were maintained around fisheries, inadvertently leading to the potential protection of introduced and hatchery fishes over native species. This legislated change, and its associated regulations, also undermined Canada's obligations to protect species at risk and the associated need to protect their habitat (Ecojustice, 2013).

The new version of this legislation even enabled the delegation of decisions about habitat and species protections to the industries the Act was originally designed to help regulate. Other legislation from which environmental protections were substantially stripped included the Canadian Environmental Assessment Act and the Navigable Waters Protection Act (renamed as the "Navigation Protection Act"). Far from creating "red tape," such legislation helped Canada maintain a reputation as an environmental leader (Schindler, 2015) on the world stage (though I am not certain that this reputation was well deserved: Canada's environmental performance has historically depended strongly on its low population and large area, but the concentration of land-use changes around high biodiversity areas has imperiled a large proportion of Canadian species (Kerr and Cihlar, 2004) and hinders their recovery (Kerr and Deguise, 2004; Robillard and Kerr, 2017; Fig. 4.1).

The absence of detectable scientific integrity in decision-making exerts greater impacts than simply leading to bad policy: it can fatally undermine public support for both policy and in governmental and industrial decision-making in general. The loss of social license for various kinds of economic activities is an

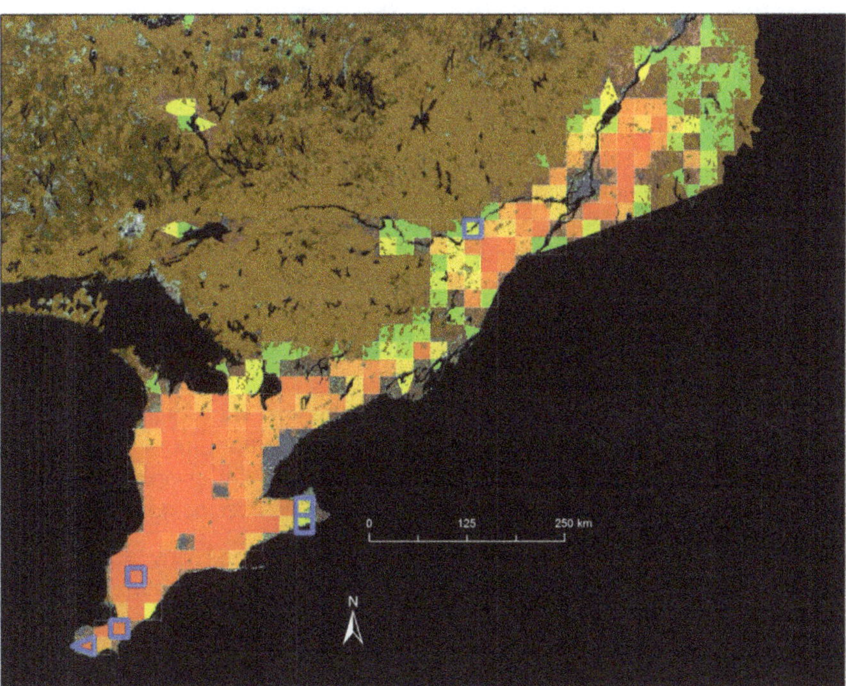

Figure 4.1 Gradients of land-use intensity for agriculture across southern Ontario, Canada, the region in which species richness and numbers of species at risk is highest. Thick squares outline areas where, if recovery of species at risk was implemented, the costs would be minimized and the potential benefits (in terms of numbers of species potentially recovered) maximized. See Robillard and Kerr (2017) for details.

indicator that members of the public have lost confidence in governance. This is not merely risky in a democratic society, it is profoundly dangerous.

The last stand: in defense of Canada's species at risk act

The environmental onslaught during the so-called Harper years was relentless, but one piece of federal environmental legislation survived that time intact: the Species At Risk Act (SARA). This legislation is the Canadian counterpart to the US Endangered Species Act (ESA). For a variety of reasons, it is weaker in its capacity to protect species at risk than the ESA, most notably in not mandating habitat protection for terrestrial species at risk in most areas. This difference is a consequence of the constitutional division of responsibilities in Canada: provinces manage resources, which includes land, so the federal government has limited authority to protect habitats. Despite the relative weakness of this legislation in terms of its capacity to mitigate known causes of extinction or to effect species recovery, the Conservative Government initiated a process to limit SARA effectiveness still further (Duncan, 2013). One of the specific ways in which the Harper government sought to limit SARA was to add a socioeconomic litmus test to it. In assessments of whether to list a species, the opportunity costs of protection would be weighed. It is unclear on how that process would have worked, but there was profound concern among scientists that the Government's record on conservation spoke for itself.

Efforts to dismantle the SARA followed numerous previous failures to enforce it, which led to embarrassing legal losses for the Harper Government in Federal Court. Prominent among these was the Environment Minister's refusal to declare critical habitat for species even in rare cases where that habitat was known definitively from scientific evidence. The federal government infamously lost a 2009 case in which it had refused to declare critical habitat for the Nooksack Dace (*Rhinichthys cataractae*), a small fish found in British Columbia streams. The court ruling was damning, and stated, "This is a story about the creation and application of policy by the Minister in clear contravention of the law, and a reluctance to be held accountable for failure to follow the law. Therefore, this is a case about the rule of law…" (Campbell, 2009). This case demonstrated yet again that efforts to undermine scientific integrity and attacks on the environment were combined, undermining fundamental requirements of democratic governance.

The attempt to dismantle SARA was one attack too many, and it galvanized opposition from within the scientific community, nongovernmental groups, and the public. This defense was led substantially by Canada's ecology researchers, who had advocated for endangered species legislation in the first place and had been contributing scientific evidence to the process of protecting and recovering at risk wildlife continuously, both before and after the passage of this legislation in 2003.

As with the US ESA, SARA has a mixed track record of conservation success, particularly in terms of recovering listed species. Habitat loss is the primary cause of decline (Kerr and Cihlar, 2004) for most species at risk, reflecting particularly the conversion of natural areas across much of southern Canada to agricultural and urban land uses (Kerr and Cihlar, 2003). Yet, habitat protection measures in SARA are weakest in precisely these areas, where federal lands are very limited and protected areas are small and distributed in ways that include species at risk no better than areas chosen at random and sometimes worse (Deguise and Kerr, 2006). Recovery of species at risk also clearly requires efforts to manage and potentially restore habitat for the majority of species affected by habitat loss (Kerr and Deguise, 2004). This SARA limitation led to the creation of the associated Habitat Stewardship Program, which seeks to find cooperative ways to work with landowners to improve habitats for species at risk. Tools like these were underused, but the fundamental challenge appeared to be that any conflict, perceived or real, between conservation and industry was consistently resolved in favor of maintaining industrial activities. This led to indefinite delays in listing decisions and refusals to identify critical habitat for species that had previously been listed. These were not problems with SARA, which emphasizes a cooperative approach to conservation. They were problems of implementation.

A small group of us prepared a letter that outlined these issues, pointing out in clear terms that there was no evidence that the SARA was either broken or ineffective. In fact, the evidence showed clearly and repeatedly that it had simply not been fully implemented. We sought to make the scientific evidence in this policy debate as clear as possible. Moreover, we communicated this letter broadly within the scientific community and asked researchers with expertise in relevant areas (e.g., ecology and conservation) to consider signing on to the letter. The result was a clearly worded statement that the Government would not be able to dismantle SARA under the false pretense that it could defend such an action using evidence that SARA was somehow fundamentally flawed, a statement that was signed by over 1000 academics, including >400 with doctorates. If released, this letter would have been another embarrassment for the Government, which had claimed to have consulted "wildlife experts" in proposing to change SARA.

Having assembled this letter, we sought to meet with the Minister of Environment, who agreed, then repeatedly delayed, and finally refused to meet. Our group held a meeting at the Prime Minister's Office instead. The goal of this meeting was to convince the Government to change course away from weakening SARA. Failing that, we planned to release the letter to an array of media contacts and to begin a public campaign to demonstrate the absence of evidence in the Government's assertions around SARA shortcomings.

Our efforts were successful. The Government did not alter SARA legislation. In response, we did not release the science letter publicly. This was a rare victory for scientific integrity and environmental protection prior to 2015 (I believe Canadians owe a debt—not yet publicly recognized—to Drs Jeannette Whitton and Scott Findlay for successfully navigating a tricky meeting at the

Prime Minister's Office). That meeting and the broader effort to galvanize the scientific community demonstrated again the power of clearly communicated consensus around scientific evidence that affected the trajectory of a vital policy debate, an approach I had used in the past when I helped lead provincial efforts to identify key scientific principles that should be in the Ontario ESA. Efforts by the research community to make the evidence clear were matched by a broad coalition of environmental groups and advocates, whose statements were similarly unequivocal, referencing the scientific perspectives we presented, but also drawing on legal arguments presented by groups like Ecojustice. In any event, by 2013, the Government had no residual credibility on environmental issues: their track record had become so unequivocally pro-development by this point, that their claims to be making SARA more efficient were viewed as an effort to gut the last and most cherished piece of federal environmental legislation (Editorial Board of Toronto Star, 2012).

Ignoring the species at risk act

Defending SARA did not prevent the Government from continuing to do mischief to species at risk protections in Canada. In addition to slow-walking every part of the recovery planning process for species at risk (McCune et al., 2013)—including refusing to designate critical habitat for listed species—they found and exploited a bizarre loophole in the SARA that made it possible for the Government to halt the listing of nearly all newly recognized species at risk.

The scientific body that is responsible for identifying which species are at risk is the Committee on the Status of Endangered Wildlife in Canada (COSEWIC), whose role is explicitly described in SARA. By law, COSEWIC deliberations must be independent of influence from any external organization, including the Government, which is one of the rare instances where scientific integrity in a policy process is explicitly protected in legislation ("Each member of COSEWIC shall exercise his or her discretion in an independent manner") (Species at Risk Act, 2002). COSEWIC's work continued, as required, throughout the Harper Government's tenure, but listing recommendations were halted internally in the public service, so that the Minister was not formally notified of COSEWIC decisions and consequently did not have to act on them. The Government created something we called "listing purgatory" (Mooers et al., 2017) by exploiting a loophole in the legislation. The SARA requires that the Government act on a recommendation it receives from COSEWIC within 9 months. To avoid having to start that clock, the Minister stopped receiving the listing recommendations. It is hard not to imagine these files accumulating over several years on a political staffer's desk, waiting to be physically rolled into the Minister's office. But this just never happened: by the time this Government was defeated in 2015, >100 species had entered this listing purgatory. This meant that the evidence to justify listing recommendations was sometimes outdated in ways that could alter the original listing recommendation. For example, a species COSEWIC recommended should be listed as threatened in 2008 might have become endangered by 2015. The absence of protection afforded to species at risk obviously cannot mitigate the threats that caused their decline in the first place.

While we had succeeded in efforts to prevent SARA from being weakened, this proved to be a partial victory. Government efforts to undermine the listing process by refusing to receive new listing recommendations meant that SARA was limited to protecting and recovering historically recognized species at risk. While the scientific integrity of COSEWIC was not undermined directly, it was effectively rendered useless because its recommendations were effectively ignored for the better part of a decade. Endangered species legislation resembles a hospital emergency room (Benson, 2012). In Canada the Harper Government was signaling that no new "patients" would be treated, regardless of how urgent their needs or modest the costs. This was another failure of governance.

Recovering scientific integrity and environmental governance

Struggles to defend environmental legislation, such as the Fisheries and SARAs, mirrored more fundamental battles over scientific integrity. While we fought an effective defensive action against Government efforts to undermine environmental legislation, at least for SARA, there had been no progress in recovering scientific integrity. In 2011 a number of colleagues and I visited a well-known pub near the University of Ottawa to consider ways that scientists could engage with this issue. From this discussion, the "Death of Evidence" rally was born, which kicked off national demonstrations in defense of scientific integrity. These were organized in part by Evidence for Democracy, a nongovernmental organization formed in recognition that decision-making in government had become extremely ideological and was demonstrably ignoring or suppressing scientific evidence. Ideology was trumping evidence, and this was undermining public confidence in governance in Canada, exacerbated by documented links between oil and gas lobby groups and government decision-making. For example, on December 12, 2011, shortly after the Harper government was elected to its majority mandate, lobbyists from the oil and gas sector wrote to then-Minister of Environment, Peter Kent, and Minister of Natural Resources, Joe Oliver, requesting that federal environmental legislation (such as the Fisheries Act and SARA) be modified to enable "shovel-ready" projects to proceed (Kenny et al., 2011). Lobbyists were writing government policy evidence that these policies were filtered through an Orwellian communications filter that meant federal government communications on such issues were fundamentally untrustworthy.

Scientific integrity and passionate advocacy are not mutually exclusive

Scientists were about to shift from communicating to policy-makers about their traditional areas of expertise to a public defense of scientific integrity itself. This transition was challenging and came with significant uncertainties. Where is the boundary between speaking publicly and perhaps passionately as a scientist versus expressing ethical views and personal opinion? These are different activities and distinctions between them need to be recognized.

Excellence as a scientist can provide a platform for communication, but it is a contingent one that is based on trust. Communicating as a scientist—using that scientific platform to make a point—must not compromise scientific integrity. Communications must be honest (i.e., have integrity) and reflect diligent efforts to understand the issues and examples being discussed thoroughly. Failures to maintain that standard are hypocritical and potentially deceptive, blurring the boundaries between personal opinion (which scientists are also entitled to have) and views that can be defended on the basis of expert knowledge and experience.

Speaking out about problems does not compromise scientific integrity. Ensuring scientific integrity in communication, even while contributing passionately at a demonstration, can be readily accomplished. In 2013, "Stand up for Science" rallies were organized in communities across Canada. Along with an inclusive group of colleagues from across the research community, I spoke at the large rally on the front steps of the Parliament of Canada. The purpose of presentations like these is to speak memorably and with impact, contributing to broader awareness of the issues through various forms of media coverage (Canadian Broadcasting Corporation, 2013). I approach speeches of this kind from the perspective of peer review: would the factual points I present stand up to peer review? Can I cite my sources and are those sources reputable? If so, then the material could be fair game for inclusion in a speech. If I think a point is not correct, I exclude it, but the style of communication must be appropriate for a public demonstration, as with this excerpt from my speech:

> *"Species are disappearing and we can't reverse this trend for most of them unless we protect their habitat. Yet, the previous Minister of the Environment was missing in action when it came time to take the advice of scientists on the government's own panel, the Committee on the Status of Endangered Wildlife in Canada. The Minister has started interfering with the membership on the panel, removing some members and refusing the appoint others. Over, and over again, we've seen this disregard for evidence create policy disasters."*

It was all factual. Speaking publicly and passionately does not imply a compromise with principles of scientific integrity, particularly when integrity itself has become a partisan issue. Conversely, however, silence can become complicity, and this too can betray principles of scientific integrity.

Scientific integrity: the renaissance

Efforts to defend environmental legislation and causes led to limited success, though these were sometimes bloody-minded affairs that were only settled after litigation, but the massive work some scientists did to restore scientific integrity in the federal government changed no practice or policy. The government never wavered in its Orwellian approach to many issues, like climate change mitigation, that absolutely needed strong scientific input. In a way, this made it easier to keep organizing and demonstrating. The lack of progress in changing any federal policy related to scientific integrity was frustrating but motivated us to keep speaking forcefully to the issue, raising awareness among Canadians (and globally), and defense of scientific integrity grew into an election issue. A small number of us began working with Members of Parliament, who were often just as appalled at the politicization of basic issues as university-based scientists were.

One such Member of Parliament was then-Opposition Member, Dr. Kirsty Duncan. A medical geographer, she had done pioneering (and, it turns out, prescient) work to find samples of Spanish Flu in the hopes of staving off the next global pandemic (Duncan, 2006). Dr. Duncan made the jump to politics in 2008 and had been one of the most consistent voices speaking out against degradation of scientific integrity, the climate change denying antics of various Government ministers and MPs, and a champion of equity, diversity, and inclusion. She was also one of the lead authors of the IPCC Fourth Assessment Report, so shared in its Nobel recognition in 2007.

I had long known of Dr. Duncan's work, and we met with her to discuss issues around scientific integrity. There were three items to discuss in the meeting: Canada should re-establish the long defunct role of Chief Science Advisor in government who would improve the flow of scientific advice to Cabinet, unmuzzle scientists so federal researchers could speak publicly about their research (recognizing that some kinds of research would be classified, exceptions were built into this position) and make scientific information and data public by default. The motivation for these changes, which would have created a 180 degrees about-face for Canada's Government at that time, was that scientific integrity needed to be integrated into decision-making, and scientific understanding should be as accessible as possible.

I had worked with the United Kingdom's Chief Scientist, Sir Robert May, during my postdoctoral research in conservation biology, and we had discussed his meetings with the Prime Minister in general terms regularly, so the role of Chief Science Advisor was as close to my heart as the more widely recognized imperative around unmuzzling. One of the reasons for this was that having an effective Chief Science Advisor can catalyze creation of similar positions within individual government departments, which clearly needed stronger scientific leadership in the Canadian context. This is the UK model: the Government has a Chief Scientist who advises Cabinet, but individual Departments also have their own lead scientists whose job is to advise their Ministers. Another reason is that strong scientific advice during emergencies can be absolutely vital and having a small group of world class scientists working on a problem can give decision makers strong policy options and also eliminate bad ones. The UK model for this is the Science Advisory Group for Emergencies. Both Dr. Duncan and I were also aware of the work of star researchers in the United States, like Dr. Jane Lubchenco, who had helped recover scientific integrity policy at NOAA after the election of President Obama, and who had established a clear a science integrity policy for the agency. We recognized that policies of this kind would serve as vital models for restoring scientific integrity in government departments and agencies in the future.

The discussion with Dr. Duncan was converted into a Motion to Parliament, which was constructed carefully and consultatively. I remember running out of a scientific meeting of the Canadian Society for Ecology and Evolution to discuss the Motion just before I had to give a talk on a separate piece of global change biology research that was soon to be published (Kerr et al., 2015) and that would occupy most of the rest of my summer. When finalized, the Motion read:

> *"That, in the opinion of the House:*
> *(a) the government has constrained the ability of federal scientists to share their research and to collaborate with their peers;*
> *(b) federal scientists have been muzzled and prevented from speaking to the media about their work;*
> *(c) research is paid for by taxpayers and must be done in the public interest in order to protect the environment and the health and safety of Canadians; and, therefore,*
> *(d) the government should immediately rescind all rules and regulations that muzzle government scientists, consolidate government-funded or -created science so that it is easily available to the public at large through a central portal, create a Chief Science Officer whose mandate would include ensuring that government science is freely available to those who are paying for it, namely, the public, and allow scientists to be able to speak freely on their work with limited and publicly stated exceptions."* (Hansard, 2015a)

Why fight for scientific integrity?

Throughout these struggles, it was vital to remain focused on the problems that needed to be solved and remember that the benefits of solving those problems would flow to Canadians. In other words, engagement with policy work, whether it is around recovering scientific integrity or improving environmental protection, is service *to others*, not to oneself. The goal is not about advancing a career, seeking attention, and holding meetings with prominent leaders. To think (or act) otherwise is the pathway toward hypocrisy and away from scientific integrity.

The Motion to Parliament was brought forward by the opposition Liberals on May 26, 2015, and was debated in the House of Commons for much of the day. Sitting in the gallery above the Commons, I watched as the Motion was introduced. As Dr. Duncan spoke eloquently in Canada's Parliament to the need to restore scientific integrity, I reflected on the knowledge that this Motion, addressing issues I had worked continuously and publicly on for years, was doomed to be voted down by the Conservative majority. However, I wondered if this might be an even better outcome in a strategic sense. This was the last major vote held in Parliament before the federal election scheduled for autumn, 2015. If the Conservative Government was going to continue with its continuous attacks on both scientific integrity and environmental protections, as seemed utterly certain, then this Motion would be a catalyst for ensuring that scientific integrity would be a true election issue for the first time in Canadian history (Fig. 4.2).

This is exactly what happened. The Motion was defeated in a vote along partisan lines, with all opposition parties voting in favor and Members of Parliament on the other side voting against. The Liberals translated the contents of the Motion to Parliament into their science platform for the election, the election was called, and the Liberals were elected to a majority Government. Prime Minister Trudeau appointed Dr. Duncan, a scientist, as Canada's Minister of Science, the first time this had been an independent Ministerial portfolio in Government. The Honorable Dr. Duncan's first words in the new Parliament and as Minister literally sent shivers down my spine: "I am proud to say the war on science is now over. This government respects research and science and the important work it does. We will work with the scientific community to ensure openness now and in the future" (Hansard, 2015b). Everyone in the House of Commons, except defeated MPs from the former government, rose and cheered. Over the next 2 years, the role of Chief Scientific Advisor was crafted, despite significant cultural resistance from the public service, and then created in 2017 (Fig. 4.3).

Recovering scientific integrity: obstacles to progress

Policies around scientific integrity and environmental protection have changed profoundly since 2015. Two scientific integrity policies changed at the end of the very first meeting of the new Federal Cabinet after the election: the long-form census from Statistics Canada was reinstated, and scientists in the federal

Figure 4.2
The introduction of the new role of Chief Scientific Advisor (CSA) in the Government of Canada, 2017, by Minister of Science Kirsty Duncan in Center Block of Parliament. Prime Minister Trudeau is standing beside the pillar, and Chief Scientific Advisor Dr. Mona Nemer (yellow jacket) is just behind Minister Duncan. *Photo by Jeremy Kerr.*

public service were unmuzzled, officially enabling them to communicate to the public about their work without interference. Yet, despite clear and repeated direction from the Minister of Science that scientists must be allowed to communicate freely about their research and did not need to ask permission to do so, implementing this policy proved challenging.

Policies can change quickly, but culture changes slowly (De Souza, 2018). This was a vital lesson in the transition to a new Government that had campaigned, in part, on scientific integrity. The Public Service had been through nearly a decade of intense message control and many found it difficult to understand or implement the *absence* of invasive controls on how scientists could communicate. Simply explaining that scientists could speak freely led to uneven results for the simple reason that the culture of the Public Service had been warped by

Figure 4.3 A group photo of architects and advocates of the newly announced Chief Scientific Advisor role in Government moments after Minister of Science Duncan (front row center, white jacket) and the Prime Minister introduced the first CSA, Dr. Mona Nemer (front row center, cream jacket), in Parliament. Dr. Jeremy Kerr is in the center, second row, directly behind Minister Duncan and CSA Nemer.

years of intense pressure to suppress and control communication of even the most trivial issues. Yet, scientific integrity policy was soon added to a new collective agreement with the Professional Institute of the Public Service, making it very difficult to reverse should the Government change in the future.

The long fight to recover scientific integrity in Canada has created numerous changes in Canada, but the constant concern is that casually crafted policies could easily be reversed by an antiscience, antienvironment Government in the future. Changes need to be "sticky" and difficult to reverse. The Chief Science Advisor role is one such example: it is closely integrated with science policy across Government, is supported by staff, and a growing group of chief scientists from the SBDAs. While it would be possible for an ideologically minded Government in the future to eliminate the role of the Chief Science Advisor (as the previous Government did in 2008 to a previous, far less effectively integrated, version of this position), the effects of such a change would ramify broadly.

The recovery of scientific integrity has mirrored rapid progress on environmental legislation and policy. Environmental protections have been restored to legislation such as the Fisheries Act and Canadian Navigable Waters Act (changed from the Navigation Protection Act), and there has been progress in reducing the backlog of listing decisions under SARA. The advisory group COSEWIC's recommendations are now transmitted to the Minister. There has been rapid progress in expanding protected areas networks through an open, inclusive process that explicitly sought to incorporate strong scientific evidence and indigenous knowledge.

It is important to note that Government decisions still do not simply mirror the desires of environmental scientists. Public policy is not dictated by scientific evidence alone and must obviously balance potentially competing interests as well as account for subjective, ethical perspectives, among a host of other constraints. Nevertheless, recovery of environmental protections and stewardship has been accompanied by recognition that scientific evidence is necessary for decision-making and that this evidence should be accessible to the public.

Speaking truth to power: a fragile renaissance for scientific integrity

Science integrity in the federal domain was under concerted attack in Canada from 2006 to 2015. Open communication of scientific insights from federal scientists was ended and replaced by Orwellian communications that were routinely and profoundly misleading. This violation of basic principles of integrity (and scientific integrity) led to organized opposition among Canadian researchers and the public, who marshaled scientific evidence to defend environmental policy and began the Herculean task of restoring scientific integrity federally. While we were successful in defending a single piece of federal environmental legislation against the depredations of an ideological Government—namely, the SARA—we accomplished little in terms of policy change around scientific integrity. Nevertheless, we established trust with Members of Parliament who had independently been working on similar issues. Our longstanding efforts to speak publicly and passionately (but truthfully) about the degradation of scientific integrity enabled the creation of a movement that influenced the outcome of the Canada's federal election in 2015. This progress was most obviously evident in our Motion to Parliament on scientific integrity that was debated and defeated by Conservatives just prior to the 2015 election. Subsequent policy changes, following the election of a progressive Government meant that the contents of that Motion were implemented comprehensively, rapidly, and in ways that were make it more difficult to reverse.

Scientific integrity has undergone a renaissance in Canada. Nevertheless, recovering scientific integrity has been slow, as rapid policy changes cannot undo years of cultural shifts toward authoritarian practices. This serves as a reminder that it is easier to break policies than to create them and the renaissance of scientific integrity in Canada, or anywhere else, must always be viewed as fragile and provisional.

Just as the degradation of scientific integrity previously were accompanied by near-comprehensive losses of environmental protections, the recovery of scientific integrity and environmental legislation were also linked. The Canadian experience of the loss and restoration of scientific integrity can inform other jurisdictions that have undergone similar (or even more egregious) failures of integrity at the federal level.

References

Barnosky, A.D., Matzke, N., Tomiya, S., Wogan, G.O.U., Swartz, B., Quental, T.B., et al., 2011. Has the earth's sixth mass extinction already arrived? Nature 471 (7336), 51–57. Available from: https://doi.org/10.1038/nature09678.

Benson, M.H., 2012. Intelligent tinkering: the endangered species act and resilience. Ecol. Soc. 17 (4), 28.

Bernstein, R., 2015. HIV researcher found guilty of research misconduct sentenced to prison. Science. Available from: https://doi.org/10.1126/science.caredit.a1500171.

Campbell, J. Environmental defence Canada, Georgia Strait Alliance, Western Canada Wilderness Committee and David Suzuki Foundation Versus Minister of Fisheries and Oceans. 2009.

Canadian Broadcasting Corporation, Stand up for Science Rallies. *The National*. Canada. 2013.

Carroll, C., Hartl, B., Goldman, G.T., Rohlf, D.J., Treves, A., Kerr, J.T., et al., 2017. Defending the scientific integrity of conservation-policy processes. Conserv. Biol. 31 (5). Available from: https://doi.org/10.1111/cobi.12958.

Chase, S., T. Grant, Statistics Canada Chief Falls on Sword over Census. *The Globe and Mail*, 2010.

Colbert, S., 2006 White House Correspondent's Dinner. 2006.

Crutzen, P.J., 2002. The 'Anthropocene'. J. Phys. IV Fr. 12 (10), 1–5. Available from: https://doi.org/10.1051/jp4:20020447.

Davidson, J., Trump Doesn't Get It. Civil Servants Shield Taxpayers from a Politicized Government. *Washington Post*, 2020.

Deguise, I.E., Kerr, J.T., 2006. Protected areas and prospects for endangered species conservation in Canada. Conserv. Biol. 20 (1), 48–55. Available from: https://doi.org/10.1111/j.1523-1739.2005.00274.x.

Desikan, A., T. MacKinney, and G. Goldman, Let the Scientists Speak: How CDC Experts Have Been Sidelined During the COVID-19 Pandemic. 2020.

De Souza, M., Senior Bureaucrats 'clinging' to Harper Government Rules and Muzzling Scientists, Says Survey. *National Observer*, 2018.

Díaz, S., Fargione, J., Chapin III, F.S., Tilman, D., 2006. Biodiversity loss threatens human well-being. PLoS Biol. 4 (8), e277. Available from: https://doi.org/10.1371/journal.pbio.0040277.

Duncan, K., 2006. Hunting the 1918 Flu: One Scientist's Search for a Killer Virus. University of Toronto Press, Toronto, Canada.

Duncan, K., 2013. Is the government planning to gut the species at risk act? IPolitics 2013.

Duncan, K., Scientists Must Be Free to Learn, to Speak and to Challenge. 2018. https://www.ted.com/speakers/kirsty_duncan.

Ecojustice, Fisheries Act Backgrounder. 2013.

Editorial Board of Toronto Star, Do We Want a More 'Efficient' Species at Risk Act? *Toronto Star*, 2012.

Gaieck, W., Lawrence, J.P., Montchal, M., Pandori, W., Valdez-Ward, E., 2020. Opinion: science policy for scientists: a simple task for great effect. Proc. Natl Acad. Sci. 117 (35), 20977 LP – 20981. Available from: https://doi.org/10.1073/pnas.2012824117.

Gatehouse, J., When Science Goes Silent. *Macleans*, May. 2013.

George, S.L., Buyse, M., 2015. Data fraud in clinical trials. Clin. Investigation 5 (2), 161–173. Available from: https://doi.org/10.4155/cli.14.116.

Hansard, House of Commons Debates, May 26, 2015a.

Hansard, House of Commons Debates, October 28, 2015b.

Hussain, Y., The Environmental Movement Has Lost Its Way. *Financial Post*, 2012.

Hutchings, J.A., Post, J.R., 2013. Gutting Canada's fisheries act: no fishery, no fish habitat protection. Fisheries 38 (11), 497–501. Available from: https://doi.org/10.1080/03632415.2013.848345.

Office of the Information Commissioner of Canada (institution), Result of Investigation into 'Muzzling' of Scientists. 2018.

Integrity, 2020. https://en.wikipedia.org/wiki/Integrity.

Kenny, B., T.M. Egan, P. Boag, and D. Collyer, Letter to Minister of Natural Resources. 2011.

Kerr, J.T., 2017. A cocktail of poisons. Science 356 (6345). Available from: https://doi.org/10.1126/science.aan6713.

Kerr, J.T., 2020. Racing against change: understanding dispersal and persistence to improve species' conservation prospects. Proc. R. Soc. B: Biol. Sci. 287 (1939), 20202061. Available from: https://doi.org/10.1098/rspb.2020.2061.

Kerr, J.T., Cihlar, J., 2003. Land use and cover with intensity of agriculture for Canada from satellite and census data. Glob. Ecol. Biogeography 12 (2). Available from: https://doi.org/10.1046/j.1466-822X.2003.00017.x.

Kerr, J.T., Cihlar, J., 2004. Patterns and causes of species endangerment in Canada. Ecol. Appl. 14 (3). Available from: https://doi.org/10.1890/02-5117.

Kerr, J.T., and I. Côté, How Justin Trudeau Can Bring Science Back to Ottawa. *Toronto Star*, 2015.

Kerr, J.T., Currie, D.J., 1995. Effects of human activity on global extinction risk. Conserv. Biol. 9 (6). Available from: https://doi.org/10.1046/j.1523-1739.1995.09061528.x.

Kerr, J.T., Deguise, I., 2004. Habitat loss and the limits to endangered species recovery. Ecol. Lett. 7 (12). Available from: https://doi.org/10.1111/j.1461-0248.2004.00676.x.

Kerr, J.T., Pindar, A., Galpern, P., Packer, L., Potts, S.G., Roberts, S.M., et al., 2015. Climate change impacts on bumblebees converge across continents. Science 349 (6244). Available from: https://doi.org/10.1126/science.aaa7031.

Learn, J.R., Canadian Scientists Explain Exactly How Their Government Silenced Science. *Smithsonian Magazine*, 2017.

Linnitt, C., Harper's Attack on Science: No Science, No Evidence, No Truth, No Democracy, no. May 2013.

Makuch, B., Stop Muzzling Scientists, Protesters Tell Tories. *Toronto Star*, 2013.

McCarthy, S., Ottawa's New Anti-Terrorism Strategy Lists Eco-Extremists as Threats. *The Globe and Mail*, 2012.

McCune, J.L., Harrower, W.L., Avery-Gomm, S., Brogan, J.M., Csergő, A.-M., Davidson, L.N.K., et al., 2013. Threats to Canadian species at risk: an analysis of finalized recovery strategies. Biol. Conserv. 166, 254–265. Available from: https://doi.org/10.1016/j.biocon.2013.07.006.

Minister of Public Works and Government Services, Values and Ethics Code for the Public Service. 2003.

Mooers, A., J. Hutchings, J.T. Kerr, J. Whitton, S. Findlay, and S. Otto, Recovering the Species at Risk Act. *Policy Options*, no. February 2017.

National Academies of Science, 2019. Reproducibility and Replicability in Science. Reproducibility and Replicability in Science. National Academies Press (US). Available from: https://www.ncbi.nlm.nih.gov/books/NBK547546/.

National Round Table on the Environment and the Economy, Framing the Future: Embracing the Low-Carbon Economy. 2012.

Nelson, J., The Harper Government's War on Science. *Canada Centre for Policy Alternatives: The Monitor*, no. June 2013.

Oreskes, N., 2019. Why Trust Science? Princeton University Press, Princeton, NJ.

Oreskes, N., Conway, E.M., 2011. Merchants of Doubt: How a Handful of Scientists Obscured the Truth on Issues from Tobacco Smoke to Climate Change. Bloomsbury Press, New York, NY.

Orwell, G., 1949. 1984. Signet Classics, London.

Palen, W., When Canadian Scientists Were Muzzled by Their Government. *New York Times*, 2017.

Phelps, J., Trump's New Political Strategy: Attack Biden by Attacking Fauci, Scientists: ANALYSIS. ABC News. 2020.

Popper, K., 1963. Conjectures and Refutations: The Growth of Scientific Knowledge. Basic Books, New York, NY.

President of the Treasury Board, Values and Ethics Code for the Public Sector. 2011. <https://www.tbs-sct.gc.ca/pol/doc-eng.aspx?id = 25049>.

Reasoner, D.J., Monitoring Heterotrophic Bacteria in Potable Water BT—Drinking Water Microbiology: Progress and Recent Developments. In, G.A. McFeters (Ed.), 452–477. 1990. New York, NY: Springer New York. Available from: https://doi.org/10.1007/978-1-4612-4464-6_22.

Robillard, C.M., Kerr, J.T., 2017. Assessing the shelf-life of cost-efficient conservation plans in Canada's farmland. Conserv. Biol. 31, 837–847. Available from: https://doi.org/10.1111/cobi.12886.

Schindler, D.W., 1974. Eutrophication and recovery in experimental lakes: implications for lake management. Science 184 (4139), 897–899. Available from: https://doi.org/10.1126/science.184.4139.897.

Schindler, D.W., The Harper Decade. 2015.

Schindler, D.W., Bayley, S.E., Parker, B.R., Beaty, K.G., Cruikshank, D.R., Fee, E.J., et al., 1996. The effects of climatic warming on the properties of Boreal lakes and streams at the experimental lakes area, Northwestern Ontario. Limnol. Oceanography 41 (5), 1004–1017. Available from: https://doi.org/10.4319/lo.1996.41.5.1004.

Scoffield, H., and J. Ditchburn, Tories Admit to Closing Enviro Research Group Because They Disliked Results. *The Tyee*, 2012.

Solomon, E., and K. Everson, 7 Environmental Charities Face Canada Revenue Agency Audits. 2014.

Soroye, P., Newbold, T., Kerr, J.T., 2020. Climate change contributes to widespread declines among bumble bees across continents. Science 367 (6478), 685–688. Available from: https://doi.org/10.1126/science.aax8591.

Species at Risk Act, Canada. 2002.

Staff, An Open Letter from the Honourable Joe Oliver, Minister of Natural Resources. *The Globe and Mail*, 2012.

Union of Concerned Scientists, Timeline: Legal Harassment of Climate Scientist Michael Mann. 2014. <https://www.ucsusa.org/resources/legal-harassment-michael-mann>.

Vitousek, P.M., D'Antonio, C.M., Loope, L.L., Reymanek, M., Westbrooks, R., 1997. Introduced species: a significant component of human-caused global change. N. Zealand J. Ecol. 21 (1), 1–16. <http://www.jstor.org/stable/24054520>.

Watts, J., Katharine Hayhoe: 'A Thermometer Is Not Liberal or Conservative.' *The Guardian*, 2019. <https://www.theguardian.com/science/2019/jan/06/katharine-hayhoe-interview-climate-change-scientist-crisis-hope>.

Weiland, N., 'Like a Hand Grasping': Trump Appointees Describe the Crushing of the C.D.C. *New York Times*, 2020. <https://www.nytimes.com/2020/12/16/us/politics/cdc-trump.html>.

Wherry, A., The Cost of Scrapping the Long-Form Census. *Macleans*, 2013.

Wong, E., Trump Has Called Climate Change a Chinese Hoax. Beijing Says It Is Anything But. *New York Times*, 2016. <https://www.nytimes.com/2016/11/19/world/asia/china-trump-climate-change.html>.

Woo, A., Three Scientists on the Research They Couldn't Discuss with Media under Harper. *The Globe and Mail*. 2015.

Blowing the whistle on political interference: the Northern Spotted Owl

Dominick A. DellaSala
Wild Heritage, A Project of Earth Island Institute, Berkeley, CA, United States

In the beginning, there was the owl

As a newly minted post-doctoral research fellow in the late 1980s, I cut my intellectual teeth in Oregon's old-growth forests studying the dietary preferences of my test subject—the Northern Spotted Owl (*Strix occidentalis caurina*). During the field season, I was equipped with over 50 pounds of "tomahawk" small mammal traps strapped tightly on my backpack for positioning the traps systematically along sampling grids within spotted owl territories. To reach the owl's clandestine haunts, I stepped gingerly over massive logs strewn across steep slopes while occasionally stirring up ground-nesting wasps that took their revenge out on me with a blitzkrieg of merciless stingers. Despite all the perils of fieldwork, I relished every minute of it, even when I was skunked by the occasional stinky mustelid caught in one of my live traps. This was the world of the owl. I was a mere spectator dwarfed by rainforest giants towering >70 m to the forest skyline.

On a good day, I would capture one of the owls' favorite meals—the northern flying squirrel (*Glaucomys sabrinus*). After identifying my captured animals' gender and taking body measurements, the treat of the day was opening the trap door (much better than releasing a skunk!). I watched in awe as the adroit squirrel scurried up a massive tree trunk, poised as if to give thanks for its freedom, and then extending its underbelly "wing" membrane as it glided effortlessly through small breaks in the dense forest canopy. And while it took me years to uncover the habitat needs of my captives, nothing prepared me for the day when the owl would set the halls of Congress ablaze in controversy over the Endangered Species Act (ESA).

During the George W. Bush two-term "reign of terra" (January 20, 2001–09), the environment was under siege. One of the biggest mistakes the administration

Conservation Science and Advocacy for a Planet in Peril.
DOI: https://doi.org/10.1016/B978-0-12-812988-3.00014-4

made at the time was putting me on the U.S. Fish & Wildlife Service's (Service) spotted owl recovery team toward the end of the second Bush term from 2006 to 2008. Reluctant to participate at first, because the team was stacked with timber industry executives and political appointees and not top owl experts, I became an unheralded whistleblower of political interference in the ESA. My status was procured from an insider's perspective with access to internal communications from top administrative officials during recovery team planning sessions. Seeking an advantage to throw sand-in-the gears of an anti-environmental agenda, I used the flow of privileged information as evidence in congressional hearings on how political appointees were instructing the recovery team to downplay the owls' mature forest habitat needs in favor of ramped-up logging.

This chapter is about how speaking truth to power had a successful outcome in restoring scientific integrity to the ESA. It's a lesson to scientists willing to step forward as whistleblowers even when faced with political reprisals. In this case, the owl scored a major victory for scientific integrity, at least temporarily, as we shall see.

What the owl needs

> Note: There are three subspecies of spotted owls: Northern (this chapter), California (*Strix occidentalis occidentalis*), and Mexican (*Strix occidentalis lucida*). General information about all three subspecies is available on The Cornell Lab of Ornithology website—https://www.allaboutbirds.org/guide/Spotted_Owl/overview.

Scientific interest in the owl has supplied more biologists with research opportunities than any raptor to date. While there is more to learn about this magnificent bird, we do know some basic ecology. The Northern Spotted Owl is a medium-sized (~46-cm) owl (see Fig. 5.1A) of the Pacific Northwest found in low- to mid-elevation forests along the Cascade Mountains and Coastal Ranges from southwestern British Columbia (although it is now functionally extinct in BC in the wild) to Marin County, California (U.S. Fish & Wildlife Service, 2011).

In the northern part of its range and in coastal temperate rainforests, the spotted owl reaches its highest densities in structurally complex, older forests—owls in coast redwoods can use younger forests because those forests can develop the advanced structure quickly that supports both nesting and owl prey. Older forests used by the spotted owl are generally characterized by multilayered tree canopies (i.e., emergent trees, mid story, lower story, understory), abundant large snags and logs, and diverse understory vegetation (Fig. 5.2A). In these towering forests, flying squirrels and red tree voles (e.g., *Arborimus longicaudus*) occupy the forest-penthouse

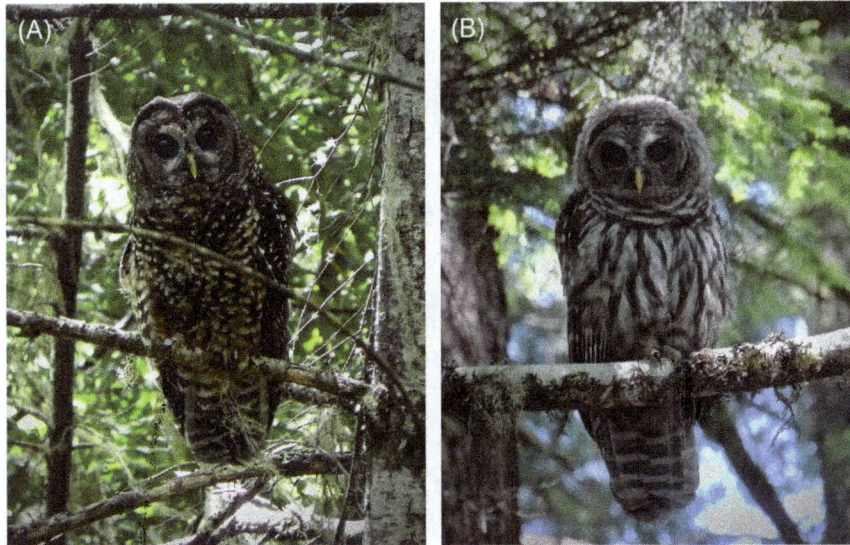

Figure 5.1 The federally threatened Northern Spotted Owl (A) and its competitor, the Barred Owl (*Strix varia*) (B). *Photos with permission from Francis Etherington.*

where the owl hunts (Forsman et al., 2004). Spotted owls do best when the amount of older forest habitat at the provincial home-range scale (nearly 6000 ha in places) is maxed out, allowing for optimal foraging (Forsman et al., 2004) and nesting opportunities (Anthony et al., 2006; Carroll and Johnson, 2008). The more and better the quality of old-forest habitat there is, the greater the chances for spotted owls to resist invasion by Barred Owls (Fig. 5.2B), a superior competitor that only recently strong-winged itself into the same penthouse space of the spotted owl due to expansion of the Barred Owls' geographic range (Wiens et al. 2014; Dugger et al., 2016; Dunk et al., 2019) enabled by human-caused habitat alterations (Livezy, 2009). Normally, one would celebrate the appearance of any raptor in the forest, but in this case, the Barred Owl spells competitive doom for the spotted owl.

Further south (e.g., Klamath providence of California) and in drier mixed conifer forests along the eastern slopes of the Cascades in Washington and Oregon, the spotted owl nests in older forests juxtaposed with dense shrubs occupied by its favorite meal—woodrats (*Neotoma* spp.) (Forsman et al., 2004). Here, fire is Nature's architect that periodically sculptures a mosaic of burn severity habitat patches (e.g., low, moderate, and severe fire effects on tree mortality, Fig. 5.2B) that the owl does best in (Franklin et al., 2000; Dugger et al., 2005; Lee, 2018). Reoccurring wildfires produce a "bed-and-breakfast" like effect where older forest patches that survived fire serve as the owls' "bedroom," and severely burned patches where most trees were killed, the "breakfast room." Just how much of each the owl needs is the subject of intense debate (see Jones et al., 2016 vs Lee, 2018, see below) with important recovery implications (see below).

Figure 5.2 (A) Classic old-growth forest used by spotted owls in the northern and coastal portions of its range versus (B) the mixed-severity "bed-and-breakfast" effect produced by periodic wildfires in dry mixed conifer forests in the southern and eastern portion of the owls' range. Wildfires are assumed to cause habitat loss by federal agencies (Davis et al., 2015), some conservation groups (e.g., Metlen et al., 2018), and some scientists (e.g., Jones et al., 2016), despite the lack of scientific agreement (see Hanson et al., 2009; Odion et al., 2014; Lee, 2018; Hanson et al., 2018).

How a 46-cm bird stopped the timber industry in its tracks

The owl gets listed and all hell breaks out

The Northern Spotted Owl was listed as threatened under the ESA in 1990 due to destruction and adverse modification of its old-forest habitat and inadequacy of regulatory mechanisms (i.e., listing factors, U.S. Fish & Wildlife Service, 2011). That listing ushered in a chronology of legal and policy decisions that reflect in many ways how politics was used to bypass the ESA (Box 5.1). Despite the ESA listing, however, it was actually another law—the National Forest Management Act (NFMA)—that silenced chainsaws throughout the Pacific Northwest when the Honorable Judge Dwyer in 1990 first enjoined federal timber sales because the US Forest Service forest plans were deemed inadequate to maintain viable populations of the owl and hundreds of other old-forest associated species well-distributed across the owls' range (https://casetext.com/case/seattle-audubon-soc-v-evans). The timber wars had reached a climax over the viability ruling (which was later weakened by the Obama administration's 2012 forest plan rule), so something had to give.

Box 5.1 Brief chronology of events regarding Northern Spotted Owl and the Northwest Forest Plan (NWFP). Not all events are shown,[1] the focus here is on significant developments pertinent to the owl and political interference as covered in this chapter. Chronology was developed in consultation with Attorney Susan Jane Brown of the Western Environmental Law Center.

- ESA listing of the owl in 1990—destruction and adverse modification of habitat, inadequacy of regulatory mechanisms as listing factors.
- Judge Dwyer 1990 decision enjoins the first of several timber sales on federal lands due to inadequate protections for hundreds of species under National Forest Management Act (NFMA) viability regulations (other legal decisions would follow).
- Draft owl recovery plan 1992 (U.S. Fish & Wildlife Service, 1992a) and critical habitat (U.S. Fish & Wildlife Service, 1992b) ruling. Note—there have been ongoing loopholes in critical habitat determinations that allow for substantial

(Continued)

Box 5.1 (Continued)

"take" (e.g., killing) of owls and degradation/downgrading of critical habitat so long as it does not trigger a "jeopardy decision" (the criteria for reaching a jeopardy decision also has been weakened over several administrations).

- President Clinton holds Portland forest summit in 1993 directing federal agencies to work with scientists in drafting an ecosystem management and biodiversity plan for >10 million ha of federal lands (all federal agencies) that would meet the standard of "scientifically sound, ecologically credible, and legally responsible."
- Northwest Forest Plan (NWFP) Record of Decision 1994—de facto recovery plan for the owl via coarse- (reserves) and fine- (survey and protect rare species) filter approaches along with other measures like the aquatic conservation strategy and watershed analysis goes into effect for hundreds of old-forest associated species.
- George W. Bush administration recovery plan and critical habitat (2006–08) forced by Seattle Audubon Society 2005 lawsuit.
- Congressional hearings on political interference in the ESA (2007–08)
- Office of Inspector General (OIG) (2008) and Government Accountability Office (GAO) (2008) find evidence of widespread political interference in several ESA decisions, including the Bush administration's owl recovery plan.
- President Obama issues 2009 scientific integrity memo rescinding 20 of the illicit ESA decisions made under the Bush administration, announcing a redo of the owl recovery plan and critical habitat decisions.
- Obama White House memo instructs federal agencies to reject "hands-off approach" and allow logging (active management) in owl habitat.
- Obama recovery plan (U.S. Fish & Wildlife Service, 2007, 2011) and revised critical habitat ruling recommend logging to reduce presumed wildfire risks to owls despite criticism from owl scientists and scientific societies.
- Implementation of Obama recovery plan and critical habitat determination by agencies and land managers (e.g., The Nature Conservancy) allows logging/thinning in owl habitat, much of which has been questioned by new studies and challenged by scientists.

[1]Legal cases used relevant provisions of the National Environmental Policy Act, NFMA, and the ESA to force the federal government to develop a plan for protecting the owl and hundreds of old-growth associated species. Relevant legal cases included Seattle Audubon Soc. v. Robertson, 1991; U.S. Dist. LEXIS 10131 (W.D. Wash. Mar. 7, 1991); Seattle Audubon Soc. v. Evans, 771F. Supp. 1081 (W.D. Wash, 1991); Seattle Audubon Soc'y v. Moseley, 798F. Supp. 1473 (W.D. Wash, 1992); Lane County Audubon Soc. v. Jamison, 958 F.2d 290 (9th Cir. Or. 1992).

Enter the Northwest Forest Plan

In an effort to break the "log jam" in the Pacific Northwest, President Bill Clinton and Vice President Al Gore along with several presidential cabinet members convened a historic forest summit in Portland, Oregon, in April 1993 that would set the stage for the landmark NWFP over a year later. At the time, the president ordered federal agencies to work with scientists in developing a region-wide framework that was *"scientifically sound, ecologically credible, and legally responsible"* (emphasis added). An attempt at an earlier draft owl recovery plan in 1992 was never finalized because it became moot when the NWFP was approved. For over a decade, the NWFP would stand as the de facto spotted owl recovery plan governing the viability of the owl and hundreds of old-growth associated species on some 10 million hectares of federal forests. It was science-based (a unique example of scientists calling the shots), was hailed as a global conservation model (DellaSala and Williams, 2006), scorned by the timber industry for not delivering on anticipated timber targets (i.e., "broken promises"), and thus far has withstood numerous attempts, which continue to this day, to "chip away" at its scientific and protective foundations.

The framers of the NWFP created 10 options for maintaining the viability of old-forest species across the owls' range before the plan was finalized. They built the selected "Option 9" on a coarse- (reserves) and fine-filter (localized habitat protections for rare species) approach rooted in conservation biology and developed to up the ante on population viability for hundreds of species from dire to a >80% viability probability outcome (DellaSala et al., 2015). The ~10 million ha federal forest land base was partitioned into Late-Successional Reserves (LSRs, 30%), Riparian Reserves (11%), and other designations (13%) that were added to the existing Congressionally Reserved Areas (30%), with 11% available to log within the Matrix (Fig. 5.3; note—~40% of LSRs were actually recovering clear-cuts that would need to be restored over time, Strittholt et al., 2006). An Aquatic Conservation Strategy and watershed analyses (coarse level; see Reeves et al., 2006) were added to address habitat and restoration needs of salmonids and other aquatic species. In addition, a fine-filter approach was designed to survey for rare species (outside the reserve network) in timber sale areas and, if any were found, a small protective buffer prohibiting logging would maintain some habitat (Molina et al., 2004). It was assumed that with these measures in place there would still be a "sustained" volume of timber that could be logged [known as "probable sale quantity" (PSQ) or "allowable sale quantity"] from the Matrix. However, in practice the PSQ was never met because of continued controversy over old-growth timber sales, the complexities of the plan itself (especially the fine filter, Molina et al., 2004), and a diversifying regional economy that reduced timber demand on federal lands (Power, 2006).

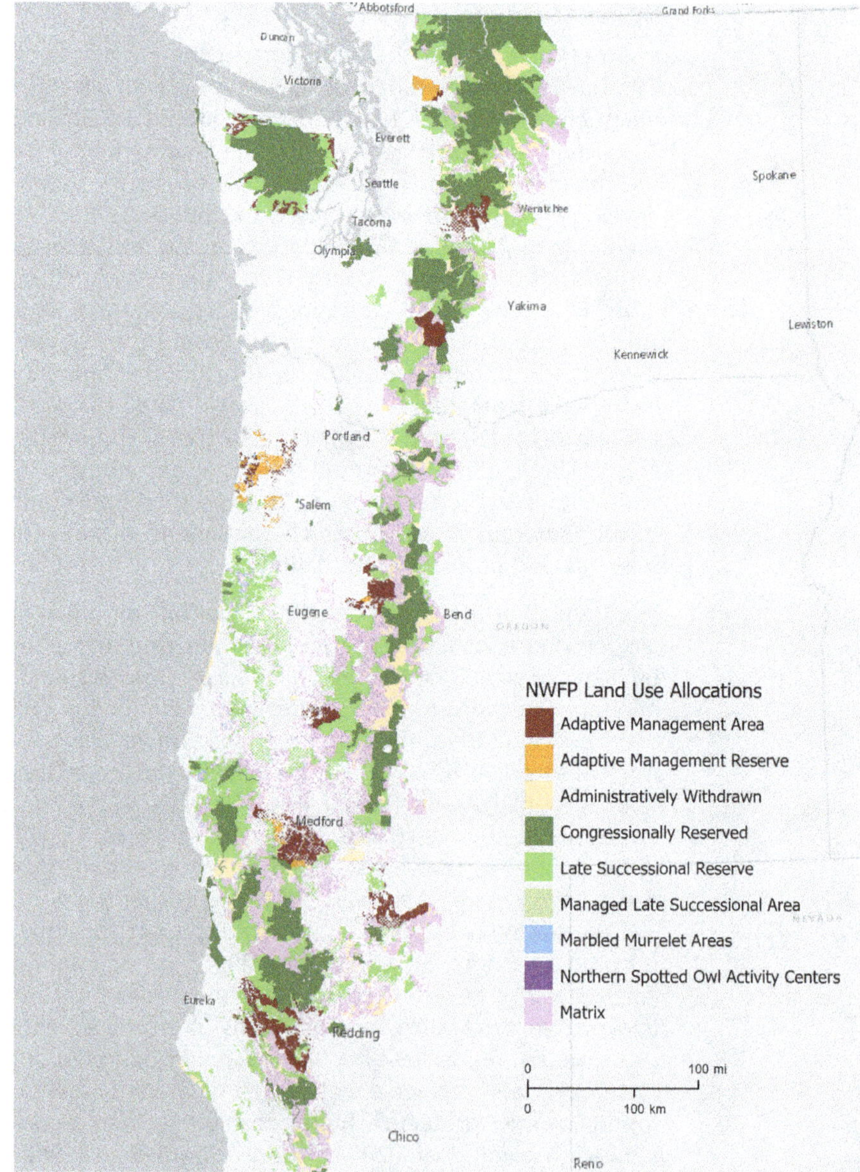

Figure 5.3 Land-use allocations, owl activity centers, and Marbled Murrelet (*Brachyramphus marmoratus*) areas designated within the NWFP area: Congressionally reserved; Late-Successional Reserves; Managed Late-Successional Reserves; Adaptive Management Areas; Administratively Withdrawn; Riparian Reserves; and Matrix. *NWFP*, Northwest Forest Plan. *Figure created using Data Basin (www.databasin.org) and NWFP data layers. Discussion in this chapter is mainly focused on the reserves and matrix areas. Figure credit: J. Leonard.*

Enter the owl recovery plan

In recognition of the collapsing owl populations in Washington, Seattle Audubon Society successfully sued the Service in 2005 to develop a recovery plan aimed specifically at the Northern Spotted Owl (http://wflc.org/cases/

Box 5.2 Best available science and recovery plans (developed in consultation with Attorney Susan Jane Brown).

According to the ESA, "information used to develop and implement recovery plans is reliable, credible, and represents the best scientific and commercial data available" [ESA, 59 Fed. Reg. 34,271 (June 1, 1994)]. Typically, recovery plans are developed by *recognized experts* in the ecology and management of the listed species to ensure that recovery objectives and delisting criteria are based on the best available science (Department of Interior and Department of Commerce, 1994, emphasis added). Under the ESA, the purpose of recovery plans is to get listed species to the point where delisting is warranted and protection is no longer needed. In order for a listed species to move from the "intensive care unit" to a self-sustaining population, recovery plans must be based on the best available science. However, the courts do not recognize recovery plans as legally binding, instead, they only carry the weight of recommendations that the Service can make to other agencies [e.g., Forest Service, Bureau of Land Management (BLM)] and landowners.

docket/seattleaudubonvnorton). While an important legal victory, in hindsight taking on a recovery plan during the hostile George W. Bush administration was risky business especially considering that the administration was also involved in sue-and-settle agreements with the timber industry to escalate logging. Controversy erupted when the Service decided to assemble a recovery team in 2006 that excluded top owl scientists and was made up of a multistakeholder team of federal and state agencies, timber industry, Seattle Audubon Society and myself that were tasked with developing a recovery plan to comply with the Audubon lawsuit. The Service's charter document under which the recovery team made decisions emphasized "recommendations for recovery actions from the Team will be made in a collaborative manner, striving for the highest level of consensus possible." While this gave the appearance of playing nice across interest groups, in reality recovery plans are supposed to be based on the best available science (Box 5.2), not a popularity contest to achieve consensus (e.g., nothing in the ESA specifies a consensus requirement). Furthermore, the two conservation groups represented on the team were flanked by timber, political appointees, and agency representatives.

Despite the stacked deck, in September 2006, the recovery team forwarded its draft plan to the Service's headquarters in Washington DC for final approvals. The team recommended a recovery strategy that was anchored mostly in the existing reserve network of the NWFP (see below). We reached consensus because it was the most scientifically credible way to recover the owl even though the timber industry resisted.

The scientific rationale for using NWFP reserves for conserving spotted owls and other old-growth dependent species has been reaffirmed many times (e.g., Noon et al., 2003; Courtney and Franklin, 2004; DellaSala and Williams, 2006; Noon and Blakesley, 2006; Strittholt et al., 2006; Carroll et al., 2009; DellaSala et al., 2015) and still is the best science of the day, despite ongoing proposals to that question the efficacy of reserves in favor of "active management" (Spies et al., 2018, see below). For instance, in the Service's commissioned 5-year "status review" of the spotted owl in 2004, Courtney and Franklin (2004) concluded that:

- "the Reserve and Matrix strategy of the NWFP has been successful and is performing as expected;" and
- "the NWFP has made important contributions to protect and recover the endangered owl and without the plan the situation of Northern Spotted Owls would be far bleaker."

Further, a demography study of spotted owls (Anthony et al., 2006) found that owls were reproducing and surviving better on federal land managed under the NWFP reserves than on nonfederal lands where logging levels were much higher (i.e., the annual rate of owl population declines on nonfederal lands was more than twice that on federal lands).

Although the recovery team agreed that a network of protected reserves linked to the NWFP would remain the foundation of the recovery strategy, we did not reach consensus on specific reserve management recommendations, particularly in the dry, fire-dependent forests. The team agreed to forward its recommendations to the Service's Washington DC headquarters on the condition that the draft plan undergo rigorous scientific peer review with substantive revisions made pending review. The Service initially rejected this request, citing insufficient time as a constraint, even though they had more than 5 months to initiate peer review before finalizing the plan. With pressure from conservation groups, the Service finally agreed to peer review the draft just two days before public release. Panels of scientific experts were brought in to provide the cover of best available science that was anything but that as political interference was about to become the administration's quid pro quo with the timber industry.

The politics of the Northwest Forest Plan and owl recovery

Political interference

Although Option 9 of the NWFP had scientific underpinnings, it was based on a political calculus handed down to the science team (known as

FEMAT—Federal Ecosystem Management Assessment Team) to placate the timber industry and decision makers into accepting ~1 billion board feet (~2,359,737 m^3 or roughly 40,000 fully loaded log trucks per year!) of timber that theoretically could be extracted annually from the Matrix even though team members apparently knew these volumes would never see the light of day. In my personal communications with FEMAT scientist Dr. Robert Anthony (now deceased), most of the team wanted the more restrictive Option 1 because it best met the coarse- and fine-filter approaches even though there would be much less timber produced. However, President Clinton ostensibly put pressure on the FEMAT team leader who pushed for Option 9 to meet higher timber volume projections. Years later, the timber industry, their allies in Congress, and the Bush administration would declare the plan's missed PSQ a "broken promise" and a failure to meet assumed timber targets (Thomas et al., 2006; Charnley, 2006) even though that view was challenged by other economists as overblown (Niemi et al., 1999; Power, 2006) because of an adaptive regional economy that no longer was dependent on federal logs. Nevertheless, the pressure was on to remove habitat protections to meet unrealistic timber volume projections.

"Delinking" (separating) Northwest Forest Plan reserves from owl recovery

Sensing an opportunity to ramp up logging volumes and under multiple sue-and-settle legal agreements with the timber industry (Box 5.3), the Bush administration realized that a weakened owl recovery plan would be the raison d'être for toppling the NWFP's protective reserve network. In late September 2007 the Pacific Regional Director of the U.S. Fish & Wildlife Service notified the recovery team of a shadow "Washington [DC] Oversight Committee" (OC), consisting of high-ranking officials from the Departments of Agriculture and Interior that would oversee recovery, essentially pushing aside the recovery team's reserves and NWFP endorsements. At the time, the OC included the infamous Julie MacDonald (Deputy Assistant Secretary for Fish, Wildlife, and Parks) who was under investigation by Congress for widespread political interference in the ESA (e.g., see https://www.doioig.gov/sites/doioig.gov/files/Macdonald.pdf). On October 17 the recovery team was told that the OC rejected the September stakeholder draft because it was based on a version of the NWFP's reserves and therefore did not meet the "flexibility" (code for more logging, see below) requirements of the Forest Service and BLM. The OC instead directed the recovery team and federal agency staff to rewrite the plan by including an alternative that did not rely on habitat protections and the NWFP reserves. The recovery team received communications from then Interior Deputy Secretary Lynn Scarlett (a political appointee) to greatly limit the discussion of owl habitat in the recovery plan (summarized below).

Box 5.3 The sue and settle (quid pro quo) with industry. Based on personal communication with environmental attorney Susan Jane Brown who provided this timeline of sue-and-settle agreements that was used by the Bush administration to usher in proposed logging increases.

In 2003, the timber industry files a legal challenge to Northwest Forest Plan (NWFP) alleging that the plan violates the O&C Act, Endangered Species Act (ESA), National Environmental Policy Act (NEPA), Federal Lands Policy and Management Act, and other laws [American Forest Resources Council v. Clarke (D. DC 2003)]. Federal Defendants and timber Plaintiffs enter "global settlement agreement" to (1) amend the NWFP aquatic conservation strategy (ACS); (2) eliminate Survey and Manage fine filter; (3) reduce ESA protections for the spotted owl and marbled murrelet; (4) amend BLM resource management plans to better reflect O&C Act provisions promoting timber volume. In 2004, the Forest Service and BLM issued a supplemental Environmental Impact Statement (EIS) to remove and/or modify the Survey and Manage program. Conservation groups successfully challenge this on NEPA grounds (Northwest Ecosystem Alliance v. Rey, 380F. Supp. 2d 1175) (W.D. Wash, 2005) (S&M II). Federal Defendants, conservation Plaintiffs, and timber Defendant-Intervenors settle the relief in that case, allowing for some aspects of the 2001 Supplemental EIS and Record of Decision (ROD), and 2004 Supplemental EIS and ROD, to move forward, known as "the Pechman exemptions." In 2004, the Forest Service and BLM issue a supplemental EIS and ROD to modify the ACS to eliminate the requirement that the ACS applies at all spatial and temporal levels as required. In July 2006, the Forest Service and BLM issue supplementals to remove and/or modify the Survey and Manage program again.

Congressional action and government oversight

During a July 2007 Congressional hearing on ESA interference that I testified at (DellaSala, 2007), then Congressman Jay Inslee read an email that I forwarded to his staff from then Deputy Secretary Scarlett with instructions to the recovery team to downplay habitat in the recovery plan. In a flurry of verbal exchanges, the congressman called for Scarlet's resignation (featured on the Jim Lehrer News Hour). Although she did not resign, her incredulous response to the accusations, and those of other OC officials, lead Congress to call on the investigative branch of government, the OIG and GAO, to look into political interference in multiple ESA decisions.

Here's some of what the OC told the recovery team in 2007 that I uncovered at the congressional hearing and helped to trigger government oversight (as summarized from congressional testimony presented in DellaSala, 2007).

- *"De-emphasize past science and rely on new science."* Already compromised by not having top owl experts on the team, the owl recovery team was told to base habitat recommendations mainly on studies in the southern part of the owl's range that conflicted with decades of old-growth studies further north. Two of those studies aptly pointed to the owl's reliance on a mixture of forest age classes in dry forested regions maintained by periodic wildfire (Franklin et al., 2000—northern California Klamath province, Olson et al., 2004—Oregon Coast Range). However, and to their credit, the authors of those studies repeatedly warned the recovery team not to use their results to nix old-growth protections. A third study (Dugger et al., 2005), also in the southern range near Roseburg, Oregon could not confirm these findings and also informed the recovery team that the owl still needed old-growth forest for its survival. Regardless, under pressure from the OC, the Service rewrote the recovery plan by cherry-picking the science in devaluing the old-growth habitat to accomodate "flexibility" (detailed below).
- *"Flip and switch the presentation of threats to the spotted owl in the draft plan" by minimizing the importance of habitat loss and placing more emphasis on Barred Owls as the primary threat.* An October 25 memo from the OC directed the recovery team to "indicate [the Barred Owl] was [the] only threat given priority number 1…and summarize the habitat threats discussion *into less than a page*" (emphasis added). The flip and switch part of the direction was a shell game to prop up Barred Owls as the main culprit of spotted owl declines so that habitat protections could be deemed nonessential. An untitled document dated October 27 and distributed to the recovery team at a meeting in Portland by the Service's recovery team leader, contained direction from then Deputy Secretary Scarlett telling the recovery team to make the new option "less focused on habitat preservation." Although Barred Owls emerged as a recent threat to spotted owls (Kelly et al., 2003; Crozier et al., 2006), the science of conservation biology and endangered species management is clear on this point—when a species is faced with multiple threats it is best to conserve more habitat, not less (see Anthony et al., 2006; Carroll and Johnson, 2008; Dugger et al., 2016; Dunk et al., 2019).
- *"De-link the recovery plan from the NWFP."* On October 18 the recovery team received notice from the Service following direction from the OC to "de-link the owl plan from the NWFP" to provide the Forest Service and BLM the "flexibility" it wanted. On October 26 the Service (i.e., the regulatory branch of government responsible for ESA implementation and federal agency compliance)

> candidly admitted to the recovery team that the Forest Service and BLM were actually calling the shots on recovery actions. This was a fox-in-the hen house strategy whereby the same agencies responsible for habitat destruction were deciding what type, how much, and where spotted owl habitat could be protected. Under questioning from recovery team members (myself and the Seattle Audubon Society rep Tim Cullinan), both the recovery team leader, and Forest Service and BLM representatives, explained that "flexibility" meant giving the Forest Service and BLM carte blanche discretion to alter or eliminate Owl Habitat Conservation Areas (a proposed revision of the NWFP reserve network). At the time, the BLM was revising its forest plans on ~1 million hectares in western Oregon to up the cut and the Forest Service was proposing to exempt forest plan revisions from NEPA (Federal Register Vol. 71, No. 241, Friday, December 15, 2006, pp. 75481–75495). It should be noted that one of the primary factors for listing the owl in the first place was "inadequacy of regulatory mechanisms," the exact problem now unfolding via political interference in the recovery plan. By adopting a "flexible" system where reserves could be moved around or eliminated at will to increase timber yields, forest plan revisions would reach a new low around the nebulously deceptive concept of "flexibility."

During the hearing on political interference in the ESA, congressional members included intimidation tactics to swear me in as a witness whereby if any of my statements noted above were contradicted by government witnesses, I would be charged with contempt (this is most unusual at congressional hearings of this type). The Forest Service recovery team representative at the time submitted a counter-deposition that challenged my declaration about political interference. When I was pressed on the witness stand and under the potential of contempt charges, I opened a briefcase full of supporting emails (hundreds of them), recovery team notes, and related documents with exact quotes from OC officials that were hard to dismiss even for contrarian congressional and agency members trying to muzzle me.

The outcome of this and other hearings on ESA interference at the time was both the GAO (2008) and OIG (2008) concluded there was widespread political interference in the ESA by political appointees. Most of the criticism was aimed at Julie MacDonald; however, the tentacles of corruption had sucked the life out of the Service that was supposed to enforce the nation's premier environmental law. The interference came from the top (as high as Vice President Dick Cheney in some cases and not just MacDonald and Scarlett) and it placed career Service employees that wanted to do the right thing in a retaliatory position from higher ups. I know this to be true because the *Jim Lehrer News Hour* interviewed me and featured the hearing exchange between Inslee and Scarlett on national TV. At the time, I was worried of being ousted from the recovery team as a traitor. But instead of a scolding, I was approached by career employees with expressions of gratitude for taking a stand for the owl that they could not for fear of reprisal. The moral of the story is there are honest professional employees at the Service that even in the darkest of administrations still believe in the agencies' noble endangered species mission.

In the end, the information presented on political interference provided by multiple witnesses was so overwhelming that it became grounds for President Obama to overturn the Bush administration's politically motivated owl "recovery plan," and restore, at least partially, scientific integrity to many of the 20 or so illicit ESA decisions of the Bush administration's reign of terra (there were more interference decisions that never made it into the investigative reports; https://www.theguardian.com/environment/2009/mar/03/obama-bush-endangered-species-act-us). However, as we will see, even the revised Obama plan was fraught with political interference and confirmation biases carried forward to this day. A new excuse—wildfire prevention—would become the next flexibility argument that would once again call into questions the reserves (e.g., see Spies et al., 2018 for challenges to the NWFP reserves versus criticisms of his views—http://forestlegacies.org/press-room/latest-news/northwest-forest-plan-under-review-2/).

Wildfire as the new flexibility excuse

When President Obama came into office in January 2009, there was renewed optimism for the return of scientific integrity (https://obamawhitehouse.archives.gov/the-press-office/memorandum-heads-executive-departments-and-agencies-3-9-09). The president delivered immediately on his promise by overturning last-minute interference decisions by the Bush administration orchestrated by Julie MacDonald and the infamous OC. In doing so, the president cited the OIG and GAO findings along with scientific peer review by leading owl experts and three of the nation's top scientific societies (Society for Conservation Biology, American Ornithologists' Union, The Wildlife Society). Soon after, John Holdren, Office of Science and Technology Policy, issued a memorandum to the heads of Executive Departments and Agencies to restore scientific integrity to agency decisions (https://obamawhitehouse.archives.gov/administration/eop/ostp/library/scientificintegrity). Apparently not everyone got or read that memo, as a series of contradictory follow-up memos from the White House and Service made stunning and unfounded revelations that once again pushed for logging in owl habitat and building on "flexibility" double-speak (Boxes 5.4 and 5.5). Unfortunately, this level of interference, not immune to any single administration, was coupled with further ESA rollbacks through the final days of the Obama administration (https://www.biologicaldiversity.org/news/press_releases/2016/endangered-species-02-05-2016.html).

The statements emphasized in the boxes below warrant careful scrutiny, as they are now inculcated in agency decision documents allowing timber

Box 5.4 From the Obama White House: the proposed rule (critical habitat) recommends, on the basis of *extensive scientific analysis*, that areas identified as critical habitat should be subject to *active management*, to produce the variety of stands of trees required for *healthy forests* (emphasis added).

The proposal rejects the traditional view that land managers should take a "*hands-off*" approach to forest habitat to promote species health; ongoing logging activity may be needed to enhance *forest resilience* (emphasis added) (White House Office Press Secretary Memo dated February 2012).

Box 5.5 Flexibility language adopted by the Bush White House: these management approaches are intended to be consistent with the principles of Executive Order 13563, which, as noted, directs agencies to consider regulatory approaches that *reduce burdens and maintain flexibility and freedom of choice* for the public (emphasis added)"

(U.S. Fish & Wildlife Service, 2007).

sales to proceed in owl habitat often with minimal environmental review. The Forest Service, BLM, and even organizations like The Nature Conservancy (pers. comm. K. Metlen, TNC, regarding BLM thinning projects in the Rogue River basin, southwest Oregon) repeatedly cite the Obama recovery plan and critical habitat rule that was built on those infamous White House memos and questionable assumptions about wildfire impacts that have been challenged by scientists (e.g., Lee, 2018).

Debunking bogus claims about "extensive scientific analysis"

The Obama White House memo refers to logging needed for owl critical habitat maintenance without citing a single scientific study. If anything, extensive scientific analysis of owl habitat assembled painstakingly by researchers over many decades and the listing factors for which the owl was placed under ESA protections repeatedly show how logging is still a main culprit of population declines (see Dugger et al., 2016 for review of habitat needs of the owl that point to more habitat protection needed, not less, and Dunk et al., 2019 that point to Barred Owls and *ongoing* habitat losses working synergistically against the spotted owl). This latest active management argument is a recycled attempt by federal agencies to knock habitat protections out and to set the stage for future revisions of the NWFP that reduce or eliminate reserves (e.g., Spies et al., 2018).

Debunking active management versus hands off rhetoric

In congressional hearings and media coverage (e.g., http://forestlegacies.org/programs/fire-ecology/geos-scientist-testifies-in-congress-on-climate-change-and-forest-fires/), timber representatives, congressional allies, and agency scientists often refer to unspecified "active management" to save the owl from wildfires. Truth be told it is code for just about any form of "management," including postfire "salvage" logging, expansive "thinning," road building, and logging with minimal standards (e.g., misusing the categorical exclusion provisions under the NEPA to limit environmental review). This specious argument is framed as an either-or situation—active management or "hands off." In reality, the recovery of a listed species is never completely hands off. Recovery actions rooted in the science of conservation biology can in fact include captive breeding as in the case of the extinct owl population in British Columbia, minimizing competition between species as in the case of Barred Owl experimental removal from spotted owl territories (Dugger et al., 2016; Wiens et al., 2014), forest restoration in clear-cuts within reserves as in the case of the NWFP standards (DellaSala et al., 2015), and cessation of logging in owl habitat (see Hanson et al., 2009). To proactively address listing factors, multiple active and passive measures are needed that respond directly to the listing factors instead of this overly simplistic either-or dichotomy and disingenuous message framing.

Debunking healthy forests framing

This is another term that can mean just about anything (see Johns and DellaSala, 2017 for discussion of commonly used euphemisms). At congressional hearings and in agency forest plan revisions, forests with trees killed by wildfires and insect outbreaks are deemed "unhealthy" with the cure-all being more logging with minimal environmental review (e.g., see https://forestlegacies.org/?s=Westerman; also see Chapter 1, The Nuts and Bolts of Science-Based Advocacy). This is a recycled message from the days when old-growth forests were assumed by foresters to be "decadent," a phase where natural tree death is somewhat prominent and thus it is assumed that the forest needs to be logged and replanted with young vigorously growing nursery stock that bears no resemblance to a natural forest (DellaSala, 2020). A more holistic definition linked to ecological integrity (see Karr, 2000)—a quantifiable measure based on native species richness and functional ecosystem processes—needs to replace subjective forest health terminology.

Debunking forest resilience framing

This term is simply defined as the ability of a system to return to its predisturbed state. However, it is subjectively used in practice to mean just about anything, including proposals to log or commercially "thin" natural forests to "save them" from wildfires and insect outbreaks, replace mature forests with vigorously growing seedlings manufactured in nurseries, the use of pesticides and herbicides to reduce competition with commercial conifers, and suppression of natural disturbances all in the name of "resilience." Many of my conservation colleagues also use this term indiscriminately and without tying it to ecological integrity.

Burdens, flexibility, and freedom of choice messaging is code for more logging

This rather ominous Orwellian terminology is message framing for eliminating habitat protections, lifting environmental regulations, and selecting whatever is deemed necessary with little if any checks and balances (e.g., misapplied categorical exclusions of environmental review that allow logging over thousands of hectares).

These two decisions (Boxes 5.4 and 5.5) set the stage for the "flexible" proposals aimed at potentially gutting the NWFP by reducing the need for reserves in favor of "active management" (e.g., Spies et al., 2018). It set the stage for the Obama owl recovery plan and critical habitat determination to be routinely cited by land managers, TNC, and agency scientists that promote logging in owl territories, all the while ignoring or discounting scientific criticisms and dissenting views (see Chapter 2, When Scientists Are Attacked: Strategies for Dissident Scientists and Whistleblowers). In sum, the accumulation of over a decade of political interference and cherry-picked science, even when blunted by the change in administrations from Bush to Obama, have become indoctrinated in the federal agency mindset to log first and ask questions later. Herein lies the disingenuous doctrine of *active management* held up as the "new science" with conservation biology and reserves dismissed as the "old science."

Distinguishing cause from effect: is wildfire a threat or an excuse to log?

Logging as the cause, nest-site abandonment as the effect

Like so many researchers in the 1980s, I calibrated my search image of owl habitat as old-growth forest (see Fig. 5.2A). Clearly, this is the case in the northern range of the owl along the western slopes of the Cascades and in coastal rainforests of the Olympics where fire is an infrequent visitor (fire rotation intervals (landscape scale) are centuries long). Further south and along the dry eastern slopes of the Cascades, owls have adapted to more frequent and sometimes intense wildfires in mixed conifer forests (Fig. 5.2B). How else would the owl be able to successfully colonize these areas without surviving the postfire landscape?

All scientists are taught to distinguish cause from effect. It's drummed into our heads in introductory statistics classes before we even set foot in the field. In the case of the owl and wildfires a large fire that kills mature trees within owl territories can certainly be alarming to the untrained eye as *cause* of owl nest-site abandonment. Importantly, to gain clarity on what was driving owl nest-site abandonment in fire-adapted regions, Lee (2018) conducted a systematic evidence review of all available scientific literature on effects of wildfire on the three subspecies of spotted owl. Using empirical data, Lee (2018) set out to answer the question: do mixed-severity fires with patches of high-severity (all trees killed) within owl home ranges affect foraging habitat selection, demography, and site occupancy? What he found sent shock waves through the owl research community, resulting in career-threatening reprisals and claims of "agenda-driven science" aimed at Lee and fellow researchers (Peery et al., 2019, see Chapter 2, When Scientists Are Attacked: Strategies for Dissident Scientists and Whistleblowers).

Lee (2018) found that wildfires of mixed severity actually had mostly positive effects on owl recruitment, owl reproduction, and owl foraging in low- and moderate-severity burns with the inclusion of high-severity patches (also see Baker, 2014). His results were based on the compilation of dozens of studies but we all know that correlation does not mean causality. However, for simplicity, here's where cause and effect can be inferred. In the majority of cases where owls abandoned nesting territories there was clear evidence that unoccupied sites were associated with logging rather than wildfires per se. Despite these findings, wildfire is routinely considered the penultimate *cause* of habitat loss in recovery actions even though fire effects are in dispute (e.g., Jones et al., 2016 vs Lee, 2018, also see Hanson et al., 2018 for similar findings). In wildlife ecology studies, there are often multiple effects that statisticians must tease out from complex data sets. In the case of fire, however, the main effect on owl nest abandonment is actually logging and not necessarily the fire per se.

In addition to Lee's findings on postfire logging, an earlier habitat simulation study by Odion et al. (2014) tested whether the forest thinning recommendations in unburned owl habitat rooted in the Obama recovery plan constituted a short-term impact to avoid the longer term effect of high-severity fires as required in the spotted owl recovery plan. Instead, Odion et al. (2014) found that rotations of severe fire (the time required for high-severity fire to burn an area equal to the area of interest once) in spotted owl territories were ~362 and 913 years for the Klamath and dry Cascades provinces, respectively—more than adequate to sustain old-growth forests in fire-dominated regions. They projected that over a 40-year period, thinning would remove ~3–6 times more-dense, late-successional forests than it presumably "saved" from high-severity fire. Even if rates of high-severity fire increased under climate change, the recovery plan requirement that the long-term benefits of commercial thinning clearly outweigh adverse short-term impacts was summarily rejected. The researchers also concluded that exclusion of high-severity fire may not benefit spotted owls in areas where owls evolved with reoccurring fires due to owl foraging preferences (also see Baker, 2014). Not surprisingly, this study was met with swift condemnation by the Service recovery team supervisor Paul Henson that continues to this day (also see Chapter 6, Overcoming the Politics of Endangered Species Listings, for similar ESA decisions made by supervisor Henson).

In another study, Raphael et al. (2013) conducted a coarse-scale simulation of forest succession, wildfire effects, and thinning treatments on spotted owl habitat in Oregon and Washington projected over a 100-year time series that despite the fact that it was a government-issued report never saw the light of day with the Service. Active "fuel reduction" was anticipated to cause substantial short-term (simulation years 0–30) owl population declines.

Despite my sending of the Odion et al. (2014) (peer reviewed) and Raphael et al. (2013) (government publication) findings to the Service, commercial thinning in owl habitat continues all the while citing the recovery plan's questionable assumptions. For instance, whenever I raise these and other concerns about thinning in owl habitat being a much bigger threat to owls than fire, I am met with swift opposition from land managers adhering to the recovery plan. This ongoing tiering to unresolved claims about wildfire underscores problems with scientific integrity that span administrations (passing the political baton from Bush to Obama to Trump, etc.). Federal agencies need to be honest in reporting levels of uncertainty and disagreement in the scientific community that can best be settled by evidence reviews and further hypothesis testing along with an adherence to the Planetary Hippocratic Oath and precautionary principle as noted in Chapter 1, The Nuts and Bolts of Science-Based Advocacy.

"Thinning" is not benign habitat alterations

Every gardener knows that thinning of dense plantings allows proper maintenance by reducing competition for nutrients and available space. But forests are not gardens, they are living organisms and plucking trees individually or in mass is not some benign gardening activity especially when done within the habitat of an endangered species originally listed because of logging-related habitat losses. I have personally witnessed aggressive thinning projects within owl habitat deemed by the Service as a necessary but temporary habitat loss to avoid a bigger presumed loss from wildfires.

For instance, logging (with the approval of TNC) within owl critical habitat in the Ashland Watershed, Oregon reduced the dominant overstory from nearly continuous canopy coverage to <60% cover, below the threshold needed by nesting owls (Fig. 5.4A). For the Shasta-Trinity, California owl sites (Fig. 5.4B and C), the Forest Service claims that logging is needed to provide a "fire-resistant stand," ignoring the evidence that it is actually harmful to owls. All the while, the Service, citing the recovery plan, grants incidental take permits and allows habitat to be degraded by commercial thinning projects as a "short-term" habitat reduction.

What's next for the owl and the Northwest Forest Plan

The mantra of "active management" and "flexibility" may soon be the fatalistic ending to the owls' remarkable evolutionary chapter. Owl populations are tanking across the range and the species has been proposed for uplisting to endangered. (Note: The Service was petitioned by conservation groups in 2015 to uplist the owl to endangered, which it recently deemed as warranted but "precluded" —see Chapter 6, Overcoming the Politics of Endangered Species Listings, for many other examples of Service ESA inactions.) The Forest Service in its recent assessment of the NWFP (USDA Forest Service, 2020) knows owl extinction is imminent; however, the agency now claims they need a much heavier "hands-on, all-lands" approach to reduce fires even though owls are doing best within the reserves where logging is disallowed and are adaptable to fires if owl territories are not logged (Lee, 2018).

For instance, there are now proposals to dismantle existing protections for large trees, allow logging within LSRs, and even the possible elimination of the reserve network in dry forests, all the while using fire as the flexibility double-speak messaging (see Spies et al., 2018, USDA Forest Service, 2020). New Orwellian terminology has been introduced like "disturbance restoration" to usher in a build-up of thinning projects and a preference for low-moderate intensity fires, even though high-intensity fires produce biological diverse complex early seral forests that the owl needs for the bed-and-breakfast effect. As Abraham Maslow once said *"if the only tool you have is a hammer, to treat everything as if it were a nail,"* to which one could substitute chainsaw given that the Forest Service continues to find ways to get logs out using the latest message frame derived from the pressures to log first and ask questions later, all the while ignoring the contrarian evidence.

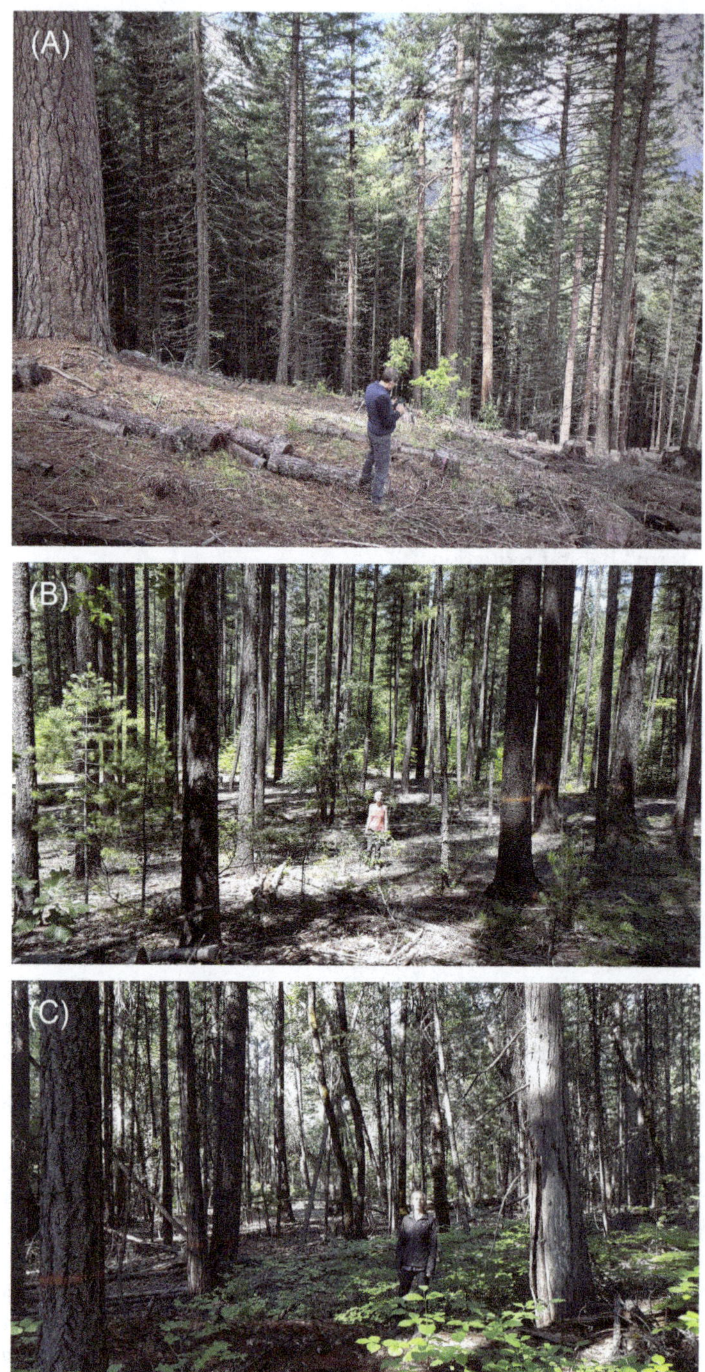

Figure 5.4 (A) Forest canopy "thinning" in owl critical habitat conducted by the Forest Service, supported by TNC, and the Southern Oregon Forest Restoration Collaborative in the Ashland Watershed of the Rogue-Siskiyou National Forest, Oregon showing tree (large gap) removals. (B and C) Proposed logging within a productive (6 of 7 years with successful nesting) on the Shasta-Trinity National Forest, California. The Forest Service wants to log all but the orange marked trees. *Photos by the author and with permission from Monica Bond.*

What's needed for the NWFP is a 21st century conservation vision that builds on—rather than tears down—the success story of the NWFP. Notably, streams are improving, owls are much better off within reserves (even if they are still declining), fire is not the penultimate cause of owl declines or significant old-growth losses, and the reserves were widely distributed to ensure their persistence using redundancy principles in conservation biology. That is - the loss of any single reserve or cluster of reserves would not topple the overall network. Rather than dismantle the NWFP, build upon it.

My vision for the NWFP and the owl going forward is this:

> The magnificent mature forests of the Pacific Northwest are once again thriving, severely burned areas are a celebration of Nature's resilience and are protected along with mature forests, spotted owl populations have recovered and no longer require endangered species protections, carbon is being drawn from the atmosphere and stored long-term as forests age, water is clean and plentiful and protected in unroaded landscapes, salmon have returned in numbers as watersheds are de-roaded and restored, and economies have moved on from the boon-and-bust of the past (as many have already). Ecological restoration is science-based (stop the hemorrhaging—e.g., decommission roads, get the cows out of the streams!) and surgical (Hippocratic Planetary Oath) in restoring heavily degraded areas like young tree plantations. Future generations marvel in the ecological bounty and spiritual sustenance of wild areas. Decisions are made using sound, independently verified science (see Chapter 1, The Nuts and Bolts of Science-Based Advocacy).

We all lose when scientific integrity takes a back seat

Administrations come and go—some are better than others—but scientific integrity violations remain alarmingly prolific as documented in the chapters of this book. To this date, the science on spotted owls is "clear-cut" in places, unsettled in others, and most often twisted by many citing inhouse (agency driven) or cherry-picked science to achieve a desired timber outcome that has little to do with recovery needs of the owl or any other endangered species for that matter. The research substantiated part of it is this—owls need older forests and a mixture of natural disturbed forests (in places) to survive and reproduce. Protecting ALL remaining habitat in reserves is critical to avoiding extinction (see DellaSala et al., 2015; Dugger et al., 2016). Arguably, owls and the hundreds of species associated with older forests need even more habitat protected now than any time since the historic Judge Dwyer decision in 1990 because threats have been piling up, including competition with Barred Owls, chemical contamination from herbicides and rodenticides (Wiens et al., 2019), and climate change (Dugger et al., 2016) that have contributed to genetic population bottlenecks (Funk et al., 2010) and ongoing range-wide declines (Forsman et al., 2011). It should be noted that when the initial findings of the peer-reviewed Funk et al. (2010) genetic study on owl population bottlenecks were presented at recovery team meetings to call attention to a potential range-wide collapse, the Service was so adamant about the negative repercussions of this study on "flexibility" and "active management" that it contracted with an outside firm to refute the findings even though that firm's report was never peer reviewed (see U.S. Fish & Wildlife Service, 2011: Appendix B). A clear example of inhouse science over peer-reviewed science to cherry-pick findings (also see Chapter 2, When Scientists Are Attacked: Strategies for Dissident Scientists and Whistleblowers, and Chapter 8, Why Advocate—and How?).

Where the science is unsettled is on wildfires. While recent studies show the owl is robust to wildfire and impacted mainly by logging and less so by wildfire, there is uncertainty in how much high-severity wildfire at the territory scale is tolerable to the owl (contrast Jones et al., 2016 vs Lee and Bond, 2015). Like a pistol duel between archrivals, scientists settle debates like this by each side taking tried-and-true research steps to falsify competing hypotheses while examining alternative explanations diplomatically and in honor of the scientific method. Rather than resorting to character assignations and unfounded accusations of "agenda-driven science" by those that advocate their own findings and receive government research funds (Peery et al., 2019), teams should cross-examine the evidence, define levels of uncertainty and risks, and do what is best for the integrity of science and the species of interest.

At the very least, researchers should avoiding inflammatory and unprofessional character assignations (Chapter 2, When Scientists Are Attacked: Strategies for Dissident Scientists and Whistleblowers, and Chapter 8, Why Advocate—and How?).

Speaking Truth to Power

Even with the best available science in place there are no guarantees that it will see the light of day. We have seen how political interference overrides independent science but when that is happening scientists need to speak out so as to avoid silent complicity (e.g., Chapter 1, The Nuts and Bolts of Science-Based Advocacy, and Chapter 8, Why Advocate—and How?). Decision makers and land managers that choose to ignore independent science must at least be transparent (accountable) about it, instead of bending the rules to fit politically motivated outcomes and confirmation biases. To do otherwise makes a mockery out of the scientific method and assures exposure and public mistrust to follow. Scientific integrity should never depend on which administration is in power, blowing in the political winds like the pungent scent of a skunk trapped between nefarious (Bush) or less-obvious distortions (Obama) of the truth.

So, what began for me decades ago in Oregon's old-growth rainforest giants continues today with scores of scientists willing to expose a litany of mistruths passed on from administration to administration and treated as gospel by decision makers, some scientists, and even some conservation groups, regardless of new discoveries and unsettled debates (see cognitive dissonance and confirmation bias discussions in Chapter 2, When Scientists Are Attacked: Strategies for Dissident Scientists and Whistleblowers, and Chapter 8, Why Advocate—and How?). While it may seem futile at times, we have no other choice but to defend scientific integrity as Nature-caring citizens with a Planetary Hippocratic Oath and an adherence to the precautionary principle (Chapter 1, The Nuts and Bolts of Science-Based Advocacy).

Finally, because many times I been branded as agenda driven by those that seek to discredit me as a scientist, know this about me: my quest for the truth and love of Nature drives my passion, I am a smasher of dominant paradigms and policies when compelled by the evidence (at times a contrarian or dissident scientist, Chapter 2, When Scientists Are Attacked: Strategies for Dissident Scientists and Whistleblowers), and I seek to daylight new discoveries respectfully and diligently via the scientific method. I value biodiversity as intrinsically important and worthy of existence and I am still a scientist—so please, get the criticism right!

A word of caution to new administrations: if you ever put me on another recovery team, you must be willing to accept and use the facts even if it means letting go of long-held beliefs. Science is about unlocking the mysteries of the universe—from atom to owl to planetary forces and beyond. We stand on the shoulders of those that came before willing to speak truth to power. Keeping up with the science is a tireless obligation in an age of misinformation but scientists need to talk to decision makers and decision makers need to listen with intent and admit when they are headed in the wrong direction.

Authors note: at the time of this chapter's publication, U.S. Fish & Wildlife Service's Paul Henson issued a determination of "warranted but precluded" on a petition made by conservation groups to "uplist" the owl to endangered, meaning according to the Service, no further habitat protections are necessary even though the owl is facing imminent extinction. Wildfires were again erroneously blamed as a causative factor in ongoing owl declines. The misinformation drumbeat goes on....!

References

Anthony, R.G., et al., 2006. Status and trends in demography of Northern Spotted Owls. Wildl. Monogr. 163, 48.

Baker, W.L., 2014. Historical Northern Spotted Owl habitat and old-growth dry forests maintained by mixed-severity wildfires. Landsc. Ecol. 30. Available from: https://doi.org/10.1007/s10980-014-0144-6.

Carroll, C., Johnson, D.S., 2008. The importance of being spatial (and reserved): assessing Northern Spotted Owl habitat relationships with hierarchical Bayesian models. Conserv. Biol. 22. Available from: https://doi.org/10.1111/j.1523-1739.2008.00931.x.

Carroll, C., Odion, D.C., Frisell, C.A., DellaSala, D.A., Noon, B.R., Noss, R., 2009. Conservation Implications of Coarse-Scale Versus Fine-Scale Management of Forest Ecosystems: Are Reserves Still Relevant? Klamath Center for Conservation Research, Orleans, CA, <http://www.klamathconservation.org/docs/ForestPolicyReport.pdf>.

Charnley, S., 2006. The Northwest Forest Plan as a model for broad-scale ecosystem management: a social perspective. Conserv. Biol. 20, 330–340.

Courtney, S., Franklin, J.F., 2004. Chapter 9: Conservation strategy. In: Courtney, S.P., Blakesley, J.A., Bigley, R.E., Cody, M.L., Dumbacher, J.P., Fleischer, R.L., Franklin, A.B., Franklin, J.F., Gutierrez, R.J., Marzluff, J.M., Sztukowski, L. (Eds.), Scientific Evaluation of the Status of the Northern Spotted Owl. Sustainable Ecosystems Institute, Portland, OR.

Crozier, M.L., et al., 2006. Does the presence of Barred Owls suppress the calling behavior of spotted owls? Condor 108, 760–769.

Davis, R.J., Hollen, B., Hobson, J., Gower, J.E., Keenum, D., 2015. Northwest Forest Plan—the first 20 years (1994–2013): status and trends of Northern Spotted Owl habitats. In: Gen. Tech. Rep. PNW-GTR-xxx. U.S. Department of Agriculture, Forest Service, Pacific Northwest Research Station, Portland, OR, xx p.

DellaSala, D.A., 2007. Testimony to the U.S. House Committee on Natural Resources for the Hearing on "Endangered species implementation: science or politics?". <https://forestlegacies.org/wp-content/uploads/2010/08/DDStestimonyHNRCconfidencehearing-July31-07.pdf>.

DellaSala, D.A., Williams, J., 2006. Northwest Forest Plan Ten Years Later – how far have we come and where are we going. Conserv. Biol. 20, 274–276.

DellaSala, D.A., 2020. Real vs. fake forests. Forest Biome: Trees of Life. In: Goldstein, M.I., DellaSala, D.A. (Eds.), Encyclopedia of the World's Biomes, vol. 3. Elsevier, pp. 1–15.

DellaSala, D.A., et al., 2015. Building on two decades of ecosystem management and biodiversity conservation under the Northwest Forest Plan, USA. Forests 6, 3326–3352.

Department of Interior and Department of Commerce, 1994. Endangered and Threatened Wildlife and Plants: Notice of Interagency Cooperative Policy on Information Standards Under the Endangered Species Act. Friday, July 1, 1994 (34271).

Dugger, K.M., et al., 2016. The effects of habitat, climate, and barred owls on long-term demography of northern spotted owls. Condor Ornithologic. Appl 118, 57–116.

Dugger, K.M., Wagner, F., Anthony, R.G., Olson, G.S., 2005. The relationship between habitat characteristics and demographic performance of Northern Spotted Owls in Southern Oregon. Condor 107, 863–878.

Dunk, J., et al., 2019. Conservation planning for species recovery under the Endangered Species Act: a case study with the Northern Spotted Owl. PLoS One 14, 45. Available from: https://doi.org/10.1371/journal.pone.0210643.

Forsman, E.D., Anthony, R.G., Meslow, E.C., Zabel, C.J., 2004. Diets and foraging behaviour of Northern Spotted Owls in Oregon. J. Raptor Res. 38, 214–230.

Forsman, E.D., et al., 2011. Population Demography of Northern Spotted Owls. Published for the Cooper Ornithological Society. University of California Press, Retrieved 6 March 2020, from. <http://www.jstor.org/stable/10.1525/j.ctt1pnr7z>.

Franklin, A.B., Anderson, D.R., Gutierrez, R.J., Burnham, K.P., 2000. Climate, habitat quality, and fitness in Northern Spotted Owl populations in northwestern California. Ecol. Monogr. 70, 539–590.

Funk, W.C., Forsman, E.D., Johnson, M., Mullins, T.D., Haig, S.M., 2010. Evidence for recent population bottlenecks in Northern Spotted Owls (*Strix occidentalis caurina*). Conserv. Genet. 11. Available from: https://doi.org/10.1007/s10592-009-9946-5.

Government Accountability Office (GAO), 2008. Testimony before the Committee on Natural Resources, House of Representatives. In: U.S. Fish & Wildlife Service Endangered Species Act Decision Making. GAO-08–688T.

Hanson, C.T., Odion, D.C., DellaSala, D.A., Baker, W.L., 2009. Overestimation of fire risk in the Northern Spotted Owl recovery plan. Conserv. Biol. 23, 1314–1319.

Hanson, C.T., Bond, M.L., Lee, D.E., 2018. Effects of post-fire logging on California spotted owl occupancy. Nat. Conserv. 23, 93–105.

Johns, D., DellaSala, D.A., 2017. Caring, killing, euphemism and George Orwell: how language choice undercuts our mission. Biol. Conserv. 211, 174–176.

Jones, G.M., et al., 2016. Megafires: an emerging threat to old-forest species. Biology 14. Available from: https://doi.org/10.1002/fee.1298.

Karr, J. 2000. Ecological integrity and ecological health are not the same: the folly of the status quo. pp. 97–169 In: D. Pimentel, L. Westra, and R.F. Noss (Eds.). Ecological Integrity: Integrating Environment, Conservation, and Health. Island Press: Washington, DC. 448 pp.

Kelly, E.G., Forsman, E.D., Anthony, R.G., 2003. Are Barred Owls displacing spotted owls? Condor 105, 45–53.

Lee, D.E., 2018. Spotted Owls and forest fire: a systematic review and meta-analysis of the evidence. Ecosphere 9 (7), 22. Available from: https://doi.org/10.1002/ecs2.2354.

Lee, D.E., Bond, M.L., 2015. Occupancy of California Spotted Owl sites following a large fire in the Sierra Nevada, California. Condor 117, 228–236. Available from: https://academic.oup.com/condor/article/117/2/228/5153140.

Livezy, K., 2009. Range expansion of Barred Owls, Part II: Facilitating ecological changes. Am. Midl. Nat. 161, 323–349.

Metlen, K., et al., 2018. Regional and local controls on historical fire regimes in the Rogue River Basin, Oregon, USA. For. Ecol. Manage. 430, 43–58.

Molina, R., Marcot, B.G., Leshner, R., 2004. Protecting rare, old-growth, forest-associated species under the survey and manage program guidelines of the Northwest Forest Plan. Conserv. Biol. 20, 306–318.

Niemi, E., Whitelaw, E., Johnston, A., 1999. The sky did not fall: the Pacific Northwest's response to logging restrictions. <https://pdfs.semanticscholar.org/5b26/07f392b92c0951fc00da459a1039158b06b3.pdf?_ga=2.14802550.1795502661.1583783406-285058161.1583783406>.

Noon, B.R., Blakesley, J.A., 2006. Conservation of the Northern Spotted Owl under the Northwest Forest Plan. Conserv. Biol. 20, 288–296.

Noon, B.R., et al., 2003. Conservation planning for US National Forests: conducting comprehensive biodiversity assessments. Bioscience 53, 1217–1220.

Odion, D.C., Hanson, C.T., DellaSala, D.A., Baker, W.L., Bond, M.L., 2014. Effects of fire and commercial thinning on future habitat of the Northern Spotted Owl. Open Ecol. J. 7, 37–51.

Office of Inspector General (OIG), 2008. Report of Investigation: The Endangered Species Act and the Conflict Between Science and Policy. Memorandum Dated December 15, 2008 From Inspector General Earl E. Devaney.

Olson, G.S., et al., 2004. Modeling demographic performance of Northern Spotted Owls relative to forest habitat in Oregon. J. Wildl. Manage. 68, 1039–1053.

Peery, M.Z., et al., 2019. The conundrum of agenda-driven science in conservation. Front. Ecol. Environ. 17. Available from: https://doi.org/10.1002/fee.2006.

Power, T.M., 2006. Public timber supply, market adjustments, and local economic assumptions of the Northwest Forest Plan. Conserv. Biol. 20, 341–350.

Raphael, M.G., et al., 2013. Assessing the compatibility of fuel treatments, wildfire risk, and conservation of Northern Spotted Owl habitats and populations in the Eastern Cascades: a multi-scale analysis. In: Final Report: JFSP Project 09-1-08-31 Project. <https://sites.google.com/a/pdx.edu/vegetation-fire-owl/home>.

Reeves, G.H., Williams, J.E., Burnett, K.M., Gallo, K., 2006. The aquatic conservation strategy of the Northwest Forest Plan. Conserv. Biol. 20, 319–329.

Spies, T.A., Stine, P.A., Gravenmier, R., Long, J.W., Reilly, M.J., 2018. Synthesis of Science to Inform Land Management Within the Northwest Forest Plan Area, vol. 1. USDA Forest Service Pacific Northwest Research Station, Portland, OR, PNW-GTR 966. <https://www.fs.fed.us/pnw/pubs/pnw_gtr966_vol1.pdf>.

Strittholt, J.R., DellaSala, D.A., Jiang, H., 2006. Status of mature and old-growth forests in the Pacific Northwest, USA. Conserv. Biol. 20, 363–374.

Thomas, J.W., Franklin, J.F., Gordon, J., Johnson, K.N., 2006. The Northwest Forest Plan: origins, components, implementation experience, and suggestions for change. Conserv. Biol. 20, 277–286.

U.S. Fish & Wildlife Service. 1992a. Draft Final Recovery Plan for the Northern Spotted Owl. USDI Fish and Wildlife Service, Portland, OR.

U.S. Fish & Wildlife Service, 2011. Revised Recovery Plan for the Northern Spotted Owl (*Strix occidentalis caurina*). Region 1 U.S. Fish & Wildlife Service, Portland, OR, Xvi + 258 pp. <https://www.federalregister.gov/documents/2011/07/01/2011-16456/endangered-and-threatened-wildlife-and-plants-revised-recovery-plan-for-the-northern-spotted-owl>.

U.S. Fish and Wildlife Service, 1992b. Endangered and threatened wildlife and plants; determination of critical habitat for the Northern Spotted Owl. Fed. Regist. 57, 1796–1838.

USDA Forest Service, 2020. Bioregional Assessment of Northwest Forests. USDA Forest Service, Washington, DC, <https://www.fs.usda.gov/Internet/FSE_DOCUMENTS/fseprd762764.pdf>.

U.S. Fish & Wildlife Service, 2007. 2007 Draft Recovery Plan for the Northern Spotted Owl (Strix occidentalis caurina): Merged Options 1 and 2. Region 1 U.S. Fish & Wildlife Service, Portland, OR. <https://www.fws.gov/pacific/ecoservices/endangered/recovery/documents/DraftRecoveryPlanNorthernSpottedOwlWEB_000.pdf>.

Wiens, J.D, Anthony, R.G., 2014. Competitive interactions and resource partitioning between northern spotted owls and barred owls in western Oregon. Wildl. Monogr. 185, 1–50.

II

An Imperfect Marriage: Policy and Science

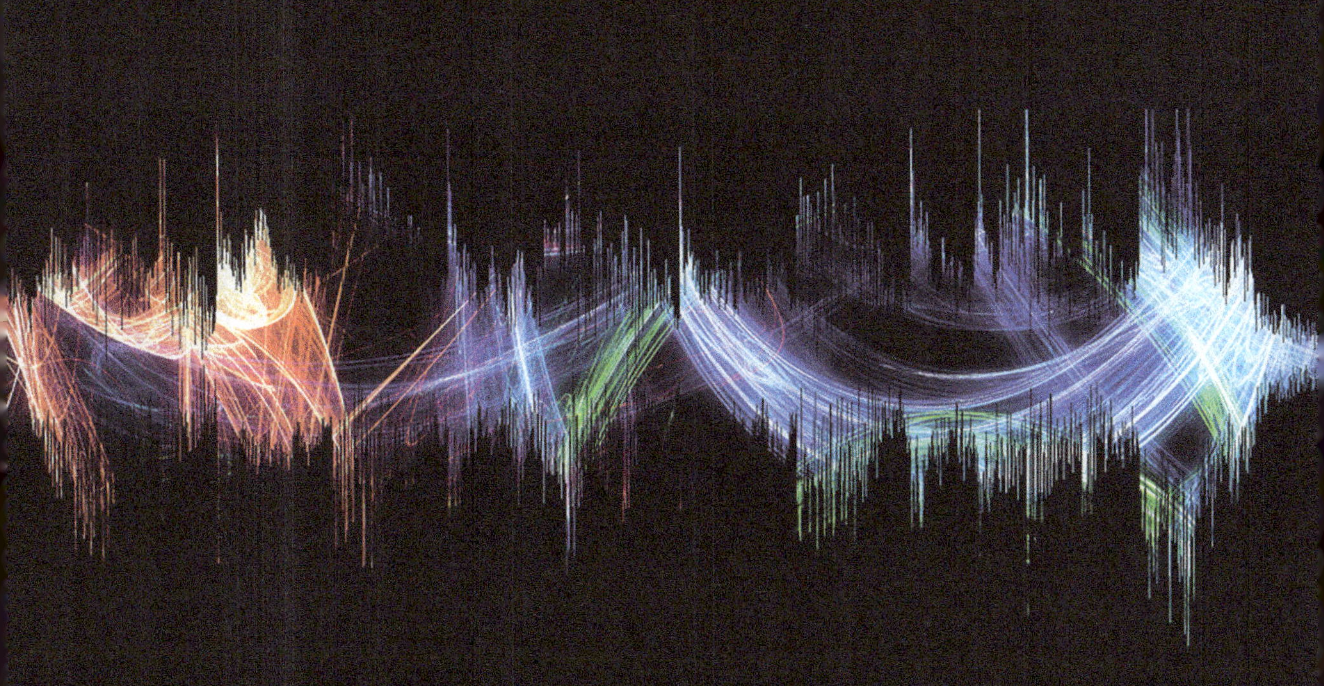

Overcoming the politics of endangered species listings

Noah Greenwald

Center for Biological Diversity, Portland, OR, United States

The US Endangered Species Act—a global model

Among environmental statutes, the Endangered Species Act (ESA) stands out for its clear wording and singular purpose. Recognizing that species "have been rendered extinct as a consequence of economic growth and development untempered by adequate concern and conservation," Congress drafted the ESA "to provide a means whereby the ecosystems upon which endangered species and threatened species depend may be conserved" (16 U.S.C. §1531b(a)(1)). To accomplish this laudable goal, the ESA provides strong protections to species, including a blanket prohibition against "take," which includes any action that causes the death, injury, or harassment of protected species, or the destruction of their habitat, and an affirmative duty for all federal agencies to avoid jeopardizing the continued existence of listed species or adversely modifying their critical habitat.

These strong protections, however, are only provided to species if—and only if—they are listed as threatened or endangered under the law. Because of this, whether a species is listed or not is in many cases a matter of life or death, literally existential. Yet too often, the U.S. Fish and Wildlife Service ("Service") has been slow to make listing decisions and denied protection to species clearly at risk of extinction (Box 6.1), bowing to pressure from politically powerful opponents of listing, such as States, federal officials, lobbyists, and Congressional members (Ando, 1997; Greenwald et al., 2006; Puckett et al., 2016; Sidle, 1998).

Under the ESA, listings should take no more than 2 years but, as one recent study found, species have waited a median of 12 years for protection, with many species waiting decades (Puckett et al., 2016). Far fewer species have been listed

Conservation Science and Advocacy for a Planet in Peril.
DOI: https://doi.org/10.1016/B978-0-12-812988-3.00011-9

Box 6.1 Listing decisions are made by the U.S. Fish and Wildlife Service ("Service," terrestrial and freshwater species) or National Marine Fisheries Service (marine and anadromous species).

The U.S. Fish and Wildlife Service oversees about 90% of the listed species, but to a much greater degree than the National Marine Fisheries Service, has long suffered the effects of overt political pressure—resulting, most pointedly, in a backlog of hundreds of imperiled species languishing while awaiting protection, often for decades.

under recent Republican administrations (62 under Bush or eight per year and 25 under Trump or six species per year), then Democratic administrations (523 under Clinton or 65 per year and 360 under Obama or 45 per year).

However, all four recent administrations, Republican and Democrat, have denied ESA protections to species when faced with political pressure, despite clear and convincing scientific evidence of ongoing declines and serious threats to their existence. Frequently, politically charged listing decisions are made in Washington, D.C. (Chapter 5, Blowing the Whistle on Political Interference: The Northern Spotted Owl), including reversing the listing recommendations of the Service's own scientists. When species are listed, it is increasingly with reduced regulatory protection. The Service justifies these actions by pointing to promises from States, industry, and landowners to help species in the future, in exchange for not listing and concordant reduced regulatory protection. Depending on your perspective, denying species protections based on promises is either a clear example of regulatory capture or, as the Service would argue, essential to saving species and perhaps the ESA itself.

Like the Clean Air Act, Clean Water Act, and National Environmental Policy Act, the ESA includes a citizen-suit provision authorizing private enforcement, an important check on political interference. The ESA also allows interested parties to petition the Service to list species and provides strict deadlines to respond. These provisions have been absolutely critical to ensuring at-risk species receive protection, identifying highly imperiled species in need of protection, reducing listing delays, and contributing to the protection of most of the more than 1700 species currently listed (Brosi and Biber, 2012; Greenwald et al., 2006; Puckett et al., 2016).

Tensions persist between public-interest groups who want the strongest possible protection for imperiled species—like the Center for Biological Diversity where I have worked for 24 years—and interests opposed to listing. In this chapter, I provide a history of the listing program, campaigns to protect species, and case studies involving species whose fate has been affected by political interests, including the Greater Sage Grouse (*Centrocercus urophasianus*), Montana arctic grayling (*Thymallus arcticus*), and Streaked Horned Lark (*Eremophila alpestris strigata*). These examples illustrate the ineffectiveness of voluntary promises, particularly when compared to the ESA's strong regulatory protections. They are not just relevant in the context of the ESA but also efforts to protect human welfare or other aspects of the environment, where similar tensions between a strong regulatory approach or voluntary promises are prevalent.

What species qualify for threatened and endangered status under the ESA

The Endangered Species Preservation Act of 1966, the predecessor to the ESA, only authorized federal protection for vertebrate species as endangered. When enacted in 1973, the ESA added plants and invertebrates and a listing classification: threatened. A U.S. House of Representatives report accompanying the ESA made clear that the threatened classification would allow the Service to list species before they declined to being in danger of extinction and could even include species that "were on the increase, so long as the Secretary was satisfied that a measurable risk to those species could be said to exist" (U.S. House of Representatives, 1973).

Under the ESA, an endangered species is defined as "any species which is in danger of extinction throughout all or a significant portion of its range," and a threatened species is "any species which is likely to become an endangered species within the foreseeable future throughout all or a significant portion of its range" (16 U.S.C. § 1532(6, 20)). Like the new threatened category, these definitions greatly expanded the universe of species that warrant federal protection. The House Report, for example, highlighted "this definition is a significant shift in the definition in existing law" because it allows the Service to protect species before they decline so far as to become "threatened with worldwide extinction" (U.S. House of Representatives, 1973).

Delays and failure to list species under the ESA start early

It did not take long following the ESA's enactment for it to become clear that thousands of species needed the new law's protection. In 1975, scientists with the Smithsonian Institution provided the Service with a list of 3187 endangered, threatened and possibly extinct plants that the Institution had identified as in need of protection. The Service accepted this as a listing petition for these species, and in 1976 proposed to list approximately 1700 species (U.S. Fish and Wildlife Service, 1975, 1976).

The Service also hired seven biologists: two mammologists, two invertebrate biologists, a botanist, ornithologist, and ichthyologist to identify hundreds of species in need of protection drawing from the IUCN Red List, Convention on International Trade, and other sources (Personal communication from retired U.S. Fish and Wildlife Service biologist, Dr. Jim Williams). In 1976, these initial seven biologists grew to 10 with two more botanists and a herpetologist added. The Service unfortunately dropped the ball on the taxa these scientists were working to save, delaying protection of most of the species for decades.

In 1978, Congress amended the ESA to require the Service to withdraw all listing proposals 2 years old or more, with a grace period of 1 year for proposals that had already been issued. The Service failed to meet this 1-year deadline for 1876 species, including 1700 plants identified by Smithsonian scientists (U.S. Fish and Wildlife Service, 1979). Instead, in 1980, the Service published the first of many "candidate notices" in the Federal Register, including the 1700 plants they had failed to list. As with many species, these plants would remain candidates for protection for years, even decades (U.S. Fish and Wildlife Service, 1980). Despite the Smithsonian's submission of its 3187 list of candidates, the Service listed just 61 plants in the 5 years following the Institution's petition (U.S. Fish and Wildlife Service, 1980).

The election of Ronald Reagan in 1980 brought significant changes to the Act. Shortly after inauguration, Reagan issued an executive order requiring economic analyses of all rules issued by the federal government, which resulted in no species listings in 1981 (Executive Order 12291, February 17, 1981). In response, Congress amended the ESA in 1982 with strict deadlines for listing decisions and clarifying that listing decisions were to be based *solely* on the best available commercial and scientific information, without consideration of economic impacts. Under these changes, the Service was to make listing decisions within 2 years of receiving a petition, with one exception—the Service could delay listing if it could show it is making "expeditious progress" in higher-priority listings. Congress emphasized, however, that this exception should not justify the "foot-dragging efforts of a delinquent agency" (U.S. House of Representatives, 1982).

Following these changes, the Service began listing species again, with the Reagan administration listing 254 species, an average of 32 species per year (Fig. 6.1). Though an improvement, the Service continued to fall short listing species identified by their own biologists, and by the end of the 1980s, there were more than 2000 candidate species not listed. Rather than work to protect more species, however, the Reagan administration instead took actions designed to slow their own biologists work to list species, and thus began what would become a pattern of stifling agency scientists that has persisted to the present day (See https://www.ucsusa.org/surveys-scientists-federal-agencies;

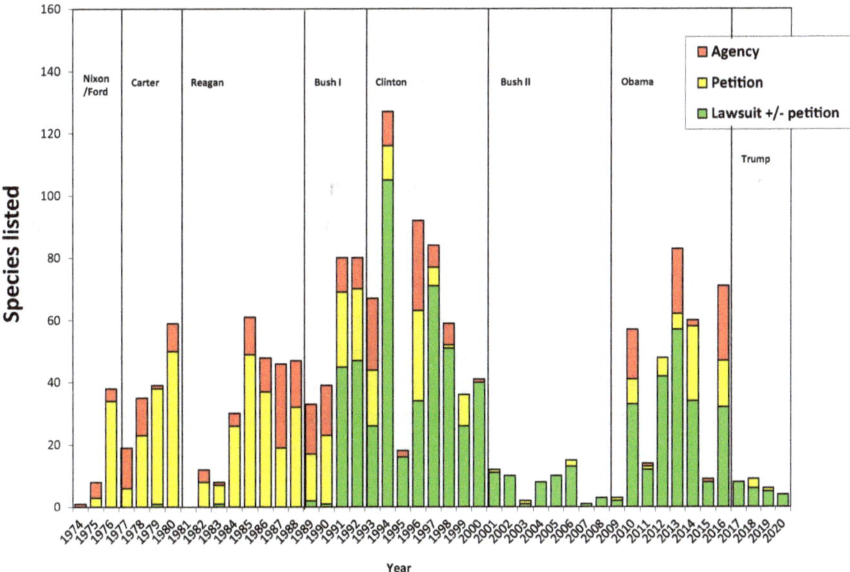

Figure 6.1 Listings per year (1973–2019) categorized by whether they were listed solely following agency action, a petition, or a lawsuit with or without a petition.

also Chapter 7, Scientific Integrity and Advocacy: Keeping the Government Honest, Chapter 8, Why Advocate—And How?).

The administration began to implement various systems for determining priorities. A 1981 guidance, for example, prioritized listings of "higher lifeforms," in order of mammals, birds, fishes, reptiles, amphibians, vascular plants, and invertebrates. This was eliminated by the 1982 amendments to the Act. As Jim Williams, one of the first biologists in the Services' listing program, describes it, the administration's efforts to develop a listing priority system was largely a make-work exercise, "an effort to keep us biologists confused, working on different things and never finishing anything." One afternoon at a D.C. bar in 1984, Service biologists expressed their shared frustration over the ever-changing priorities. After a few drinks, they developed their own satirical "schedule" of priorities. Reflecting the political mood, delistings—the removal of endangered and threatened species from the lists—would not be subject to any criteria "except ASAP," with listings priorities ordered, first, by "things that recognize their mother," followed by "things with bright colors" and "things that are hard to kill with a brick."

The administration began to isolate the biologists and apparently, to try to push them out altogether. First, the Service's headquarters transferred from downtown D.C. to across the Potomac in nearby Virginia, making it harder to interact with nongovernmental conservation organizations, the likely purpose of the move. Then, the administration began transferring biologists to far-flung locations removed from their areas of expertise. According to Ken Dodd, the herpetologist of the listing team: "We were too much trouble, knew too many people, had expertise, and were loyal to our science and principles, not a government bureaucracy. Drove them nuts." Fortunately, a memo detailing the administration's plans was leaked to Interior's Inspector General and was quashed. But most of the original biologists left the Service anyway, pursuing conservation goals in prestigious universities, natural history museums, or somewhere else.

In a harbinger of things to come, the Service's stifling of biologists did not end with the Reagan administration. In a 1998 piece in *Conservation Biology* entitled "Arbitrary and Capricious Species Conservation," former Service biologist John Sidle observed that under the Clinton administration, "[t]he Department of Interior consistently missed statutory deadlines for listing and deliberately allowed politics to influence listing decisions" (Sidle, 1998). The case studies below exemplify the agency's contempt for its own scientists and of science in general, species conservation, and even the ESA itself.

Public pressure gains protection for species

By the end of the 1980s, public concerns about dwindling wildlife populations lead to more interest in ESA implementation. The U.S. Department of the Interior Office of Inspector General (1990) audited the listing program and concluded that the Service's progress listing species was inadequate to the task. Based on the Service's promises to list 50 species per year and the standing backlog of more than 2000 candidate species, the Inspector General determined that "it may take from 38 to 48 years at current listing rates to list just those species now estimated to qualify for protection under the Act." The Inspector General concluded that "this length of time to list and protect endangered species is not indicative of the 'expeditious progress' specified in the Act and could likely result in additional extinction of certain plants and animals during the period" (U.S. Department of the Interior Office of Inspector General, 1990).

In response, conservation groups filed some of the first lawsuits to challenge the Service's failure to protect species according to the ESA's mandatory deadlines. Three lawsuits were filed that settled at the end of the George H.W. Bush administration with hundreds of enforceable deadlines for backlogged candidate species (*Conservation Council for Hawaii v. Lujan* 89–95, Haw., 1990; *California Native Plant Society v. Lujan* 91-0038, Cal. 1991; *Fund for Animals v. Lujan* 92–800, D.C. 1992). All told, these lawsuits dramatically increased the number of species listings from 1991 to 98, when 607 species were listed for an average of 76 per year (Fig. 6.1).

The Clinton administration, however, was not thrilled at having to list so many species with Secretary of Interior Bruce Babbitt telling the *Wall Street Journal* in

1995 that he would not have approved the settlements (Noah, 1995). In 1996, the administration greatly reduced the number of candidate species the Service tracked, apparently in an effort to limit such settlements in the future. The Service previously maintained a three-tiered system for ranking candidate species, with "C1" species for which the Service had sufficient information to list, "C2" species for which more information was needed, and "C3" species that were either apparently extinct (3A), had questionable taxonomy (3B), or were apparently abundant and not facing significant threats (3C) (U.S. Fish and Wildlife Service, 1994). From 1996 forward, the Service eliminated the C2 and C3 lists and dropped more than 2000 species from consideration (U.S. Fish and Wildlife Service, 1996). To this day, the Service does not maintain a list of species that need consideration for listing beyond those that have been petitioned by conservation groups, so it has no true understanding of how many species may warrant protection similar to the C2 list.

With the legal settlements wrapping up in the late 1990s, the Service's listing rate began to drop once again, with an average of 45 species listed per year from 1998 to 2000. But with the election of George W. Bush, the Service's rate of new listings plummeted to just 62 in total, an average of eight per year, nearly all of which followed lawsuits by conservation groups (Fig. 6.1). This was not for lack of any need for protection. The Service's 2002 candidate notice, for example, included 260 C1 species and 39 proposed listings. Rather, the drop reflected a renewed, vigorous antiregulatory agenda, as reflected in the appointment of Gale Norton as Interior Secretary. Before that, Norton had been an attorney for the Mountain States Legal Foundation, a property-rights organization. Indeed, a platoon of other industry insiders was installed at Interior and other regulatory and land management agencies.

In response, the Center sued the Bush administration under the ESA's citizen-suit provision, challenging the Service's failure to make expeditious progress in listing species and thereby unlawfully delaying protection to the backlog of candidate species. This lawsuit itself became backlogged, however, behind a large docket of pending cases in the U.S. District Court. To try to address this, we reached an agreement in principle with the Service in 2009 by which the agency agreed to process the candidate species over 5 years. But as soon as political appointees, like then-solicitor of Interior David Bernhardt, caught wind of it, it was shut down.

With lengthy delays and little progress in listings, conservation groups filed petitions to list hundreds of species. A 2010 petition, for example, sought protection for 404 southeastern United States aquatic and wetland species that NatureServe, American Fisheries Society, and IUCN had identified as imperiled. The high species richness yet severe imperilment of aquatic fauna in the American Southeast were the primary reasons for the petition. In 2011, the Service decided that 374 of these species warranted consideration (U.S. Fish and Wildlife Service, 2011), hundreds of which are still awaiting protection (U.S. Fish and Wildlife Service, 2011).

Following Barack Obama's election, conservationists had a breakthrough with two settlement agreements reached with the Service in 2011, one with the Center, and a companion agreement with WildEarth Guardians, another conservation organization. Together, these agreements required the Service to publish proposed listing rules, or withdraw these species from the candidates list—make "up-or-down findings"—for nearly 260 candidate species as well as to make initial petition findings for nearly 500 other species, including the finding for the 404 southeast species discussed above, all by 2016. These agreements would eventually contribute to the Obama administration adding 360 species to the endangered and threatened species lists, an average of 45 per year (Fig. 6.1), sharply reducing the candidate backlog.

Yet a new species backlog developed during the agreements' term—this time, of species that had received initial decisions from the Service in favor of listing, but which were awaiting the culmination of the listing process and ESA protection. Thus the Service developed a new, yet unambitious, "national listing workplan" in 2016, under which about half of the more than 500 species would receive findings by 2023.

This workplan then fell to the Trump administration for implementation.

The Trump administration brought a return to the antiregulatory agenda of the Bush and Reagan years, with a sharp reduction in species listings once again. The administration listed just 25 species in total, a rate of just six per year, and the lowest of any administration. The administration also installed David Bernhardt, the Bush administration's Interior solicitor as Interior Secretary. The same guy that nixed the 2009 agreement-in-principle with the Center over the candidates. In the interim, Bernhardt served as a lobbyist for the oil and gas industry and other powerful special interests opposed to environmental protections.

Unsurprisingly, the Service has blown even the modest timelines in the workplan missing its own deadlines for dozens of species every year since 2016. The Center brought suit over many of these missed findings, resulting in settlements that put decisions for species still further into the future. Not entirely satisfied with this result, the Center sued in 2020 to enforce the ESA as to all remaining species in the workplan, 241 in total, to ensure these species get protection. This case is currently pending.

Today, there are over 1700 listed species in the United States, most of which are in place only because of petitions from scientists and lawsuits by conservation groups. Delaying protection has real consequences for species. Delay makes recovery more difficult and expensive and increases extinction risk. Indeed, at least 47 species have gone extinct waiting for ESA protection in a bureaucratic never-never-land (Greenwald et al., 2019). The ESA, and specifically, the timely listing of species, may be humanity's best tool for preventing extinctions. ESA protection is credited with preventing the extinction of roughly 99% of species that have been listed, largely due to the ESA's substantive protections (Greenwald et al., 2019).

Nevertheless, as demonstrated in a number of recent, high-profile cases, the Service has foregone ESA protection in exchange for what are often amorphous, voluntary promises by landowners to adhere to various conservation efforts and agreements. Such voluntary conservation efforts often have more to do with politics than saving species. Yet while they may have helped to protect species, as the following case studies highlight, they are no substitute for ESA protection through listing.

Case studies of political interference in the listing process

The Service's efforts to substitute voluntary agreements for listing date at least to 1996, when the agency withdrew a proposal to list the Barton Springs salamander based on a voluntary conservation agreement with the Texas Parks and Wildlife Department and other state agencies. The petitioning conservationists challenged the withdrawal in court and won. The court found the Service's actions to be unlawful, in part because the agency had bowed to "strong political pressure." The court found that the Secretary "cannot use promises of proposed future actions as an excuse for not making a determination based on the existing record." This decision was followed by other cases involving other species, including the Oregon Coast coho salmon (*Oncorhynchus kisutch*), Queen Charlotte goshawk (*Accipiter gentilis laingi*), Yellowstone cutthroat trout (*Oncorhynchus clarkii bouvieri*), and others, as courts unanimously concluded that under the ESA, the Service could not refuse to list species that otherwise qualify based on promises of future conservation.

In response, the Service introduced two new policies encouraging voluntary conservation agreements and allowing for their consideration in listing decisions. In 1999 the Service published its "Candidate Conservation Agreements with Assurances" ("CCAA") policy, which locks in specific landowners activities that would cause "take" of candidate species, were those species protected under the ESA, such that activities covered by a CCAA, are exempted from the Act's take prohibition. Despite the stated intent of this policy, however, the Service has nearly always elected not to list species if one or more CCAAs have been signed with private landowners or states.

In 2003 the Service published the "Policy on Evaluation of Conservation Efforts When Making Listing Decisions" ("PECE") policy. The PECE policy sets criteria for the Service to find voluntary conservation agreements to be strong enough to avert ESA listings that are warranted under the statutory factors. Under the PECE policy, the Service evaluates conservation agreements based on the degree of certainty of implementation and effectiveness. PECE considers whether a given agreement incorporates sufficient resources, contains a clear schedule for conservation, sufficiently incentives implementation, and other factors. In some cases, the Service has used the PECE policy to deny ESA protection to imperiled species. In most cases, however, the Service has simply made reference to promises of voluntary conservation actions in denying species protection but has avoided expressly relying on them as a basis for the denial. This may be because the PECE policy criteria are stringent and many conservation efforts may not meet the criteria.

The below case studies examine the Service's reliance on these policies and illustrate the ineffectiveness of relying on voluntary conservation activities to ensure the survival and recovery of species.

Greater Sage Grouse

The largest grouse in North America, the Greater Sage Grouse, once numbered 1.6 million to as many as 16 million birds roaming across 13 western states and three Canadian provinces. Today, Greater Sage Grouse are lost from nearly 50% of their historic range, including two states and one province, and their abundance has declined precipitously to an estimated 500,000 birds, and probably far fewer, as the species is still in decline (U.S. Fish and Wildlife Service, 2015). Primarily, this is the result of fragmentation and destruction of habitat, as large, contiguous stands of sagebrush have been cleared for crop or pastureland, rededicated to energy development, roads and infrastructure, or lost to invasive species and altered fire regimes. The sagebrush ecosystem is unraveling before our eyes, with consequences for many species beyond its conservation flagship species, the Sage Grouse.

Concerns lead conservation interests to file three listing petitions with the Service between July 2002 and December 2003. In January 2005, the Service issued a "not warranted" decision for the grouse (U.S. Fish and Wildlife Service, 2015), a finding embroiled in controversy from the beginning. Shortly before the finding was published in the *Federal Register*, the *New York Times* broke a story about Julie MacDonald, then-deputy assistant Interior Secretary under the Bush administration, who had heavily edited and questioned the results of a 2004 status review of the grouse (Barringer, 2004; also see Chapter 5, Blowing the Whistle on Political Interference: The Northern Spotted Owl for related rollbacks). According to the Times, Ms. MacDonald's "critique of Sage Grouse biology and the biologists who work for an agency she oversees showed flashes of her strong property-rights background and her deference to industry views." MacDonald ludicrously claimed that it was "simply a fairy tale, constructed out of whole cloth" that the Sage Grouse once numbered in the millions and questioned whether Sage Grouse were truly dependent on sagebrush, stating "they will eat other stuff if it is available." After conservation groups quickly challenged the "not-warranted" finding, a court vacated the decision as "arbitrary and capricious" and in violation of the ESA, and ordered the Service to make a new decision.

Thus in 2010 the Service issued a new listing decision for the Sage Grouse—this time, finding ESA protection to be "warranted," but instead of publishing a proposed listing, deemed the Greater Sage Grouse a candidate. This "warranted-but-precluded" decision was based on the Service's frank admissions that regulatory mechanisms were insufficient to protect the grouse from multiple threats and "populations were likely to become smaller, fewer, and separated by fragmentation, placing the species at risk of extinction in the future" (U.S. Fish and Wildlife Service, 2010a,b). While an improvement from the 2004 withdrawal, this finding still extended no ESA protections to the species. It was intended to be a wake-up call to federal agencies, states, and industries to quickly put measures in place for the Sage Grouse.

In response, the Bureau of Land Management, the U.S. Forest Service, and all 11 states with Sage Grouse populations developed new conservation plans for the Sage Grouse. And predictably, the Service issued a not warranted finding for the Sage Grouse in 2015, finding that the federal plans and three of the 11 state plans sufficiently ensured the grouse's survival. Eight other states' plans were found to lack sufficient certainty of implementation and effectiveness under the PECE policy to support withdrawing the Greater Sage Grouse from the candidates list.

The BLM and Forest Service mapped 27 million hectares of Sage Grouse habitat on federal lands and divided this into two primary management zones: 14 million hectares of "priority habitat management areas" (PHMA), and 13 million hectares of "general habitat management areas" (GHMA). In PHMAs, oil and gas leasing was suspended and grazing leases were reviewed for specific habitat objectives. In GHMAs, leasing and other development were allowed, but with an overall disturbance "cap" of 3% and buffers around leks (areas where males congregate annually and conduct elaborate displays to attract females for mating) of 2–5 km, depending on the development. In the three states with sufficient plans, core areas where disturbance from oil and gas or other development was capped at less than 5% of the total area, with lek buffers of 1 km. Grazing in these zones is largely unrestricted under the state plans, although the new lek buffers could affect fencing and other infrastructure.

While these management plans were hailed for their sage grouse protections, they left open millions of hectares of habitat to development, inevitably contributing to further declines in Sage Grouse range and abundance. But in a classic example of the problems with sacrificing regulatory protection in exchange for promises, the Trump administration moved quickly to eliminate these protections on BLM lands, including >20 million ha of sage grouse habitat. These revisions reduced the conservation commitments that had averted listing, largely opening them back-up to oil and gas drilling, mining, and unrestricted livestock grazing. Conservation groups challenged these plans and as of submission of this chapter, they had been enjoined, and it is likely the Biden administration will restore protections. It is noteworthy, however, that there is nothing stopping a future administration from again opening these lands to development.

Had the Service simply listed the Greater Sage Grouse, states and federal agencies would have developed new conservation plans for the species, but it would be considerably more difficult to undo these plans later. The ESA would require the BLM and Forest Service to engage in a consultation process with the Service and show how lessening protections likely would not jeopardize the grouse's existence. Today, listing the Greater Sage Grouse under the ESA is prohibited by a Congressional rider. A rider in the omnibus appropriations bill for 2015 prohibits spending on listing the Greater Sage Grouse, and because Congress has been using continuing resolutions to fund the government, the language has been renewed annually ever since (Public Law No: 113–235, 113th Congress, December 16, 2014). Thus the Sage Grouse's future remains uncertain.

Montana arctic grayling

A member of the salmon family, the arctic grayling is a beautiful fish with a prominent dorsal fin, widely distributed across Canada and Alaska (Fig. 6.2). Historically, arctic grayling existed in only two places in the lower 48 states: Michigan and the upper Missouri River of Montana. Populations in Michigan went extinct by the 1930s, those in Montana were restricted to the Big Hole River and a few lakes by the end of the 1970s, relegating the species to roughly 5% of its historic range (Kaya, 1992). A Pleistocene relict, Montana arctic grayling are genetically distinct from other grayling, originating from a time when what we now call the Missouri River connected to the Saskatchewan River and

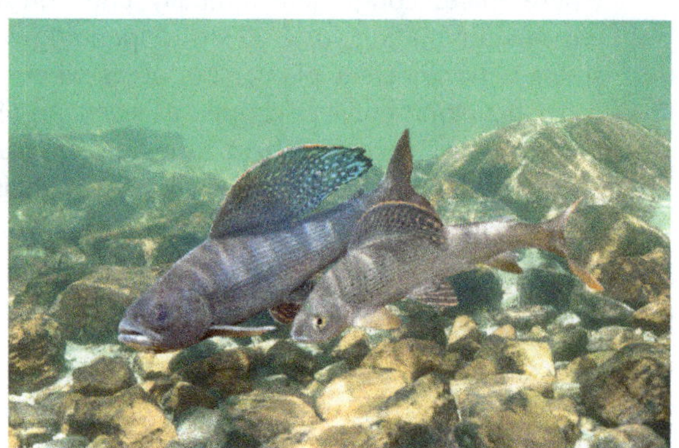

Figure 6.2 Arctic grayling in Montana where a distinct population in the upper Missouri River has declined by roughly 90% but has been denied ESA protections. *Source Photo: Pat Clayton.*

flowed into Hudson Bay rather than the Gulf of Mexico. The grayling's decline from historic levels is attributed to dams, overharvest, habitat degradation, and introduction of nonnative trout. In the Big Hole River, the primary threat since the 1980s is overdraft of the river for irrigation, a problem that is only being exacerbated by climate change (Montana Fish, Wildlife and Parks, 1995).

The Service placed Montana grayling on the candidate C2 list in 1982 and conservation groups petitioned for listing in 1991. In 1994, the Service found listing the graying to be "warranted but precluded," repeating its pattern of delaying protections for clearly imperiled species (U.S. Fish and Wildlife Service, 1994).

In response to the warranted but precluded finding, the Service and Montana Fish, Wildlife and Parks (MFWP) formed a working group that developed a conservation plan, and in 1996 a memorandum of agreement (MOA) (Montana Fish, Wildlife and Parks and Fish and Wildlife Service, 1996). The MOA established MFWP as the "lead agency for overseeing implementation," set a short-term goal of "a minimum of five Montana grayling reintroductions" by 2000 and a long-term goal of "at least five stable, viable populations distributed among at least three of the major river drainages (e.g., Big Hole; Jefferson, Beaverhead, Madison, Gallatin, Sun, Smith) within the historic range of Montana grayling in the Missouri River system upstream from Great Falls: including those upper Missouri Basin waters within Yellowstone National Park" by 2020 (Montana Fish, Wildlife and Parks and Fish and Wildlife Service, 1996). Neither of these goals has been met. To date, only one population has been restored in addition to the Big Hole population, with no other restorations in progress.

Under the MOA, the Service promised to review the grayling's status and reconsider it for ESA listing if five populations were not in progress by December 2002 or the population fell below certain criteria for 3 consecutive years (<30 adult fish per 1.6 km or age 1–2-year-old grayling $<50\%$ or $>80\%$ of sampled fish) (Montana Fish, Wildlife and Parks and Fish and Wildlife Service, 1996). The first condition was met in 2002, but the Service did not initiate a status review.

By 2003, it was clear the second criterion had also been met. Following multiple years of river flows dropping below 0.6 m^3/s in the summer (any flows below 1.7 cm trigger voluntary irrigation reductions and 0.6 cm is a mere trickle), grayling dropped to less than 5 fish per 1.6 km from 2000 to 2002 (Montana Fish, Wildlife and Parks and Fish and Wildlife Service, 2006). Although the Service still did not reinitiate listing, in a 2004 candidate notice the agency admitted that, despite conservation efforts, "there continue to be periods when flows are well below those considered 'survival' flows for grayling," "water temperatures exceed the thermal tolerance of grayling," and "in fall, 2002, the remnant grayling population in the Big Hole River apparently had declined to such a low level that not enough fish were found to estimate population density" (U.S. Fish and Wildlife Service, 2004).

With extinction of the grayling now possible, conservation groups sued for protection in 2003 and forced the Service to make a new listing decision. Once again, the Service tried to avoid ESA protection. In 2007, the Service removed Montana arctic grayling from the candidates list, this time based on the rationale that the grayling was not a "species" under the ESA (i.e., the grayling didn't qualify as a "distinct population segment"). Again, the Center brought litigation under the ESA's citizen-suit provision and in 2009, the Service agreed to reconsider, returning to its previous position—that is, that Montana grayling are a distinct population warranting protection—but once again, stopped short of an actual listing rule, instead of returning the fish to the candidates list (U.S. Fish and Wildlife Service, 2010a,b).

Forced to make another decision on the grayling's status by the 2011 settlement agreements in which the Service agreed to address the backlog of candidate species, the Service again denied the grayling protection in 2014. To justify the reversal, the Service argued that populations of grayling introduced into lakes outside their historic range would suffice to conserve the species despite having concluded the opposite in previous findings. The Service also argued that population numbers were improving, and that high temperatures and low stream flows were not as detrimental as previously thought (U.S. Fish and Wildlife Service, 2014). Again, conservationists challenged the decision, culminating in a panel of the Ninth Circuit Court of Appeals concluding the Service's most-recent denial could not be reconciled with its previous findings, or evidence in the record, and directed the Service to make a new, and this time legal, decision. It is important to recognize that in a legal challenge, the Service gets deference from the court on questions of science, so when a court overturns a decision, it can only do so by finding the Service ignored science before it, which is exactly what the court found for the grayling.

Despite the adverse court decision, the Service again denied protection for the grayling in 2020 (U.S. Fish and Wildlife Service, 2020). As in 2014, the Service relied on lake populations even though most are outside the species' historic range and have well documented differences in behavior and genetic characteristics from fluvial populations (Kaya, 1992). The Service also repeated claims that grayling numbers have improved in the Big Hole in response to voluntary conservation. The population numbers presented in the finding, however, show grayling numbers underwent serious declines from 1991 to 2006, improved to 2012, but then again declined and remained at lower numbers through the present with a population size of roughly 300 breeding fish (U.S. Fish and Wildlife Service, 2020). The relatively small size and fluctuating numbers of Big Hole grayling hardly inspire confidence that the population is secure. This combined with the fact that restoration goals for the grayling established in the 1990s have not been met and ongoing threats from low flows and high stream temperatures related to irrigation withdrawals and increasingly climate change provides a clear indication the grayling continues to warrant protection under the ESA as the Service has twice concluded.

The back and forth between warranted and not warranted findings and the agency's frequent shifting of the goalposts highlights the Service's resistance to listing a species when voluntary efforts are being carried out by a state whether or not these efforts are meeting goals or securing the species.

Streaked Horned Lark

The Streaked Horned Lark, a subspecies of the more widespread horned lark known for its feathered "horns," once nested from southern British Columbia to the Rogue Valley in Oregon (Fig. 6.3). It was so abundant around Puget Sound that it was considered a nuisance by turn-of-the-century golfers. But the widespread destruction of the lark's open prairie, beach, and floodplain habitats caused cataclysmic population declines. It has already been lost in the San Juan Islands, northern Puget Sound, Oregon's Rogue and Umpqua valleys, the Oregon Coast, and Canada. Today, as few as just 1200 Streaked Horned Larks occupy the southern Puget Sound, Washington Coast and lower Columbia River and Willamette Valley. With most native habitats gone, the lark's existence is reduced to airports, grass-seed farms, dredge spoil islands in the Columbia, and other areas where disturbance by people mimics the floods and fires that once created the lark's nesting areas.

The lark was added to the candidate species list in 2001, and in 2013 it was finally listed as a threatened species, in accordance with the 2011 settlement agreement for candidate species. In listing the lark, however, the Service included a tractor-sized loophole. Under the ESA, endangered species are

Figure 6.3 The federally threatened Streaked Horned Lark has been denied protection for its nests, which are constructed on the ground and vulnerable to plowing or mowing. *Source Photo: Andy Frank.*

automatically granted protections from take, but for threatened species, the law requires the Service to issue a rule regulating activities, when "necessary and advisable for the conservation of the species" (16 U.S.C. §1533(d)). In the case of the lark, the Service published such a rule, but rather than regulate activities to conserve the lark, the rule exempted harmful activities from all regulation—thus the rule exempts all "routine agricultural activities" from the Act's protections, including such activities which harm or kill larks and contributed to the listing in the first instance. The Service justified this all-encompassing exemption on the basis that "maintenance of extensive agricultural lands in the Willamette Valley is crucial to maintaining the population of Streaked Horned Larks in the valley" (U.S. Fish and Wildlife Service, 2013).

The problem with this, of course, is that only some agricultural activities benefit lark, but the rule exempts all "routine agricultural activities." The lark typically uses grass-seed fields because they to some degree replicate their habitat. Yet, the expansive language of the rule allows conversion of grass-seed farms to vineyards or other agricultural uses, even though such a conversion would eliminate lark habitat.

Even on grass-seed farms, the lark, a ground-nesting bird, still needs protection during the nesting season. Initially, Service biologists clarified that agricultural activities would be exempt from ESA regulation "when timed appropriately to minimize impacts to Streaked Horned Larks during the nesting season," but this language was removed by the supervisor of the Service's Oregon office, Paul Henson.

Henson rose-up the Service ranks during the Bush administration and has been a staunch opponent of ESA listings and habitat protections (see Chapter 5, Blowing the Whistle on Political Interference: The Northern Spotted Owl, for political interference in development of a recovery plan for the Northern Spotted Owl, *Strix occidentalis caurina*, which Henson also oversaw at the time). In an internal e-mail obtained through the Freedom of Information Act, Henson says of the lark language "we do not want to be perceived as 'sort of' doing a 4d exemption, but 'not really'" (Henson, 2012). In this same e-mail chain, another service employee observed this language was being deleted to make it more "palatable to the ag community."

This amounts to a textbook example of regulatory capture and is disappointing since seasonal protections limited to a small area around nests have been highly effective at protecting other ground-nesting birds, such as western snowy and piping plovers, which have seen their populations increase. Similar protections certainly would have benefited the Streaked Horned Lark with minimal impact on landowners.

The Center challenged both the decision to list the lark as threatened rather than endangered and the 4(d) rule and prevailed on the listing issue. The court did not rule on the 4(d) rule since if the lark is listed as endangered, no 4(d) rule will be possible. A new decision on the lark's status is expected in 2021.

Speaking truth to power: US Fish & Wildlife Service reformation

The ESA's strong provisions for protecting species and ecosystems have been consistently undermined by the Service's reticence to list species and enforce the law in the face of opposition from powerful special interests. The ability of the public to go to court to enforce the ESA's deadlines for listing species and to challenge politically motivated decisions to deny species' protections has provided an important check to the regulatory capture exemplified by the lark and other case studies. Even in those cases where species have been denied protection, such as with Montana arctic grayling and Greater Sage Grouse, public advocacy together with scientists and their published work has pushed States, federal agencies, and private landowners to take conservation action to avoid regulation.

Petitions from scientists, conservation groups, and members of the public to list species have also greatly contributed to species conservation by bringing relevant scientific information before the Service. Petitions have spurred further scientific research into the status and needs of literally thousands of species across the country in some cases leading to their listing under the ESA and in others not, but in all cases improving our knowledge of the natural world. As but one example, the Center's petition to list 404 aquatic and wetland species from the southeast United States has led to research on dozens of species in the decade since it was filed, including genetic research to clarify taxonomy, habitat studies, population monitoring, and distribution surveys. This research has led to some of these species finally getting listed, including the Big Sandy crayfish (*Cambarus veteranus*) and candy darter (*Etheostoma osburni*). In other cases, the Center withdrew petitions because new information indicated the targeted species were more abundant or have questionable taxonomy.

None of this is intended to suggest that petitions and lawsuits are a substitute for a vigorous regulatory agency sheltered from political influence. The Service is empowered to implement the strong provisions of the ESA and if it did so boldly and in accordance with the best available science, there is no question species would be listed more quickly and recovery enhanced. The agency's reticence to enforce the ESA for fear of political blowback goes well beyond listing. The Service has had to be sued to designate critical habitat for most species and still only roughly 40% have critical habitat (Schwartz, 2008). Likewise, the Service and other federal agencies frequently have had to be sued to consult over the impacts of federal activities on endangered species and although such consultations often result in avoidance and mitigation of impacts with considerable benefit to species, the Service almost never concludes federal actions jeopardize listed species and therefore must be stopped (Malcom and Li, 2015). The Service also rarely prosecutes private parties for destruction of habitat, preferring to limit enforcement of the take prohibition to the narrow circumstances where a dead body can be produced (see Fish and Wildlife Service, 2018). The Trump administration even went so far as to issue a guidance to Service biologists to not tell private landowners that they need a permit if their activities will harm a listed species (Fish and Wildlife Service, 2018). Here too, conservation groups with help from scientists have had to step in and sue private or municipal actors engaging in unmitigated activities that harm endangered species.

It is thus surprising that despite foot dragging in listing species, lax enforcement and chronic underfunding, the ESA has been tremendously successful with 47 species recovered and delisted to date and hundreds more on the road to recovery (Taylor et al., 2005; Schwartz, 2008; Greenwald et al., 2016; Suckling et al., 2016; Greenwald et al., 2019).

With reform and increased funding, the Service could make tremendous progress toward stemming extinction in North America and beyond. Needed reforms include leadership committed to the agency's regulatory duties, sheltering agency biologists from political interference, and restoring partnerships with scientists in academia and other federal agencies. The Service also needs substantial increases in funding for listing and recovering species. The ESA requires the development of recovery plans for listed species, including an estimate of the cost of recovery, yet the Service lacks a specific budget for implementing recovery plans, leaving many species in the dark (Greenwald et al., 2016). Perhaps the recently elected Joe Biden administration will enact these and other needed reforms, but if not, the Center for Biological Diversity and other conservation groups will be there to hold the agency's feet to the fire.

References

Ando, A.W., 1997. Waiting to be protected under the Endangered Species Act: the political economy of regulatory delay. J. Law Econ. 42, 29–60.

Barringer, F., 2004. Interior aide and biologists clashed over protecting bird. New York Times, December 5, 2004.

Brosi, B.J., Biber, E.G.N., 2012. Citizen involvement in the U.S. Endangered Species Act. Science 337, 802–803.

Greenwald, D.N., Suckling, K.F., Taylor, M., 2006. The listing record. In: Scott, J.M., Goble, D.D., Davis, F.W. (Eds.), The Endangered Species Act at Thirty: Conserving Biodiversity in Human-Dominated Landscapes. Island Press, Washington DC, pp. 51–67.

Greenwald N., Hartl B., Mehrhoff L., Pang J., 2016. Shortchanged, funding needed to save America's endangered species. <https://www.biologicaldiversity.org/programs/biodiversity/pdfs/Shortchanged.pdf.>

Greenwald, N., Suckling, K.F., Hartl, B., Mehrhoff, L.A., 2019. Extinction and the U.S. Endangered Species Act. PeerJ 7, e6803. Available from: https://doi.org/10.7717/peerj.6803.

Henson, P., 2012. Email from Paul Henson to various Service staff re: revised 4(d) rule for Streaked Horned Larks. August 27, 2012. Available upon request.

Kaya, C.M., 1992. Restoration of fluvial Arctic grayling to Montana streams: assessment of reintroduction potential of streams in the native range; the upper Missouri River drainage above Great Falls. Prepared for: Montana Chapter American Fisheries Society, Montana Dept. of Fish, Wildlife, and Parks, U.S. Fish and Wildlife Service, and U.S. Forest Service, Bozeman, MT. 102 pp.

Malcom, J.W., Li, Y., 2015. Data contradict common perceptions about a controversial provision of the US Endangered Species Act. <https://www.pnas.org/cgi/doi/10.1073/pnas.151693811>.

Montana Fish, Wildlife and Parks, 1995. Montana Fluvial Arctic Grayling Restoration Plan. November, 1995, Helena, MT.

Montana Fish, Wildlife and Parks and Fish and Wildlife Service, 1996. Memorandum of Agreement for the implementation of the fluvial arctic grayling restoration plan. February 7, 1996, corrected June 6, 1999.

Montana Fish, Wildlife and Parks and Fish and Wildlife Service, 2006. Candidate conservation agreement with assurances for fluvial artic grayling in the upper Big Hole River, FWS Tracking # TE104415-0, March 30, 2006.

Noah, T. 1995. Caught in a Trap: Democrats Get Snared By GOP Pact on List of Endangered Species—A Bush-Era 'Critter Quota' Boosts Animal Protection—And Antiregulatory Ire. Wall Street Journal, February 17, 1995.

Puckett, E.E., Kesler, D.C., Greenwald, D.N., 2016. Taxa, petitioning agency, and lawsuits affect time spent awaiting listing under the US Endangered Species Act. Biol. Conserv. 201 (2016), 220–229. Available from: http://doi.org/10.1016/j.biocon.2016.07.005.

Schwartz, M.W., 2008. The performance of the Endangered Species Act. Annu. Rev. Ecol., Evol., Syst. 39, 279–299. Available from: https://doi.org/10.1146/annurev.ecolsys.39.110707.173558.

Sidle, J.G., 1998. Arbitrary and capricious species conservation. Conserv. Biol. 12, 248–249.

Suckling K.F., Mehrhoff L.A., Beam R., Hartl B., 2016. A wild success, a systematic review of bird recovery under the Endangered Species Act. <http://www.esasuccess.org/pdfs/WildSuccess.pdf>.

Taylor, M., Suckling, K., Rachlinski, J., 2005. The effectiveness of the endangered species act: a quantitative analysis. BioScience 55 (4)), 360–367. Available from: https://doi.org/10.1641/0006-3568(2005)055[0360:TEOTES]2.0.CO;2.

U.S. Department of the Interior Office of Inspector General, 1990. Audit Report: The Endangered Species Program U.S. Fish and Wildlife Service Report No. 90-98, p. 6.

U.S. Fish and Wildlife Service, 1975. Review of status of over 3000 vascular plants and determination of critical habitat. Fed. Register 40, 27823.

U.S. Fish and Wildlife Service, 1976. Proposed endangered status for some 1,700 U.S. vascular plant taxa. Fed. Register 41, 24524.

U.S. Fish and Wildlife Service, 1979. Notice of withdrawal of five expired proposals for listing of 1,876 species, and intent to revise plant notice which includes most of these species. Fed. Register 44, 70796.

U.S. Fish and Wildlife Service, 1980. Review of plant taxa for listing as endangered or threatened species. Fed. Register 45, 82480.

U.S. Fish and Wildlife Service, 1994. Endangered and threatened wildlife and plants; animal candidate review for listing as endangered or threatened species. Fed. Register 59, 58982. Available from: https://ecos.fws.gov/docs/federal_register/fr2729.pdf.

U.S. Fish and Wildlife Service, 1996. Endangered and threatened wildlife and plants; animal candidate review for listing as endangered or threatened species. Fed. Register 61, 7595.

U.S. Fish and Wildlife Service, 2004. Review of species that are candidates or proposed for listing as endangered or threatened. Fed. Register 69, 24876. <https://www.govinfo.gov/content/pkg/FR-2004-05-04/pdf/04-9893.pdf#page = 2>.

U.S. Fish and Wildlife Service, 2010a. 12-Month findings for petitions to list the Greater Sage Grouse (*Centrocercus urophasianus*) as threatened or endangered. Fed. Register 75, 13910. <https://www.govinfo.gov/content/pkg/FR-2010-03-23/pdf/2010-5132.pdf#page = 2>.

U.S. Fish and Wildlife Service, 2010b. Revised 12-month finding to list the upper Missouri River distinct population segment of arctic grayling as endangered or threatened. Fed. Register 75, 54708. <https://www.govinfo.gov/content/pkg/FR-2010-09-08/pdf/2010-22038.pdf#page = 2>.

U.S. Fish and Wildlife Service, 2011. Partial 90-day finding on a petition to list 404 species in the Southeastern United States as endangered or threatened with critical habitat. Fed. Register 76, 59836. <https://www.govinfo.gov/content/pkg/FR-2011-09-27/pdf/2011-24633.pdf#page = 2>.

U.S. Fish and Wildlife Service, 2013. Determination of endangered status for the Taylor's checker-spot butterfly and threatened status for the Streaked Horned Lark. Fed. Register 78, 61452. <https://www.govinfo.gov/content/pkg/FR-2013-10-03/pdf/2013-23567.pdf#page = 1>.

U.S. Fish and Wildlife Service, 2014. Revised 12-month finding to list the upper Missouri River distinct population segment of arctic grayling as endangered or threatened. Fed. Register 79, 49384. <https://www.govinfo.gov/content/pkg/FR-2014-08-20/pdf/2014-19353.pdf#page = 1>.

U.S. Fish and Wildlife Service, 2015. 12-Month finding on a petition to list greater sage-grouse (*Centrocercus urophasianus*) as an endangered or threatened species. Fed. Register 80, 59858.

U.S. Fish and Wildlife Service, 2018. Guidance on trigger for an incidental take permit under section 10 (a)(1)(B) of the Endangered Species Act where occupied habitat or potentially occupied habitat is being modified. Principal Deputy Director, Greg Sheehan. April 26, 2018.

U.S. Fish and Wildlife Service, 2020. Four species not warranted for listing as endangered or threatened species. Fed. Register 85, 44478.

U.S. House of Representatives, 1973. Committee on Merchant Marine and Fisheries, House Report No. 93-412.

U.S. House of Representatives, 1982. Endangered Species Act Amendments of 1982. Conference report filed in House, House Report 97-835, September 17, 1982.

Scientific integrity and advocacy: keeping the government honest

Michael Halpern
Formerly with the Union of Concerned Scientists, Cambridge, MA, United States

When politics invades the weather forecast

It began with a tweet. Hurricane Dorian bore down on the Gulf Coast, and President Donald Trump warned Alabama residents on Twitter that they should prepare for the worst. But there was one problem: the National Weather Service (NWS), a branch of the National Oceanic and Atmospheric Administration (NOAA), did not forecast that the storm was a danger to the state.

Panicked Alabama residents quickly called the NWS office in Birmingham, Alabama looking for advice and information. NWS experts—who did not yet know about the presidential tweet—clarified in their own public message that the hurricane was not expected to pose a danger to Alabama. This is how they protect the public: share information and clear up confusion about risk in real time.

Yet some officials saw this as an attack on the president. Immediately, NWS experts were told they could not communicate with the public without approval. Quickly, both the White House and the Department of Commerce threatened NOAA political leaders with termination unless they supported President Trump's version of the hurricane forecast. The leaders subsequently buckled and issued a statement repudiating their own agency's scientists (Freeman and Samenow, 2020). A few days later, President Trump appeared in the White House with a map doctored with a felt-tipped pen suggesting that the hurricane would indeed hit Alabama, birthing the "Sharpiegate" scandal.

NOAA experts, and the greater scientific community, cried foul. Preventing scientists from communicating about severe weather—especially during a major hurricane—puts public safety at risk. Many scientists spoke publicly about the

Conservation Science and Advocacy for a Planet in Peril.
DOI: https://doi.org/10.1016/B978-0-12-812988-3.00003-X

problems that this political interference poses to public understanding of and trust in their colleagues' work.

Playing politics with an agency long known for its independence has consequences, and scientists who have long watched the intersection of science and politics knew exactly what to do. Experts inside and outside of government, including at the Union of Concerned Scientists (UCS) where I worked for seventeen years, filed complaints under NOAA's scientific integrity policy, first developed and put in place in 2010. A subsequent investigation found several officials—including Acting NOAA Administrator Neil Jacobs—had violated the policy and undermined the scientific independence of the agency (National Academy of Public Administration NAPA, 2020).

Without courageous scientists inside NOAA, however, it is unlikely the story would have been told, or much accountability would have been possible. Anonymous scientists, for example, provided emails and other background information to reporters and Congress that showed political staff clamping down on agency experts and trying to rewrite the story. They gave context to the importance of agency autonomy and independence.

The agency's top scientist, Dr. Craig McLean, even filed his own complaint under the scientific integrity policy. But these actions, like those of many government employees who speak truth to power, came at a price. Months later, McLean was relieved of his duties after asking new political appointees to simply acknowledge the agency's scientific integrity policy applied to them (Boyle, 2020).

This chapter explores the critical role that government scientists play in policymaking, the politics that get in the way, and how attacks on science have changed how the scientific community builds power and engages in public life. I discuss how scientists respond to inappropriate political pressure on or interference in their work, methods that elected and appointed officials use to sideline science they find inconvenient from decision-making, increases in the quality and quantity of advocacy in the scientific community, and ways that scientists and their organizations are pushing to create conditions that support a thriving federal scientific enterprise and more informed environmental and public health policy.

Science has always been political

Scientists who have tracked attacks on science over the years were aware that this kind of conduct, while particularly egregious in the NOAA case, was not uncommon. Politicians have long sought to highlight the evidence that supports their positions and undermine the evidence that does not (i.e., "cherry-picking" the science). Science and politics are interdependent, and many laws require decisions to be made with full consideration of the best available evidence. A tension between the two is inevitable.

On the other hand, science should not *dictate* policy; rather, science should fully *inform* policy. Most presidents have recognized the limits of evidence to solely chart our collective future. "I know few significant questions of public policy which can safely be confided to computers," said President John F. Kennedy. "In the end, the hard decisions inescapably involve imponderables of intuition, prudence, and judgment" (Kennedy, 1963).

But if science is not at the table, every decision becomes wholly political. Evidence that is never presented or considered cannot be used to cast doubt on the quality of policy decisions. Knowing this, ever since World War II, political leaders in the United States have attacked and marginalized individual government scientists and cast doubt on scientific consensus.

They have rewritten scientific documents to reflect the outcomes they prefer and undermined the independence of federal scientific advisory committees. They have clamped down on scientist communication and the ability of experts to publish research that the political leaders of the day find inconvenient, and to use words the leaders believe are politically contentious. As a result, historical researchers are able to document political pressures on government scientists during every US presidency dating back to President Eisenhower (Carter and Berman, 2018). But that does not mean that this pressure under each administration is equal.

Political and industry pressure on science ramps up

Since the 1970s, more laws that statutorily require the use of science—from the Clean Air Act to the Endangered Species Act—led to marked increases in the perceived need to sideline science from policymaking. Evidence that cannot be avoided in a regulatory context must be misrepresented to move forward with a policy decision that is not scientifically justifiable, lessening the likelihood that the decision will be overturned by the courts.

People of all political stripes say they believe that good decisions should be fully informed by the best available information. Many presidents have publicly recognized the centrality of evidence to the success of the nation. "Science, like any field of endeavor, relies on freedom of inquiry; and one of the hallmarks of that freedom is objectivity," President George H. W. Bush told the National Academy of Sciences. "Now, more than ever, on issues ranging from climate change to AIDS research to genetic engineering to food additives, government relies on the impartial perspective of science for guidance" (Bush, 1990).

But fissures in this philosophy develop quickly, especially when regulated industries that have skin in the game lean on elected and appointed officials. Industry pressure on government officials in state and federal capitals has

become a constant, renewable resource regardless of which political party controls the executive and legislative branches of government.

It gets worse when those officials previously worked for those very industries. When coal lobbyist Andrew Wheeler runs the Environmental Protection Agency, or when oil and gas lobbyist David Bernhardt runs the Department of Interior that oversees leasing on public lands and waters, targeting science becomes considerably easier (Flavelle et al., 2018; Siler, 2019).

Misrepresenting science has become big business. David Michaels served as administrator of the Occupational Safety and Health Administration in each of President Obama's terms. In his 2020 book *The Triumph of Doubt*, Michaels outlines how companies manufacture doubt and uncertainty about science that demonstrates harm caused by their products, delaying or entirely avoiding oversight and regulations or other public protections (Michaels, 2020).

Every industry has a lobbyist in Washington D.C. Even some simple chemical compounds have lobbyists. Take the Salt Institute, which argued in 2016 that reductions in sodium intake "will make our food less safe and endanger public health" (Devaney, 2016). The US Food and Drug Administration scientists had just released voluntary guidelines recommending that people consume less salt to lower their risk of heart disease and stroke.

Often the dirty work is done by trade associations who take the bad publicity that their members cannot afford. One study by UCS found that while public statements from fossil fuel companies about climate change were often accurate, the trade groups they funded were considerably more likely to misrepresent and overstate the uncertainty behind the science in official government filings (Goldman and Carlson, 2014).

Political pressures also happen completely outside of government. The National Football League attacked and stonewalled researchers, created fake scientific advisory committees, and published a series of bogus scientific papers in repeated attempts to downplay the impact of repeated blows to the head on players' brains (Fainaru-Wada and Fainaru, 2014). The NFL's actions delayed public understanding of this problem, forestalled compensation to injured players, and held up changes to league rules and protocols to make playing the game safer.

UCS describes these tactics under the "Disinformation Playbook," where some companies and industry groups try to pass off counterfeit science as legitimate, harass scientists whose results threaten their bottom line, manufacture uncertainty about established research, buy credibility through alliances with universities, and inappropriately influence government officials or decision-making processes (Union of Concerned Scientists, 2017).

Given all of this tension and pressure, scientists have the duty to be fully engaged (and speak truth to power) in pushing back against misinformation and defending the independence of government experts who are charged with evaluating scientific claims and making recommendations about scientific standards that adequately protect public health and the environment.

Still, for decades, "scientific integrity" was limited to an evaluation of whether one's own research was honest. Political interference in science was perceived as an issue of occasional anecdote and not as a systemic problem to be analyzed and addressed. That changed upon the election of George W. Bush.

Early in Bush's first term, scientists realized that something was different. A White House official attempted to edit the Environmental Protection Agency's (EPA) Report on the Environment to downplay human impacts of climate change. A report on the impact of mercury exposure on children's health was suppressed as the administration wrestled with new standards on regulating mercury emissions from coal-fired power plants. A microbiologist at the US Department of Agriculture who studied the human health impacts of airborne bacteria from farm waste was repeatedly prevented from speaking publicly about his research at scientific conferences. Too often, the effect of these actions was to sideline the science before it could even be considered by decision-makers and the public (Mooney, 2005; Union of Concerned Scientists, 2004).

Some in Congress recognized there was a problem early on. Representative Henry Waxman, then the top Democrat on the House Committee on Oversight and Government Reform, issued a report in late 2003 detailing what he saw as the president's undermining of scientific integrity in federal agencies

(House Committee on Oversight and Government Reform, 2003). He would continue to doggedly pursue the issue until he retired from Congress.

Around that time, experts from many diverse scientific backgrounds gathered in Washington, DC to talk about what they saw as a growing problem that needed a response from the scientific community. They decided to put together a statement that called on the Bush administration to restore scientific integrity to federal policymaking.

"When scientific knowledge has been found to be in conflict with its political goals, the administration has often manipulated the process through which science enters into its decisions," they wrote. "Other administrations have, on occasion, engaged in such practices, but not so systematically nor on so wide a front" (Gottfried et al., 2004).

The statement was drafted and signed by 62 prominent experts, including 20 Nobel Laureates, who advised Republican and Democratic presidents dating back fifty years. The scientists released it in February 2004, along with supporting case study evidence compiled by UCS, to an avalanche of media attention. Subsequent reporting put individual attacks on science in the context of greater scientific community concern (see Chapter 5, Blowing the Whistle on Political Interference: The Northern Spotted Owl). A new issue was born.

Building a movement to defend science

The first step was to build a constituency to advocate for reform. UCS began attending more scientific society meetings to hold discussions about the intersection of science and politics. We held forums on university campuses to explore how academic scientists could support their peers in government.

Experts around the country continued to add their names to the scientist statement on scientific integrity as abuses of science continued. Over the next 4 years, more than 15,000 scientists in every US state and territory would sign on. These scientists began meeting with their members of Congress and urging oversight. They held community discussions to educate their peers about the problem and motivate them to take action. They wrote letters to the editors of their local newspapers to raise awareness about the health and environmental consequences of attacks on science.

At the 2006 meeting of the American Association for the Advancement of Science in St. Louis, a hastily arranged session warned of the consequences of political control over government scientific work. Nobel Laureate and former California Institute of Technology President David Baltimore did not mince words.

"It's no accident that we are seeing such an extensive suppression of scientific freedom," Baltimore told a room overflowing with hundreds of scientists. "It's part of the theory of government now, and it's a theory we need to vociferously oppose." He called on the government to be "the guardian of intellectual freedom" and for scientists to speak up ardently against actions that compromise it (Dean, 2006).

We knew, although, that the issue was unlikely to swing elections. Scientific integrity is like campaign finance reform: it is incredibly important to a functioning government, but precious few people will name it as one of their top issues. That meant it was important to identify those who are most impacted and to build a broad coalition of organizations that have a stake in good science-based policymaking.

Since science pervades almost all environmental, public safety, and public health policy, pretty much any NGO working for the public interest will be impacted. So UCS helped convene environmental organizations, public health associations, transparency and government accountability groups, reproductive health advocates, and whistleblower supporters to advocate in Washington on a common agenda to create more independence for federal agency scientific work. Together, we catalyzed considerably more scrutiny and oversight during the Bush administration over actions that sidelined or suppressed science, and in doing so raised the political price for engaging in such activity.

But for all of the media and public attention this issue received, many in the science world were fearful to push back. Science funding had long enjoyed bipartisan support, and many scientific societies were fearful that defending the work of their members from political attacks would anger some policymakers and put this support at risk.

Further, many scientific organizations have for years positioned themselves as "honest brokers" who are above politics. It is hard to understate how dangerous this line of thinking can be. Merriam-Webster defines politics in part as "the total complex of relations between people living in society" (Merriam-Webster Dictionary, 2020). As science communications expert Aaron Huertas writes, politics is "about who has power to make decisions and who gets limited resources" (Huertas, 2018).

There are politics in what and who we decide should be studied. Politics are in research and experimental design. Politics in how the research results are interpreted, presented, and ultimately used. To claim that a human endeavor is above politics is to fail to understand the power dynamics of that endeavor, making it more difficult to wield that power for good and setting up inequities within scientific communities.

Scientific society members began demanding that the societies expand their policy programs beyond funding advocacy. They questioned whether societies should take funding from companies that support attacks on science or organizations that engage in science denial. Attacks on the scientific process itself created an opening for more conversation about the proper role for scientists and scientific organizations in public life.

There is now more space in the science community for conversations about everything from scientific integrity to racial equity. This evolution is both important and necessary. You cannot wield power unless you build it first.

Monitoring the federal government

Before science advocates could fully advocate for more protections for government scientists, we had to fully demonstrate the scope of the problem. A great team of researchers dug up previously known examples of political interference in science and unearthed new ones. Over the course of several years, UCS was able to document nearly 100 case studies of political interference in science during the George W. Bush administration on issues from childhood lead poisoning to right whales (Union of Concerned Scientists, 2009).

We also decided to take the temperature of the federal workforce through anonymous surveys that allowed experts to publicly share their perceptions of and experiences with political interference in science. UCS first surveyed government scientists about how politics impacts their work in 2005, starting with the US Fish and Wildlife Service (FWS) and moving onto many more federal agencies and departments. At FWS, one in five of the scientists surveyed had been "directed, for nonscientific reasons, to refrain from making jeopardy or other findings that are protective of species." More than a third said they could not openly express "concerns about the biological needs of species and habitats without fear of retaliation" (Union of Concerned Scientists, 2005).

When the Democrats took over Congress in 2007, the first hearing then-Chairman Waxman held in the House Committee on Oversight and Government Reform Waxman held was on political interference in climate science at federal agencies. UCS released the results of a survey of government climate scientists at the hearing that found widespread attempts to meddle in their work. NASA scientist Dr. Drew Shindell testified about new restrictions on the ability of researchers to publicly share their knowledge and expertise: agency press officers were required to monitor all interviews related to climate science, and any NASA climate communications must be cleared by White House political appointees (House Committee on Oversight and Government Reform, 2007a).

Later in the year, the committee released its own report detailing widespread interference. The evidence before the Committee leads to one inescapable conclusion, the investigators wrote. "The Bush Administration has engaged in a systematic effort to manipulate climate change science and mislead policymakers and the public about the dangers of global warming" (House Committee on Oversight and Government Reform, 2007b).

UCS was also involved in efforts to bring new examples of political interference to light. In 2006, the conservation organization Forest Guardians received documents from a Freedom of Information Act Request related to endangered species decisions within the US Fish and Wildlife Service. The documents showed that Interior Department Deputy Assistant Secretary Julie MacDonald had personally rewritten scientific assessments developed by agency biologists. MacDonald had a degree in civil engineering and no background in biology or other natural sciences. The edits were sufficiently substantial to prevent the protection of several species under the Endangered Species Act (Eilperin, 2006).

For example, the agency was poised to list the Gunnison's Sage Grouse (*centrocercus minimus*) as endangered and had even drafted a press release to announce the listing. MacDonald stepped in after a phone call from the Colorado governor. After delaying the announcement, MacDonald was involved in an effort to strip significant information from the listing proposal, and eventually FWS headquarters in Washington, DC took responsibility for redrafting it (Union of Concerned Scientists, 2006).

UCS worked with the conservation group to give the documents with the Washington Post. We shared them with congressional staff while the reporter was investigating the story. When the story broke a few weeks before the 2006 midterm election, Representative Nick Rahall pledged to hold hearings into Julie MacDonald's conduct should the Democrats win control of Congress (Eilperin, 2006).

When they won the election, we were ready. Members of the UCS Science Network contacted congressional representatives to encourage oversight. The committee scheduled a hearing and invited MacDonald to testify (House Committee on Natural Resources, 2007). A few days before the hearing, MacDonald resigned from her position with the reason used by many compromised officials: to spend more time with her family (Barringer, 2007). The hearing went on anyway, and those who testified focused not on MacDonald as a bad apple but on the systemic failures that made her actions possible (also see Chapter 5, Blowing the Whistle on Political Interference: The Northern Spotted Owl, for related endangered species scientific integrity hearings at the time).

A subsequent investigation by the Interior Department Inspector General found that MacDonald had "regularly bypassed managers to speak directly with field staff, often intimidating and bullying them into producing documents that had the desired affect she and the former Assistant Secretary wanted." The investigation also found that she had provided nonpublic information about internal endangered species deliberations to the Pacific Legal Foundation and the California Farm Bureau (Department of Interior Inspector General, 2006a).

Staff biologists took detailed notes throughout the process. Because of these notes, we were able to understand what edits were made by whom, and where scientists objected to those edits. The detailed administrative record was essential to the inspector general and congressional investigations that ultimately cost MacDonald her job and eventually compelled the FWS to right the wrongs.

Candidates and their commitments

The awareness raising, movement building, research, oversight, and advocacy would not lead to systemic change without pressuring candidates for federal office to commit to taking action. We needed to get presidential candidates to publicly agree to restoring scientific integrity to federal policymaking. Fortunately, we found a unique opportunity to do just that.

Public awareness of the lack of attention paid to science policy issues grew rapidly during the 2008 election. Shawn Otto and Matthew Chapman, two screenwriters idled by the Hollywood writers' strike, joined with a few scientists to propose a presidential debate on science policy. Even the idea of this kind of debate got thousands of people talking.

The organization that emerged from that effort, Science Debate, started a national conversation about the importance of science to decision-making on issues from water quality to pandemics. In written answers to a series questions posed by the organization, informed by a panel of scientific experts, both major presidential candidates recognized the problem of political interference in science and pledged reform (Science Debate, 2008).

Scientific research cannot succeed without integrity and trust, wrote Senator John McCain. "We have invested huge amounts of public funds in scientific research.... Many times our research results have identified critical problems for our country. Denial of the facts will not solve any of these problems."

Then-Senator Barack Obama's commitments were more specific. "I will restore the science integrity of government and restore transparency of decision-making by issuing an Executive Order establishing clear guidelines for the review and release of government publications, guaranteeing that results are released in a timely manner and not distorted by the ideological biases of political appointees," he wrote. "I will strengthen protection for 'whistle blowers' who report abuses of these processes."

This led President Obama to declare in his first inaugural address to "restore science to its rightful place," a pledge that many advocates would remind him of in the ensuing 8 years (Obama, 2009a). Two months later, surrounded by smiling scientists at the White House, he signed a memorandum tasking his science advisor with coming up with a plan to create strong scientific integrity standards within government (Obama, 2009b).

"Promoting science isn't just about providing resources – it is also about protecting free and open inquiry," the president said at the White House event.

"It is about letting scientists like those here today do their jobs, free from manipulation or coercion, and listening to what they tell us, even when it's inconvenient—especially when it's inconvenient."

The White House Office of Science and Technology Policy (OSTP) began drafting this plan, but soon met with significant resistance within agencies and the White House. Some officials were hesitant to give up control of information and thought of themselves as the "good guys" who could be trusted. Activists ramped up pressure on the administration to do something, and eventually, OSTP Director John Holdren released a memorandum directing agencies and departments to develop policies to create strong scientific integrity standards (Holdren, 2010).

Some agencies and departments took this charge seriously, assigning multiple staff to develop policies and implementation processes. They met with stakeholders to seek input. The EPA and the Department of Interior even sought input through a formal public comment process. These policies were more robust from the outset. Other agencies took a lighter approach, restating the principles outlined in the OSTP memorandum but not taking meaningful steps to implement their policies or otherwise integrate their principles into agency culture and practices.

Even then, the scientific integrity and related policies were a work in progress, and mistakes were inevitable. In the years since the first versions were released, several agencies and departments went back and turned mediocre policies into good ones. In the waning days of the Obama administration, the Department of Energy released a new policy that was the culmination of a 2-year process of negotiation and input. During the Trump administration, the policies remained in place and were used many times by experts both formally and informally to push back on actions that led to losses of scientific integrity. Some agencies, most notably the EPA, continued to refine and improve their policies and practices.

A major advantage of scientific integrity policies is how they can change cultural expectations and address conflict informally. Scientific integrity officers regularly help scientists solve problems before they become official complaints or public scandals. A scientist stands up to her manager and insists that her paper be published as is. An expert resists pressure from a political appointee to soften their analysis. As a result, many of the most impressive stories of government scientists speaking truth to power cannot be told. Instead, the scientists simply get back to work in service of the public interest.

Public access to government knowledge

The issue of employee rights to communicate their research and scientific opinions has generally been more contentious. Some government officials worry that scientists who are given more opportunities to speak openly about their work and expertise will abuse that authority and openly criticize policy initiatives. Scientists though are quite adept at distinguishing between statements of policy and statements of evidence. Scientists also tend to be conservative in not wanting to overstate what they say.

NOAA was the first to explicitly extend the right of scientists to communicate without interference. And for years, the policy worked well. When announcing the policy at a scientific conference, NOAA Administrator Dr. Jane Lubchenco gave concrete examples of where the policy might have led to different outcomes. "The policy is very clear about saying that NOAA scientists may speak freely to the media about their scientific findings without having to go to a public affairs office and request permission. That was not true during the Deepwater Horizon [disaster]" (Showstack, 2011).

The Deepwater Horizon was an oil drilling rig that exploded and sank in April 2010, opening a hole in the ocean floor that was not controlled for 5 months. Officials had previously been criticized for painting a rosier picture of cleanup efforts than could be supported by the evidence (Davis, 2010).

Other federal departments and agencies, such as the Department of Energy, followed suit and explicitly gave their scientists more independence. But many policies still hedge in making scientists fully available to the public and the press. President Obama pledged to be the most transparent president in history, but that transparency ended up more limited than many would have preferred. Journalists repeatedly expressed displeasure at continued roadblocks and were disappointed by the Obama administration's gatekeeping of expert speech.

"Over the past two decades, public agencies have increasingly prohibited staff from communicating with reporters unless they go through public affairs offices or through political appointees," wrote a group of journalism associations in a letter. "We consider these restrictions a form of censorship—an attempt to control what the public is allowed to see and hear. The stifling of free expression is happening despite your pledge on your first day in office to bring 'a new era of openness' to federal government—and the subsequent executive orders and directives which were supposed to bring such openness about" (Society of Professional Journalists, 2014).

A study of agency public affairs officers and reporters conducted jointly by UCS and the Society of Professional Journalists found a majority of journalists believe the public does not get the information it needs because of barriers to reporting on agency scientific work. On the other hand, agencies face constraints from resources to litigation that make transparency more difficult (Bailin et al., 2015).

Defending scientific integrity abroad

Attacking scientists for their results and scientific opinions is not just a United States phenomenon. In 2012, six Italian scientists were initially convicted of manslaughter for advice they gave prior to a deadly earthquake (Cartlidge, 2016). In 2019, government officials in Brazil prevented scientists from releasing a study on illegal drug use that might suggest that current governmental policies are ineffective (Escobar, 2019).

A September 2020 survey of Australian scientists found significant censorship and information suppression within the national government (Driscoll et al., 2020). "We are often forbidden [from] talking about the true impacts of say a threatening process...especially if the government is doing little to mitigate that threat," one survey respondent said. "In this way the public often remains in the dark about the true state and trends of many species."

But attacks on science can also have political costs. Political interference in science became a major issue in the 2015 Canadian elections because Canadian science advocates connected the attacks to direct impacts on public health and the environment.

Prime Minister Stephen Harper's dismantling of government scientific capacity and censorship of scientific work was extensive, with government scientists censored, budgets chopped, data monitoring programs eliminated, scientific libraries shuttered, and the contents thrown into dumpsters (Dupuis, 2014). The long-form census was axed, depriving public officials and businesses of critical information about the Canadian public (Maioni, 2015).

Canadian graduate student Katie Gibbs helped organize a lab coat protest in front of Parliament in Ottawa they dubbed the "Death of Evidence" (Nature, 2012). More than 2000 people showed up to observe a mock funeral. The event made for dramatic photos, and launched a movement to defend science in the federal government. The Canadians could lean on the Americans for strategy, because the parallels were remarkable.

In Canada, the Department of Fisheries shuttered its scientific library system, sold off equipment, and threw books and research documents in dumpsters. In the United States, the EPA had similarly dismantled its library system. Experts at two major environmental government departments, Fisheries and Oceans Canada and Environment Canada, were routinely muzzled. The same happened to scientists at the Department of Interior and EPA in the United States (Dupuis, 2014; Union of Concerned Scientists, 2009).

Gibbs went on to co-found Evidence for Democracy, an organization that birddogged the Harper administration for several years and then pressured the Trudeau government for reform. The organization's work was aided by surveys of scientists developed and implemented by the federal employees union, which found a demoralized workforce, numerous instances of censorship and self-censorship, and new impediments to the use of science in decision-making. Incredibly, 90% of respondents reported not being able to speak freely with the media about their scientific work (PIPSC, 2013).

One of Prime Minister Trudeau's first declarations was to restore the use of the long-form census. It is unlikely that a census form would have become such a priority without sustained public pressure and attention. Yet Canada has struggled since to fully expand public access to scientific experts; in 2018, three years into the Trudeau government, still more than half of government scientists who responded to a survey still felt they were being muzzled (Owens, 2018).

Scientific integrity and the Trump administration

Improvements made during the Obama administration were soon to be challenged by the election of Donald Trump. A loose relationship with facts and the stated desire of "deconstruction of the administrative state" did not bode well for the independence of the federal scientific enterprise. Still, scientists had laid the groundwork for aggressive oversight. Scrutiny would be high regardless of the administration's actions.

Thousands signed a letter calling on Congress and the new administration to ensure that science played a central role in protecting public health and the environment. The letter identified four areas of concern: creating a strong and open culture of science; ensuring that public safeguards remain strongly grounded in science; meeting high standards of scientific integrity; and providing adequate resources to enable federal scientists to do their work (Acrivos et al., 2017).

Yet by almost any measure, political interference in science increased markedly during the Trump presidency in both incidence and severity. The attacks came both through direct censorship and manipulation of science and, more critically, the deconstruction of systems of scientific input and advice that are designed to create both good policy and accountability.

Various organizations tracked scores of attacks on science during the 4 years President Trump was in office, including UCS and a joint project of the Sabin Center for Climate Change Law and the Climate Science Legal Defense Fund (Sabin Center, 2020; Union of Concerned Scientists, 2020). The most high-profile attacks were easy to understand: the censorship of scientists at the Centers for Disease Control and Prevention, or the firing of top experts across the government whose factual statements irked the White House. Experts at the United States Geological Survey were told to limit climate forecasts. Studies into the safety of offshore drilling and mountaintop removal mining were shelved.

The more insidious problem, however, were actions that destabilized and compromised the independence of federal agencies under the purview of the executive branch. For example, EPA sought to exclude scientists who received grants from the agency from serving on the agency's science advisory panels. Several lawsuits eventually forced the EPA to rescind the policy, but not before it stacked advisory panels with experts who had financial conflicts of interest and, often, views outside the scientific mainstream (Reilly, 2020b).

As referenced above, UCS has surveyed tens of thousands of federal government experts multiple times since 2005 to examine both the intensity of political pressure and its historical precedent. In 2018 again, UCS partnered with the Iowa State University Center for Survey Statistics and Methodology to survey experts across 16 federal agencies. More than 4200 respondents painted a disturbing picture of the state of the federal scientific enterprise, describing challenges including censorship and self-censorship, political interference in their work, and low morale. Scientific integrity in government under President Trump was at its lowest point yet (Carter et al., 2018).

Environmental Protection Agency attempts to remove science and science advice

In 2017, the EPA proposed a rule that would exclude many public health studies from informing agency decisions by suggesting that for a study to be viable, all data would need to be publicly available.

The proposed rule would significantly restrict the types of science that could be used in policymaking. It would effectively force the agency to ignore thousands of scientific studies when responding to public health threats and setting pollution standards. And it could compromise scientific assessments produced by the EPA on a wide variety of topics, including on chemicals such as ethylene oxide. The EPA cannot develop adequate public health protections without fully considering all the scientific evidence; much of the agency's work was under threat (Friedman, 2019).

The original target of the proposal was the Clean Air Act. Science continues to demonstrate massive public health benefits of stricter controls on air pollution, especially fine particulate matter pollution (Wei et al., 2019). The Clean Air Act requires pollution standards to be set solely based on the best available science. So, if you cannot manipulate the science, you manipulate the process by excluding the best available science from consideration.

Republicans in Congress tried for several years to pass legislation to hamstring EPA in this way. The legislation, variously known as the Secret Science Reform Act and the HONEST Act, was strongly opposed by scientific organizations and never made it through the Senate (Niiler, 2019).

When Scott Pruitt was appointed EPA administrator in 2017, antiscience advocates saw an opening to try to implement the failed legislation through the agency. Former House Science Committee Chair Lamar Smith, who had pushed the legislation for years and had just left Congress, went to EPA in 2020 and urged the agency to find a way to implement his bill without congressional approval (Waldman and Farah, 2018).

Chairman Smith was notorious in particular for his attacks on climate scientists. In 2015, he subpoenaed the raw data and email correspondence of a top NOAA climate scientists whose research he did not like. Urged on by science advocates, NOAA successfully resisted the subpoena as an attack on scientific inquiry (Tollefson, 2015).

Administrator Pruitt believed he would be able to get the rule through quickly. But scientific organizations came out in full force against the proposal. Scientific societies and journal editors were unanimous in opposition; the editors of *Science* and *Nature* joined several colleagues in writing that "excluding relevant studies simply because they do not meet rigid transparency standards will adversely affect decision-making processes" (Berg et al., 2018).

Hundreds of groups and thousands of experts filed public comments urging EPA to abandon the proposal. The comments were highly substantive and forced the agency to spend many months responding to them before moving forward.

The National Academies of Science also asked to review the proposal (McNutt et al., 2018). Their request was ignored. The EPA's Science Advisory Board (SAB) asked the agency to do the same and were similarly sidelined. The SAB decided to review it anyway, writing in a draft report that the rule represented a "license to politicize the scientific evaluation required under the statute based on administratively determined criteria for what is practicable" (Eilperin, 2019). This led Administrator Andrew Wheeler to curtail the SAB's authority to determine its own agenda (Reilly, 2020a).

Then, at the beginning of the COVID-19 pandemic, the EPA released a supplemental proposal that not only ignored appeals to abandon the rule but significantly expanded its reach. The agency provided thirty days for the public to comment without any public hearings just as the attention of public health experts was rightfully diverted to keeping people safe from the pandemic.

We urged the EPA, repeatedly, to hold a virtual hearing on the proposal. Each time, they refused. So UCS decided that if the agency would not do its job, someone would need to do it for them. EPA would get feedback whether they wanted it or not.

UCS scheduled a virtual public hearing on EPA's behalf and invited the public to testify. Dozens of scientists, advocates, and regular citizens showed up ready to share substantive feedback on the proposal and its impact on the EPA's ability to meet its mission, including Bill Reilly, who served as EPA administrator under George H. W. Bush from 1989 to 1993.

Reilly remembered how one of his predecessors, William Ruckelshaus, had felt limited by the insufficiency of the scientific and health information before him to set health and environmental standards. "Since that time EPA has given the highest priority to ensuring the integrity of the science on which its regulatory decisions are made," said Reilly at the hearing. "What EPA does depends on the confidence of the public, it depends on the integrity of the science basis for its decision-making. Both have been put into question by the current proposed regulation" (Halpern, 2020).

Criticism came from within the agency in the form of a dissenting scientific opinion from Tom Sinks, the official in charge of human subjects research review and former director of the agency's science advisory office. Sinks had previously been named the contact for the initial draft of the policy even though he had not yet reviewed it. "Data availability is not a measure of study quality nor is it a determinant of causation," he wrote. "This will compromise the scientific integrity of our scientists, the validity of our rulemaking, and possibly the health of the American People" (Sinks, 2020).

The fact that Sinks was able to issue a dissenting scientific opinion is notable in its own right. The right and responsibility to do so is outlined in the agency's scientific integrity policy. Without the policy in place, it is considerably less likely that he would have felt empowered to go on the record. His statement will be used in court to challenge any substantively similar proposal that is finalized by the agency.

EPA attempted to finalize the proposal just before President Trump left office. The Environmental Defense Fund subsequently successfully sued the agency to prevent the proposal from going into effect, and the new Biden EPA refused to defend it (Malakoff, 2021).

EPA leaders also removed scientific advisors whose analysis they feared would prove inconvenient to the decisions those leaders were determined to make. Particulate matter leads to respiratory and cardiovascular disease and causes premature death (Wei et al., 2019; Wu et al., 2020). The science overwhelmingly showed that current pollution standards were not sufficiently protective of public health, but the administration was hell bent on keeping the same standards in place.

For decades, under presidents of both parties, the EPA followed a well-defined, transparent process to set particulate matter standards under the Clean Air Act. EPA is required to review air pollution standards every 5 years. The agency's Clean Air Scientific Advisory Committee (CASAC) has long relied on independent panels of experts to help them understand how pollutants affect human health and welfare. EPA had regularly convened a working group comprised of the nation's best particulate matter experts to go over the agency's analysis of the science and give advice to CASAC.

Anticipating that the expert panel would agree with EPA scientific staff that current standards were inadequate, the EPA disbanded the panel in October 2018 (Friedman, 2018). Scientists were aghast. There was no way that EPA would get the independent science advice it needed without the panel. For an administration intent on avoiding strengthening air pollution standards, although, it was a great move: you cannot sideline science advice if you do not collect it in the first place.

In an unprecedented action, the dismissed scientists decided to meet anyway and conduct a full review of the EPA's scientific assessment. With logistical support from UCS, the scientists convened in a hotel conference room outside Washington DC and held a meeting using the same practices as the EPA (Goldman et al., 2020). All panelists went through ethics reviews. They livestreamed their deliberations and took comments from the public. Their subsequent analysis was highly influential in informing CASAC's deliberations and helping those outside the agency evaluate whether the decisions made by EPA leaders were scientifically supportable (Frey et al., 2019; Frey et al., 2020).

The future of scientific integrity reform

An agricultural expert at the US Department of Agriculture gives a lead to a reporter that scientists are restricted from using climate change data in their work. An analyst at the Department of Energy tells congressional staffers about suppressed studies on energy use. An epidemiologist at the Center for Disease Control and Prevention files a scientific integrity complaint about public misrepresentation of the risks of a new pandemic. These are hypothetical yet plausible scenarios, and they all depend on the courage of public servants to speak truth to power. They also depend on functional oversight mechanisms and systemic supports that protect those making allegations and holding accountable those whose actions undermine scientific integrity.

There are a number of excellent organizations with significant legal expertise that ably represent and advocate for whistleblowers. Public Employees for Environmental Responsibility has long fought for government employees who expose attacks on science. The Climate Science Legal Defense Fund similarly supports scientists both inside and outside of government who find themselves subject to attacks and harassment (also see Chapter 16, Speaking Truth to Power for the Earth).

The Government Accountability Project represents many scientist whistleblowers, including Rick Piltz, a whistleblower who exposed inappropriate political influence over the development of climate science documents during the George W. Bush administration. Piltz went onto found Climate Science Watch, which monitored interference in climate science for several years.

Yet becoming a formal whistleblower under the most ideal circumstances is a tremendous sacrifice. The process is arduous and isolating. The chances of complete success are low. And even when you do succeed, you almost undoubtedly need to find a new job (see Chapter 11, Essays from the Trenches of Science-Based Activism).

Despite these risks, more and more scientists are looking for employment where their analysis is used directly in policymaking. Over the years, hundreds of scientists have asked me how to break out of academia. Government science jobs at the state and federal levels are an excellent way to do just that—even with (and sometimes because of!) the political pressures that might come one's way.

For federal agencies to meet their missions, we need courageous government employees who are willing to advocate internally for independence and to help non-profits, scientific associations, media, and investigators inside government hold those who violate scientific integrity principles accountable.

Strong leadership can foster this courage. In one of his first official acts, President Biden issued a Memorandum on Restoring Trust in Government

The facts are coming! The facts are coming!

Through Scientific Integrity and Evidence-based Policymaking (Biden, 2021). Robust action within the executive branch will be necessary to foster and rebuild a thriving federal scientific enterprise.

But it is just as important for agencies to strengthen formal protections for agency staff so that they are not put in a position where they are forced to report abuses of power. Currently, scientific integrity policies depend on the good grace of agency leaders. Because policies could be rescinded or curtailed at any time, scientific integrity officials know they can only push so far. This is especially problematic when the agency leaders themselves—or representatives from the White House—are the ones whose actions are harming scientific integrity.

To give teeth to these policies, New York Representative Paul Tonko introduced the Scientific Integrity Act in the House of Representatives in 2017 and again in 2019 (Scientific Integrity Act, 2019). The legislation is designed to clarify basic standards of behavior and set expectations for how scientific integrity policies are developed and carried out.

At the invitation of the Democratic majority, I testified at the hearing in support of the legislation. Science and politics scholar Roger Pielke, invited by the Republican minority, also supported the codification of scientific integrity protections (Scientific Integrity in Federal Agencies, 2019). Each of us criticized both major political parties for politicizing science and scientific processes, helping to demonstrate that this is a nonpartisan issue deserving support from representatives of all ideologies. We agreed that good scientific integrity practices do not dictate policy; they simply make good, evidence-based policy more likely.

The hearing helped Representative Tonko earn bipartisan co-sponsorship for his legislation, and an amended bill passed out of the House Science Committee with a bipartisan vote. Hawaii Senator Brian Schatz has been pushing the bill in the Senate.

The legislation and related activities should be a priority for the current Congress and Biden administration. Advocates must use their power to push elected officials to take action on scientific integrity. There is always way more that an administration wants to accomplish than it has the time, resources, and political capital to do. Politicians are people, and they want easy wins that cause the least amount of pain. It's up to all of us to make scientific integrity wins as easy as possible.

Until then, scientists who work for government will do their jobs as best they can. They will continue to develop research and analysis to support their agency's mission. They will take notes and document any behavior that gets in the way. The records they keep are important both for policymaking and for court challenges to decisions that cannot be supported by evidence.

Speaking truth to power: experts as advocates

Some people believe that their civic duty ends once they vote and their candidate of choice is elected. But the election is just the beginning of creating long-lasting change. Politicians are considerably more likely to follow through on promises they make when there is an engaged constituency to advocate for those changes.

Realizing the promises that successful candidates make during campaigns requires political pressure once they take office. It is up to scientists, public interest advocates, and democracy supporters to build the political capital and space that's necessary for elected officials to act, regardless of whether elected officials are considered "friendly" to science. Successful advocates must pressure their friends and their adversaries to get things done.

The pejorative use of the term "advocate" has for years stunted the ability of scientists to fully participate in public life (Chapter 1, The Nuts and Bolts of Science-Based Advocacy). Yet the increase on attacks on science and scientists has come more willingness by the scientific community to step forward and defend the scientific enterprise.

The avalanche of misinformation put forward during the Trump and many in his administration became a wake-up call and a rallying cry. The transformative measure of the March for Science movement that began in 2017 (shortly after Trump's inauguration) is how scientists viewed themselves and their role in decision-making. What was a conversation about *whether* to engage in politics and policy became a conversation about *how* to do so most effectively.

Many more scientists, especially those early in their careers, understand that science does not speak for itself, and that an expectation of science-informed policy requires experts to cultivate and build power. This is a sea change in how the science community perceives itself. Advocacy is becoming a standard part of a scientific career, not something "nice" to do in one's spare time. And many believe that this trend can only accelerate.

Long overdue discussion and action around racial equity within scientific communities have similarly opened many eyes about the futility and danger of scientists pretending they are above politics and that their work is value neutral and apolitical. Again, politics is about power: who decides what is studied, what questions are asked, and ultimately who benefits from research results. Universities, scientific societies, NGOs, and other institutions must develop a better understanding of politics and power to effectively address the marginalization of Black scientists and scientists from other underrepresented groups. Any stigma against advocacy melts away when one sees oneself as part of a system and not outside of it.

"My hope is that this new 'political awakening' will endure and transform how scientists participate in political life," wrote former New York City Health Commissioner Mary Bassett in a viewpoint article for *Nature*. "The label 'activist' should be an honor, not a slur or reproach" (Bassett, 2020). Many experts are starting new advocacy organizations on the own campuses and in their own communities and joining together through sciencerising.org, a collaborative effort to support the ability of scientists to build power and engage in democracy and governance.

There has never been a better time for scientists to speak truth to power because the public square has eroded. A weakened media and congressional oversight infrastructure makes it more difficult for meaningful and thorough oversight. There are far fewer reporters with far less time to report on a story. Investigations that may have been afforded months in the past are now given considerably fewer resources. Legislative and federal agency staff are overwhelmed and, every year, asked to do more with less. Intentional misinformation sows confusion, creates chaos, and makes effective governance more difficult. Until these trends are reversed, subject-matter experts must step into provide context as well as research and investigative capacity.

Further, many government ethics policies depend on the good faith of those who implement them—that action will be taken when malfeasance is found. Yet there are fewer public and personal consequences for engaging in troublesome behavior. In response, we must all be the truth tellers.

For decades, UCS had engaged scientists to advocate on specific issues, from nuclear safety and security to global warming through its Science Network. UCS expanded that mandate in the early 2000s to engage scientists to defend the scientific enterprise itself. Now, the organization is helping scientists build power and apply it to the issues they care about, and to advance a culture of advocacy and engagement among scientists.

The network has evolved into an inclusive community of more than 25,000 scientists, engineers, economists, public health specialists, and other experts across the country working to educate the public and inform decisions critical to our health, safety, and environment. Complementing this work are scores of university and other science policy organizations that have sprouted over the past several years.

So many of the complex issues we face—from climate change to food security to pandemics—rely heavily on access to the best available scientific information. It is more important than ever before for scientists to build power and use their standing to advocate for the public interest and ensure that science plays a central role in both personal and governmental decisions.

Scientists and their allies can find their place in this movement at ucsusa.org/sciencenetwork, where they can explore direct engagement through the UCS and build their own skills to have an impact in their own communities.

The future of science advocacy is bright and robust. And not a moment too soon.

References

Acrivos, et. al., 2017. Science and the public interest: an open letter to President-Elect Trump and the 115th Congress. Accessed December 3, 2020 at https://www.ucsusa.org/open-letter-president-trump-and-115th-congress.

Bailin, D., Carlson, C., Halpern, M., Huertas, A., Kothari, Y., 2015. Mediated access: transparency barriers for journalists' access to scientists and scientific information at government agencies. Accessed December 3, 2020 at https://www.ucsusa.org/sites/default/files/attach/2015/07/ucs-mediated-access-report-2015.pdf.

Barringer, F., 2007. Interior official steps down over rules violation. *New York Times*. Accessed December 3, 2020 at https://www.nytimes.com/2007/05/02/washington/02interior.html.

Bassett, M.T., 2020. Tired of science being ignored? Get political. Nature 586, 337. Accessed December 3, 2020 at https://www.nature.com/articles/d41586-020-02854-9.

Berg, J., Campbell, P., Kiermer, V., Raikhel, N., Sweet, D., 2018. Joint statement on EPA proposed rule and public availability of data. Science 371, Accessed December 3, 2020 at https://science.sciencemag.org/content/360/6388/eaau0116.

Berman, E., Carter, J., 2018. Scientific integrity in federal policymaking under past and President administrations. J. Sci. Policy Gov. 13, Accessed December 6, 2020 at https://www.sciencepolicyjournal.org/uploads/5/4/3/4/5434385/berman_emily__carter_jacob.pdf.

Biden, J., 2021. Memorandum on restoring trust in government through scientific integrity and evidence-based policymaking. Accessed March 20, 2021 at https://www.whitehouse.gov/briefing-room/presidential-actions/2021/01/27/memorandum-on-restoring-trust-in-government-through-scientific-integrity-and-evidence-based-policymaking.

Boyle, L., 2020. NOAA chief scientist fired for asking new Trump hires to recognize scientific integrity policy. *The Independent*, 30 October 2020.

Bush, G.H.W., 1990. Remarks to the National Academy of Sciences. 23 April 23 1990.

Carter, J., Goldman, G., Johnson, C., 2018. Science under Trump: voices of scientists across 16 federal agencies. Union of Concerened Scientists. Accessed December 6, 2020 at https://www.ucsusa.org/resources/attacks-on-science.

Cartlidge, E., 2016. Seven-year legal saga ends as Italian official is cleared of manslaughter in earthquake trial. Science 371, Accessed December 3, 2020 at https://www.sciencemag.org/news/2016/10/seven-year-legal-saga-ends-italian-official-cleared-manslaughter-earthquake-trial.

Davis, J., 2010. NOAA on the BP Gulf oil blowout: is this any way to communicate science? Clim. Sci. Policy Watch. Accessed December 3, 2020 at http://www.climatesciencewatch.org/2010/10/27/noaa-on-the-bp-oil-blowout-is-this-any-way-to-communicate-science/.

Dean, C., 2006. At scientific gathering, U.S. policies are lamented. *New York Times*. Accessed December 3, 2020 at https://www.nytimes.com/2006/02/19/us/at-a-scientific-gathering-us-policies-are-lamented.html.

Department of Interior Inspector General, 2006a. Report of investigation—Julie MacDonald. Accessed December 3, 2020 at https://www.doioig.gov/reports/report-investigation-julie-macdonald.

Department of Interior Inspector General, 2006b. Report of investigation: Julie MacDonald. Accessed December 6, 2020 at https://www.doioig.gov/reports/report-investigation-julie-macdonald.

Devaney, T., 2016. Salt lobby warns sodium reduction will 'endanger public health.' Accessed December 3, 2020 at https://thehill.com/regulation/healthcare/281914-salt-lobby-warns-sodium-reduction-will-endanger-public-health.

Driscoll, D., Garrard, G., Kusmanoff, A., Dovers, S., Maron, M., Preece, N., et al., 2020. Consequences of information suppression in ecological and conversation sciences. Conserv. Lett. Accessed December 3, 2020 at https://conbio.onlinelibrary.wiley.com/doi/10.1111/conl.12757.

Dupuis, J., 2014. The Canadian war on science: a long, unexaggerated, devastating chronological indictment. ScienceBlogs Accessed December 3, 2020 at https://scienceblogs.com/confessions/2013/05/20/the-canadian-war-on-science-a-long-unexaggerated-devastating-chronological-indictment.

Eilperin, J., 2006. Bush appointee said to reject advice on endangered species. *Washington Post*. Accessed December 20, 2020 at https://www.washingtonpost.com/wp-dyn/content/article/2006/10/29/AR2006102900776.html.

Eilperin, J., 2019. EPA's scientific advisors warn its regulatory rollbacks clash with established science. *Washington Post*. Accessed December 3, 2020 at https://www.washingtonpost.com/climate-solutions/epas-scientific-advisers-warn-its-regulatory-rollbacks-clash-with-established-science/2019/12/31/a1994f5a-227b-11ea-a153-dce4b94e4249_story.html.

Escobar, H., 2019. Brazilian government accused of suppressing data that would call its war on drugs into question. Accessed December 3, 2020 at https://www.sciencemag.org/news/2019/06/brazilian-government-accused-suppressing-data-would-call-its-war-drugs-question.

Fainaru-Wada, M., Fainaru, S., 2014. League of Denial: The NFL, Concussions, and the Battle for Truth. Penguin Random House, New York.

Flavelle, C., Allison, B., Dlouhy, J., 2018. "New EPA chief on collision course over conflicts of interest. Accessed December 3, 2020 at https://www.bloomberg.com/news/articles/2018-07-06/new-epa-chief-can-t-avoid-conflicts-of-interest-watchdogs-warn.

Freeman, A., J. Samenow, 2020. New emails show how President Trump roiled NOAA during Hurricane Dorian. Washington Post, 1 February 2020.

Friedman, L., 2019. E.P.A. to limit science used to write public health rules. New York Times, 11 November 2019.

Friedman, L., 2018. E.P.A. to disband a key scientific review panel on air pollution. New York Times. Accessed March 23, 2021 at https://www.nytimes.com/2018/10/11/climate/epa-disbands-pollution-science-panel.html.

Frey, C., et. al., 2019. Advice from the independent particulate matter review panel on EPA's policy assessment for the review of the National Ambient Air Quality Standards for particulate matter. Accessed December 3, 2020 at https://ucs-documents.s3.amazonaws.com/science-and-democracy/IPMRP-FINAL-LETTER-ON-DRAFT-PA-191022.pdf.

Frey, C., et al., 2020. The need for a tighter particulate-matter air quality standard. N. Engl. J. Med. 383, 680–683. Accessed December 3, 2020 at https://www.nejm.org/doi/10.1056/NEJMsb2011009.

Goldman, G., Carlson C., 2014. Tricks of the trade: how companies anonymously influence climate policy through their business and trade associations. Union of Concerned Scientists. Accessed December 3, 2020 at https://www.ucsusa.org/resources/tricks-trade.

Goldman, G., Christopher Frey, H., Bachmann, J., Zarba. C., 2020. "We put science back into EPA air pollution standards, but.... Scientific American. Accessed December 3, 2020 at https://blogs.scientificamerican.com/observations/we-put-science-back-into-epa-air-pollution-standards-but/.

Gottfried, K., et. al., 2004. Scientist statement on restoring scientific integrity to federal policy making. Accessed December 3, 2020 at https://www.ucsusa.org/resources/2004-scientist-statement-scientific-integrity.

Halpern, M., 2020. EPA refused to hold a hearing on its science rule, so we held it for them. Accessed December 3, 2020 at https://blog.ucsusa.org/michael-halpern/epa-refused-to-hold-hearing-so-we-held-it-for-them.

Holdren, J., 2010. Memorandum for the Heads of Executive Departments and Agencies. Accessed December 3, 2020 at https://obamawhitehouse.archives.gov/sites/default/files/microsites/ostp/scientific-integrity-memo-12172010.pdf.

House Committee on Natural Resources, 2007. Endangered Species Act implementation: science or politics? Accessed December 3, 2020 at https://www.govinfo.gov/content/pkg/CHRG-110hhrg35221/html/CHRG-110hhrg35221.htm.

House Committee on Oversight and Government Reform, 2003. Politics and science in the Bush administration. Accessed December 3, 2020 at https://repository.library.georgetown.edu/handle/10822/998782.

House Committee on Oversight and Government Reform, 2007a. Allegations of political interference with the work of government climate change scientists. Accessed December 3, 2020 at https://www.govinfo.gov/content/pkg/CHRG-110hhrg34913/html/CHRG-110hhrg34913.htm.

House Committee on Oversight and Government Reform, 2007b. Political interference with climate change science under the Bush administration. Accessed December 3, 2020 at https://www.hsdl.org/?abstract&did = 481710.

Huertas, A., 2018. A year of protest and power. Saying no and getting to yes. Accessed December 3, 2020 at https://aaronhuertas.medium.com/a-year-of-protest-and-power-ea219955771e.

Kennedy, J.F., 1963. Address at the anniversary convocation of the National Academy of Sciences. 22 October 1963.

Maioni, A., 2015. We haven't forgotten the long-form census. Toronto Globe and Mail. Accessed December 3, 2020 at https://www.theglobeandmail.com/opinion/we-havent-forgotten-the-long-form-census/article22819338/.

Malakoff, D., 2021. Death of EPA's controversial 'censored science' rule delights researchers. Science. Accessed March 20, 2021 at https://www.sciencemag.org/news/2021/02/death-epa-s-controversial-censored-science-rule-delights-researchers.

McNutt, M., Mote, C.D., Dzau, V., 2018. "Re: strengthening transparency in regulatory science. Accessed December 2, 2020 via Internet Archive at https://tinyurl.com/y7ja7gef.

Merriam-Webster Dictionary, 2020. Definition of "politics". Accessed December 3, 2020 at https://www.merriam-webster.com/dictionary/politics.

Michaels, D., 2020. The Triumph of Doubt. Oxford University Press, New York.

Mooney, C., 2005. The Republican War on Science. Basic Books, New York, NY.

Nature, 2012. Death of evidence. Nature 487, 271–272 Accessed December 3, 2020 at. Available from: https://doi.org/10.1038/487271b.

National Academy of Public Administration (NAPA), 2020. An independent assessment of allegations of scientific misconduct. Accessed December 3, 2020 at https://nrc.noaa.gov/Scientific-Integrity-Commons.

Niiler, E., 2019. The EPA's anti-science 'transparency' rule has a long history. *Wired*. Accessed December 3, 2020 at https://www.wired.com/story/the-epas-anti-science-transparency-rule-has-a-long-history/.

Obama, B., 2009a. President Barack Obama's inaugural address. Accessed December 3, 2020 at https://obamawhitehouse.archives.gov/blog/2009/01/21/president-barack-obamas-inaugural-address.

Obama, B., 2009b. Memorandum for the Heads of Executive Departments and Agencies. Accessed December 3, 2020 at https://obamawhitehouse.archives.gov/blog/2009/01/21/president-barack-obamas-inaugural-address.

Owens, B., 2018. Half of Canada's government scientists still feel muzzled. Science 371, Accessed December 3, 2020 at https://www.sciencemag.org/news/2018/02/half-canada-s-government-scientists-still-feel-muzzled.

PIPSC, 2013. The Big Chill: silencing public interest science. Professional Institute of the Public Service of Canada. Accessed December 3, 2020 at https://www.pipsc.ca/portal/page/portal/website/issues/science/bigchill.

Reilly, S., 2020a. Wheeler sets new policy on advisory panel decisionmaking. *E&E News*. Accessed December 3, 2020 at https://www.eenews.net/eenewspm/stories/1062456481.

Reilly, S., 2020b. EPA gives up on barring grantees from science advisory panels. *E&E News*. Accessed December 3, 2020 at https://www.sciencemag.org/news/2020/06/epa-gives-barring-grantees-science-advisory-panels.

Sabin Center, 2020. Silencing science tracker. Accessed December 3, 2020 at https://climate.law.columbia.edu/Silencing-Science-Tracker.

Science Debate, 2008. Presidential election questions. Accessed December 3, 2020 at https://sciencedebate.org/sciencedebate-presidential-2008.html.

Scientific Integrity Act, 2019. H.R. 1709. 116th Cong. 2019. Accessed December 3, 2020 at https://www.congress.gov/bill/116th-congress/house-bill/1709/text.

Scientific Integrity in Federal Agencies, 2019. 116th Cong. 2019. Accessed December 3, 2020 at https://science.house.gov/hearings/scientific-integrity-in-federal-agencies.

Showstack, R., 2011. NOAA issues scientific integrity policy. EOS 92, Accessed December 3, 2020 at https://agupubs.onlinelibrary.wiley.com/doi/pdf/10.1029/2011EO500004.

Siler, W., 2019. The David Bernhardt scandal tracker. 19 April 2019. Outside. Accessed December 3, 2020 at https://www.outsideonline.com/2390596/david-bernhardt-scandal-tracker.

Sinks, T., 2020. The final strengthening transparency in regulatory science (STRS) rule differing scientific opinion. Accessed December 3, 2020 at https://int.nyt.com/data/documenttools/dissenting-scientific-opinion/8fdd7838c67f4c21/full.pdf.

Society of Professional Journalists et al., 2014. Letter urges President Obama to be more transparenrt. Accessed December 3, 2020 at https://www.spj.org/news.asp?ref = 1253.

Tollefson, J., 2015. US science agency refuses request for climate records. Nature Accessed December 3, 2020 at https://www.nature.com/news/us-science-agency-refuses-request-for-climate-records-1.18660.

Union of Concerned Scientists, 2006. Systematic interference with science at Interior Department exposed. Accessed December 3, 2020 at https://www.ucsusa.org/resources/systematic-interference-science-interior-department-exposed.

Union of Concerned Scientists, 2009. Abuses of science: case studies. Accessed December 3, 2020 at https://www.ucsusa.org/resources/abuses-science.

Union of Concerned Scientists, 2005. Survey: U.S. Fish and Wildlife Service scientists. Accessed December 3, 2020 at https://www.ucsusa.org/resources/survey-us-fish-wildlife-service-scientists.

Union of Concerned Scientists, 2017. The disinformation playbook. Accessed December 3, 2020 at https://www.ucsusa.org/resources/disinformation-playbook.

Union of Concerned Scientists, 2020. Attacks on science. Accessed March 20, 2021 at https://www.ucsusa.org/resources/attacks-on-science.

Union of Concerned Scientists, 2004. Scientific integrity in policymaking. Accessed March 20, 2021 at https://www.ucsusa.org/resources/scientific-integrity-policy-making-0.

Waldman, S., Farah, S., 2018. Smith pitched Pruitt on 'secret science.' Now it's happening. E&E News. Accessed March 20 at https://www.eenews.net/stories/1060079655.

Wei, Y., Wang, Y., Di, Q., Choirat, C., Wang, Y., Koutrakis, P., et al., 2019. Short term exposure to fine particulate matter and hospital admission risks and costs in the Medicare population: time stratified, case crossover study. BMJ 367, l6258. Available from: https://doi.org/10.1136/bmj.l6258.

Wu, X., Braun, D., Schwartz, J., Kioumourtzoglou, M.A., Dominici, F., 2020. Evaluating the impact of long-term exposure to fine particulate matter on mortality among the elderly. Science Advances. EABA5692. Available from https://advances.sciencemag.org/content/6/29/eaba5692.

Why advocate—and how?

Robert M. Hughes[1,2], Robert L. Vadas, Jr.[3], J. Hal Michael, Jr.[4], Beverly E. Law[5], Arthur C. Knutson, Jr.[6], Dominick A. DellaSala[7], Jim Burroughs[8] and Hal Beecher[9]

[1]Amnis Opes Institute, Corvallis, OR, United States [2]Oregon State University, Corvallis, OR, United States, [3]Unaffiliated, [4]Ecologists Without Borders, Olympia, WA, United States [5]Oregon State University (Professor Emeritus), Corvallis, OR, United States [6]California Department of Fish and Wildlife (Retired), Sacramento, CA, United States [7]Wild Heritage, A Project of Earth Island Institute, Berkeley, CA, United States [8]Oklahoma Department of Wildlife Conservation, Oklahoma City, OK, United States [9]Washington Department of Fish and Wildlife (Retired), Olympia, WA, United States

Why advocate

Citizens, scientists, and managers have an ethical obligation to advocate for improved natural resource management, particularly when science is being ignored or altered (Karr, 2006; Nelson and Vucetich, 2009; Wood, 2014; Ripple et al., 2017). This is especially true for natural resource scientists and managers during this time of unprecedented planetary-scale changes (Crutzen and Stoermer, 2000) in which human activities have markedly transformed Earth's climate (Masson-Delmotte et al., 2019), hydrology (Poff et al., 2007), biodiversity and ecosystem services (Sullivan et al., 2006; Grossman and Parker, 2012; Diaz et al., 2019), and earth-moving processes (Hooke, 2000). Speaking truth to power is needed when mythology and profit contests with science in decision-making (Scheufele and Krause, 2019), as is the case when politics largely ignore scientific information (Lepore, 2018). And it should be our professional mantra whenever cut-backs in federal and state agency budgets and staffing are taking place (Kelderman et al., 2019). However, there are often professional consequences for questioning dominant paradigms, policies, or programs that scientists need to prepare for as discussed in this chapter (also see Chapter 2, When Scientists are Attacked: Strategies for Dissident Scientists and Whistleblowers).

For example, scientists affiliated with federal agencies have made personal attacks on nongovernmental organization (NGO) scientists who have published research questioning federal land management policies pertaining to wildfire effects on wildlife (Peery et al., 2019; Chapter 2, When Scientists are Attacked: Strategies for Dissident Scientists and Whistleblowers). Indeed, some argue that scientists and

Conservation Science and Advocacy for a Planet in Peril.
DOI: https://doi.org/10.1016/B978-0-12-812988-3.00015-6

managers should not engage in policy development because such activities impugn their impartiality, instead believing that knowledgeable people should simply stick to the science and leave such decisions to political decision makers (e.g., Cohn, 2005; Lackey, 2007; Wilhere, 2012; Boreman, 2013; Reiser, 2017; Chapter 1, The Nuts and Bolts of Science-Based Advocacy). We disagree with this position because the current state of the planet requires individuals with scientific training to provide expertise in outcomes affecting nature and humans. To do otherwise is professionally irresponsible, unethical, and contradictory to our public missions (Nelson and Vucetich, 2009). In other words, just as epidemiologists, virologists, geneticists, doctors, and nurses have obligations to broadcast their concerns about the coronavirus pandemic and other human health threats, natural resource professionals have comparable obligations to voice and publish their concerns about the effects of anthropogenic pressures and stressors on local landscapes and the planet writlarge. In addition, the Trump administration's appointments of corporate lobbyists and attorneys to agency leadership positions and judges has led to increasing environmental and social justice losses and long-term threats to both (Lerner, 2020). Remaining silent in the face of such attacks makes us complicit.

Given widespread obstacles to science-based advocacy, we convened a special symposium at the joint 2019 conference of the American Fisheries Society and The Wildlife Society to discuss options for ethical environmental practitioners. The objective of that symposium and this chapter is to offer advice from professionals working in different institutions concerning how to successfully contend with one's ethical obligations to advocate for what is right while remaining gainfully employed. We define advocacy as:

> "the act or process of supporting a cause or proposal" (Merriam Webster).

In this context, we propose that the underlying advocacy rationale is the perpetuation of nature in an ever more human-dominated world (Harari, 2015). In particular, we are concerned with how scientists can engage in policies affecting the natural world while maintaining their credibility and standing within their respective institutions whether they work for a natural resource agency, industry, university, or NGO. In this chapter, we sprinkle in *Key Takeaways* and short stories to call attention to concepts from our collective experiences as practitioners and science-based advocates.

> *Key Takeaway*: *Be Clear and Honest*. The public greatly appreciates clear and honest statements by health care professionals when their science supports changes in health care policies. They should appreciate similar statements from natural resource professionals.

Although this may seem like a straightforward obligation of trained professionals, we begin with the rhetorical question: Why is science-based advocacy such a touchy subject in natural resource science and management? We think that there are several fundamental reasons why many scientists avoid science-based advocacy.

Ideas are routinely dismissed that conflict with those in charge

At least since Copernicus and Galileo, institutions have actively or implicitly discouraged ideas, comments, or questions that run counter to business as usual from moving up management lines to top decision makers (Bella, 1992). Notably, institutional leaders often have different values than their scientific employees and therefore prioritize political gains or losses over scientific information.

> *Key Takeaway: Learn How to Overcome Obstacles.* The US Environmental Protection Agency (EPA) scientists proposed incorporating biological and physical habitat indicators with landscape conditions into local and national water body monitoring and assessment programs. The EPA did not begin national surface water assessments until a decade or more elapsed. Why? Although such programs are expensive and complex, EPA had the money. The major resistance came from middle and upper level bureaucrats in EPA and state agencies who felt threatened by new ideas and the prospect that some states and regions would appear to be in much worse condition than others in national assessments. Those obstacles were overcome by persistent publication of research results, service on technical committees, and presentations at professional society meetings and state and regional workshops.

Another example of dismissing science is when agency scientists recommend endangered species listings or environmentally proper stream flows. Often, their science is scrubbed, and their advice is overruled by decision makers for socioeconomic or political reasons (Vadas and Weigmann, 1993; Cohn, 2005; Sullivan et al., 2006; Carroll et al., 2012; Chapter 6, Overcoming the Politics of Endangered Species Listings). Unfortunately, such science dismissal has resulted in multiple catastrophic failures as follows:

- NASA decided to launch the Challenger shuttle despite warnings from engineers that cold weather might lead to O-ring failure (Dalal, 2016).
- Exxon denied decades of climate change threats despite its own studies and others to the contrary (Banerjee et al., 2015).
- The CIA misused psychological findings for its torture program to obtain questionable information from suspected terrorists, disregarding warnings from professional psychologists (Singal, 2014).

- Boeing decided to sell 737 Max aircraft despite warnings from its own staff that its antistall software overrode pilot controls (Rivers, 2019).
- Governments have failed to address climate change, human population growth, and overconsumption of natural resources despite decades of warnings from scientists (Vitousek et al., 1997; Mckee et al., 2004; Sullivan et al., 2006; Ripple et al., 2017; Diaz et al., 2019; Masson-Delmotte et al., 2019).
- Many people in the United States widely denied the potential national and global threats of inadequate health insurance (Stack, 2018), pandemics and climate change (Krugman, 2020; Lancet, 2020; Zakaria, 2020) on human health, mortality, and economies—despite multiple warnings to the contrary. The United States leads the world in Covid-19 deaths and infection rates (Sanger et al., 2020) as a result of a science-denying president (Baker, 2020) who has promoted a dangerous, unscientific approach called "herd immunity" (Alwan et al., 2020).
- The US Department of Energy blocked presentations, publications, and implementation of a proposal to upgrade the US electrical grid system to make electricity less expensive and more reliable because it would favor green energy sources over coal (Fairley, 2020). Subsequently, California experienced rolling blackouts in a heat wave and Texas experienced a multiday power outage during a winter snowstorm.
- Scientists have been warning for years that many western United States towns were ill-prepared for increased climate-induced wildfires (Grossman and Parker, 2012), yet local, state, and federal policies promote logging in the backcountry and housing in the wildland-urban interface that do nothing to protect homes or communities from wildfire (DellaSala and Hanson, 2015).

Fear of "Rocking the Boat"

A key road to success in almost any institution is simply to do our jobs as well as possible, stay focused only on our work, avoid considering its greater repercussions, and do not rock-the-boat by challenging status quo thinking (Bella, 2006). This process has led to otherwise moral physicians developing biological warfare weapons (Bella, 2006), military psychologists developing a torture program for suspected terrorists (Singal, 2014), software engineers producing election disinformation (Barrett, 2019), robotic engineers displacing the middle class through automation (Harris et al., 2018), parochial compartmentalization of natural resource research and management instead of an ecosystem approach (Bigford, 2015; Michael, 2017; Callisto et al., 2019), widespread natural resource mismanagement (Sullivan et al., 2006; Wood, 2014), and destruction of nature globally (Leopold, 1949; Ripple et al., 2017).

> *Key Takeaway: Consider the Ethical Consequences of Just Going with the Flow When the Science Says Otherwise.* Think about what you would tell your
>
> *(Continued)*

(Continued)

grandchildren if they were to ask sometime in the future why you allowed such degradation of the resources that you were being paid to conserve—thereby limiting their enjoyment of those resources.

Decision makers often prioritize inhouse science over objective scientific approaches

Most natural resource agencies fund inhouse and academic research oriented toward current management practices rather than testing alternative hypotheses and practices, which would be more appropriate in a scientifically oriented adaptive management approach (e.g., Sullivan et al., 2006; Maas-Hebner et al., 2016; Hanson et al., 2018). As a result, staff and grant funding from such agencies can involve implicit or explicit political expectations of affiliated researchers, meaning that much agency and academic research is focused on supporting current management paradigms rather than regularly questioning them (i.e., confirmation bias, Vadas, 1994; Chapter 1, The Nuts and Bolts of Science-Based Advocacy, Chapter 2, When Scientists are Attacked: Strategies for Dissident Scientists and Whistleblowers). Staff are then rewarded with opportunities to present at professional meetings if the presentation advances the mission of the agency, as defined by managers. For such situations, effective advocacy is critical for avoiding scientific distortion (Nelson and Vucetich, 2009).

Key Takeaway: Make Sure You Include All the Relevant Science and Not Just What's Inhouse. Current search engines and professional societies offer multiple options for discovering alternative insights.

Going against the status quo has career consequences

People naturally resist change and different ways of thinking (i.e., cognitive dissonance) whether they are scientists (Kuhn, 1962; Vadas, 1994; Nelson and Vucetich, 2009) or humans in general, preferring the status quo (Lazer et al., 2018) and prevailing mythologies (Lepore, 2018). This makes questioning of those dogmas and practices unbelievable, if not heretical. For example, Galileo's scientific defense of heliocentrism versus geocentrism led to his trial by the Inquisition, forced denial of his theory, and a house arrest life-sentence (DellaSala, Preface). Ruch (2017) documents several cases where federal employee whistleblowers were suspended or fired when they protested scientific mismanagement by supervisors (who were not reprimanded, also see

Chapter 7, Scientific Integrity and Advocacy: Keeping the Government Honest).

> *Key Takeaway*: *Challenge the Status Quo When It Conflicts with the Science.* Be careful, tactful, and prepared for the potential consequences when doing so. Have a Plan B.

Decision makers have difficulty acknowledging the problem even exists

Both individuals and institutions (from agencies to nations) have difficulty in acknowledging there is a problem, accepting responsibility for that problem (rather than blaming others), identifying what needs changing, and selecting effective ways of changing it (Lepore, 2018; Diamond, 2019). A classic example of this denial was the perspective of Donald Trump (former President) and several state governors and city mayors regarding the seriousness of the coronavirus pandemic. Months after the virus was initially detected and control efforts implemented in multiple nations, the numbers of the US cases and deaths continued rising. Science denial and pushing the virus problem aside, the modus operandi of the Trump administration, was at the core of an explosion of exposure and mortality cases.

> *Key Takeaway*: *Decision Makers Need to Admit When They are Wrong and Change Direction Accordingly.* You can help them by continuing to clearly and tactfully present your science, as well as its policy and potential political implications, in multiple outlets.

Scientists have poor communication skills that limit their ability to influence change

Scientists often do a poor job of communicating their findings with the public in a manner that engenders understanding of an issue (Chapter 1, The Nuts and Bolts of Science-Based Advocacy), let alone the confrontational nature of science whereby ideas are continually being tested, retested, and revised—unlike myths that are often accepted as fact (Sullivan et al., 2006). The appearance of dueling science does little to help solve this communication problem (Peterson, 2017; Chapter 1, The Nuts and Bolts of Science-Based Advocacy). Scientists are trained to speak and write for other scientists. Far too few of us learn how to clarify and condense our science for presentations in public schools, community meetings, and legislative hearings (Callisto et al., 2019; França et al., 2019).

Key Takeaway: Learn How to Communicate Beyond Preaching to the Choir. Practice by giving short presentations in public schools and before local civic clubs that welcome speakers.

Others casting doubt blunts the effect of your work in enacting change

Corporations fund research and publications that are deliberately designed to cast doubt on the science that would lead to governmental regulation (Michaels, 2020). Their lobbyists further that doubt in governmental legislative, judicial, and executive bodies (Wood, 2014; Oreskes and Conway, 2010; MacLean, 2017; Mayer, 2016; Michaels, 2020), which makes it difficult for unbiased science to be separated from biased science or mythology in a timely and unconfounded manner.

Key Takeaway: Document whether the science was independently obtained without ties to political motives or preconceived outcomes. Determine where the science was published, whether it was peer-reviewed, and its funding sources (Chapter 1, The Nuts and Bolts of Science-Based Advocacy, Chapter 2, When Scientists are Attacked: Strategies for Dissident Scientists and Whistleblowers).

A desire for fame and fortune gets in the way of doing the right thing

Scientists themselves are sometimes their worst enemies. Scientific and management advances would likely occur more rapidly if one's desire for fame and fortune did not deter cooperative research. Although scientific debates are healthy components of the scientific process, when scientists spend time publicly debating minor shortcomings of others' work it can lead to distrust of all scientists by the general public and managers (as in Chapter 2, When Scientists are Attacked: Strategies for Dissident Scientists and Whistleblowers).

Key Takeaway: *Check your ego at the door.* Learn to work as a collaborative team member and be tactful and respectful when questioning others' research and perspectives. Avoid becoming offended when your research and viewpoints are questioned.

How to advocate

What follows in this section are experiential events and ways to get involved from the collection of different perspectives (federal agencies, state agencies, universities, NGOs, and environmental consultants) of chapter authors.

Insights from federal agency employees

Implementing new scientific concepts typically takes decades to become acceptable practice at many institutions and is resisted when supervisors have different goals than scientists (Cohn, 2005; Sullivan et al., 2006). This concept, referred to as prematurity in scientific discovery, is well laid out in Hook (2002). Effective measures for accelerating acceptance of such concepts include: collaborating on research with scientists working for other agencies, universities, private consultants, industries, and NGOs; presenting research results at professional meetings; publishing in scientific journals; serving on state, federal, and international technical committees (Cohn, 2005); and assuming leadership roles in professional societies (Boreman, 2013). Choose your journal carefully before submitting so as to increase its publication probability because editors and reviewers also have professional biases (Vadas, 1994; Chapter 2, When Scientists are Attacked: Strategies for Dissident Scientists and Whistleblowers). It is also important to clearly distinguish scientific from policy recommendations, to be tactful and not surprise supervisors with presentation and publication contents, and to spend time resetting oneself in nature. However, leadership perspectives often change fundamentally during one's career at an agency, meaning that once-appropriate publications, professional presentations, and media interviews may eventually run counter to the altered agency viewpoint. This can lead to employee gag orders, reprimands, suspensions, narrowing of duties, substandard office facilities, and requiring prior permission for all outside communication (Sullivan et al., 2006; Chapter 7, Out of the Ivory Tower: Campaign-Based Science Messaging for the Public). Employees must then endure those restrictions until reaching retirement status (Boreman, 2013), seek employment elsewhere, or resist such restrictions with potential risks to continued employment. Because a colleague may eventually become your supervisor, try not to burn bridges unnecessarily during your career.

> *Key Takeaway*: Avoid spending agency time and equipment on activities that are only tangentially related to your official job duties. Instead, conduct these activities before and after work, during lunch and coffee breaks, and on weekends and holidays. Seriously ramp up those activities during your retirement. They help keep your mind fit.

Insights from state agency employees

Because outside influences (political, stakeholder, interagency conflicts, funding) affect agency decisions, be constructive and tactful in recommendations to supervisors and be willing to heed constructive criticism despite their often-limited ecological knowledge. Avoid angry outbursts and wise cracks that can lead to disciplinary action (Mottram, 1983) and be prepared for senior biologists and managers altering your conclusions given political pressures (Schuyler, 1975). Include management and conservation implications of research findings in publications, but obtain supervisor clearance before doing so, and document interpersonal conflicts in a logbook in case of future disciplinary or legal proceedings. Become a mentor for fellow colleagues as well as younger scientists and staff but be prepared for someone to turn on you when political winds change. Focus on being a problem solver versus a constant complainer. If you cannot present the issues that most concern you as an employee, do so as a member or leader in an environmental or professional organization—and continue doing so when you retire (it is good for the mind). Be willing to pay for publication, to attend meetings on your own funds, and attend on you own (documented as leave) time. Clearly separate when you are speaking as a representative of the agency and when you are speaking as a private citizen, especially when dealing with NGOs, industry, decision makers, and the public that know who employs you.

Key Takeaway: Avoid Surprising Supervisors. A state employee was invited to give a presentation on salmon recovery. His topic indicated that recovery of salmon populations could also result in substantial increases in nutrients, fecal coliforms, dog-poisoning flukes, and movement of salmon pathogens from strays. The presentation was challenging the audience to imagine that if we restored a salmon-based ecosystem, these were some of the consequences (all of which were documented in the literature or had been encountered in the field). He intended to present on his own time, without reference to his agency. Agency management demanded to see the presentation first. The next day he was told that someone else would make a different presentation and that he must restrict his work to his specific District. The Agency did not want to have people even consider alternatives to established policy. Two years later, wholly on his own time, he organized a session at a local professional society meeting where he delivered that prior presentation. At that meeting, he received an award for publishing his research results and was chastised by management for having received it because the award mentioned the agency for which he worked. A year later he retired. The lessons learned are to avoid embarrassing the employer, establish a network of friends and colleagues, be seen talking to like-minded Commissioners, and work very hard to have a widely known reputation for honesty.

It is especially important to share papers, articles, and opinion pieces with colleagues inside your agency and worldwide. Simply provide information and then follow the discussion, but use your personal email and home computer to do so. However, avoid sending constant barrages of emails because they may eventually be dismissed *en masse*. Science-based advocacy involves more than challenging management decisions; it involves educating people inside and outside your work unit. Know when to call for outside peer review or drop out of scientifically biased, interagency projects, to preserve your scientific integrity and sanity. Finally, serve as a peer reviewer for journal and book chapter manuscripts.

> *Key Takeaway: Do Not Advocate on Employer Time.* A state agency employee improperly called the White House on a state phone line because of concerns about human overpopulation, which resulted in a year-long pay-grade demotion. Subsequently, he found better ways to get involved. These included starting a county Planned Parenthood affiliate, serving on its board for 18 years, contributing many letters to the editor and guest editorials in his local newspaper, and writing an article on the effects of population growth on ecosystems for *Outdoor California* that the agency distributed widely to the public. The agency indicated that the article was a personal, not an agency, view, however.

Insights from university employees

Universities are the last bastions for unfettered research and sharing of results with the public in addition to doing so at professional meetings. Yet, as state and federal support for universities has waned, corporations have filled the gap. Corporate decisions are driven by short-term profits, and they lobby to control policy to suit their needs. When scientific research on natural resources is in the news and corporate leaders feel it may affect their spin or bottom-line, attacks begin by them and special interest groups (Nelson and Vucetich, 2009), particularly when your work is published in high-profile journals. Read Oreskes and Conway (2010) to help you better understand how and why this occurs.

When research is on topics that are likely to have policy and decision-making relevance, it is important to assemble a research team of coauthors who have strong expertise in the various areas to be covered in publications (e.g., Hossain et al., 2018). Know the relevant scientific literature, be thorough, and leave no stone unturned. Scientists are usually their own worst critics, yet additional reviews before submission to a journal are helpful. Once a publication is accepted, authors are encouraged to write a press release for public dissemination with clearly written findings and importance. Some universities have press offices, and your draft can be their starting point. Others require that administrators, from department chairs to presidents, review the press release. It is your responsibility to ensure the press release is accurate. You will have the opportunity to clarify with the press when they contact you. Prepare for the kinds of questions the public, corporations, and the press are likely to ask.

If your findings are challenged by industries or other institutions, bury them in the well-established scientific knowledge that preceded and supports your analysis. Take the moral high ground, be calm and professional, and know that it is your responsibility to communicate the science to help humanity make wise decisions to (1) minimize further impacts on ecosystems and (2) restore function for life on Earth.

> *Key Takeaway*: Become an active member of professional organizations, build your network of colleagues, and give presentations in a variety of venues, including to the public and public schools. Professional organizations, like the American Geophysical Union and the Ecological Society of America, provide training programs for communicating science. Take advantage of them early in, and throughout, your career. The US Carbon Cycle Science Program also aims to identify a viable way for carbon cycle science to have a more direct and useful contribution to policy and decision-making in the context of climate change. This goal was one of the central tenets of the community-based US Carbon Cycle Science Plan. Many academic scientists in the United States are part of this program. Become part of it. The American Fisheries Society (AFS) and 110 other aquatic scientific societies (2020) published a climate action statement. Read it and become part of the solution.

Insights from environmental consulting firm and nongovernmental organization employees

Politically motivated suppression of science by federal and state agencies as well as by universities is not new (although it may be more evident currently in the United States). This may result from distorting, starving, totally defunding, or colonizing the science with politicized ideas and individuals (Cohn, 2005; Sullivan et al., 2006). To minimize and expose science suppression, it is important to establish open communication links between government and independent scientists (e.g., NGOs and private consultants) so that both are aware of the best available science, as well as its limitations, which may not be evident to the agencies or the public (Sullivan et al., 2006; Dasgupta, 2017; Reiser, 2017; Chapter 1, The Nuts and Bolts of Science-Based Advocacy). Although management agencies tend to collaborate more closely with the industries that they manage (Hughes, 2014a), it is imperative for independent scientists to develop trusting working relationships with those same agencies and industries to the degree possible. It is equally important to clearly enunciate areas of scientific and policy consensus and differences to facilitate resolution of environmental problems by sharing unpublished research results, collaborating in joint workshops, and working as interinstitutional teams. Recognize that science is not the only essential factor in managing natural resources, given the importance of socioeconomic evaluation (Vadas and Weigmann, 1993; Dasgupta, 2017), but when communicating science to decision makers and the public be aware of your audience and keep it simple and short (Olson, 2020, Chapter 1, The Nuts and Bolts of Science-Based Advocacy).

Acknowledge where scientists may have apparent political or financial conflicts of interest and discuss such topics as institutional challenges, while avoiding personal attacks on other scientists in workshops, peer-review journals, or in social media (e.g., Vadas, 1994; Peery et al., 2019; Chapter 2, When Scientists are Attacked: Strategies for Dissident Scientists and Whistleblowers). Recognize that one's funding sources and employers may be perceived as sources of potential conflict of interest because of that institution's values, regardless of the quality of the science (Chapter 1, The Nuts and Bolts of Science-Based Advocacy). When working in markedly different cultures or nations, learn how the inherent local constraints, values, and desires differ from your own, and be flexible and embrace involvement and adaptations of your science by locals (Grossman and Parker, 2012). NGOs should avoid "evidence complacency," for which ideologic/intuitive, trial-and-error, or opportunistic (socio-politically and/or fund-wise) approaches are used, rather than available scientific evidence, effectiveness monitoring, and statistically rigorous analysis (Dasgupta, 2017). Expert judgement and anecdotes can be considered evidence-based if used transparently and systematically. Finally, be prepared for (1) institutions that overpromise the environmental benefits of their actions, when their mandates and expectations do not adequately address species and ecosystem limiting factors; and (2) character assassinations by decision makers that will disagree with your findings (Chapter 2, When Scientists are Attacked: Strategies for Dissident Scientists and Whistleblowers, Chapter 5, Blowing the Whistle on Political Interference: The Northern Spotted Owl).

Key Takeaway: Scientists should clearly state personal perspectives and values when delivering scientific findings in policy-sensitive decisions. It is best to state values and assumptions upfront to avoid any misunderstandings or mischaracterizations. For instance, if you support biodiversity conservation or climate change mitigation, state those values consistent with your findings as an objective scientist (Nelson and Vucetich, 2009; Chapter 1, The Nuts and Bolts of Science-Based Advocacy).

Science-based advocacy implications

Natural resource management agencies are some of our most complex governmental institutions because they incorporate the complexities of ecosystems, the socioeconomic and political (e.g., license-sale and legislative-funding) pressures altering both those ecosystems and their management institutions, the group dynamics of the agencies themselves, missions and underlying legal requirements, and ethical differences regarding how humans should relate to nature (Leopold, 1949; Hughes, 1997; Cohn, 2005; Sullivan et al., 2006; Lackey, 2007). Thus, attempts to separate natural and social sciences, as well as diverse natural resource disciplines, leads to over-simplification of critical issues and continued degradation of biodiversity and ecosystem services (e.g., Hossain et al., 2018; Callisto et al., 2019). Further, the information available to natural resource scientists is rarely clean or complete, which then brings into play how one weighs competing information. Likewise, ignoring employees that see things differently, rather than encouraging diverse perspectives, results in a calcified agency ill-suited to manage its staff, let alone the natural resources for which it is responsible (Lichatowich, 1992). Such scientific pluralism and inclusiveness (Paine, 1994; Vadas, 1994; Sullivan et al., 2006; Nelson and Vucetich, 2009) is particularly important today given current concerns with human diversification programs meant to better respect various cultural, racial, and gender perspectives (Potok, 2010; Wilkerson, 2020).

Natural resource management decisions are nearly always based on more than one line of evidence (Sullivan et al., 2006). Decision makers must weigh laws, regulations, treaties, politics, human health, fiscal and staff resources, and many other factors. Each of these has its own factual basis, benefits, and costs. As such, any given decision is neither purely wrong nor purely right from a management perspective (there are levels of uncertainty). Therefore, the impacts of some decisions are mostly borne by the natural resource, some by the resource exploiters, and some by competing stakeholders. Nonetheless, as natural resource scientists, it is mandatory that we clearly describe the likely outcomes that any given decision will have on the resources under our purview. Therefore, we must be allowed—if not mandated—to publicly present the scientifically defensible facts and implications that are indicated by our research.

Key Takeaway: Like physicians, our first obligations should be to our patients—the natural resources. If we fail to sustain them, current and future citizens, as well as natural resource management agencies, will suffer (Nelson and Vucetich, 2009; Hughes, 2014b).

(*Continued*)

(Continued)

Key Takeaway: Place Science-Based Advocacy in the Proper Context. An agency assistant director told employees that they should not advocate. An employee politely stated that their jobs actually mandated conditional science-based advocacy. For example, if an agency goal or mission is to fulfill (to the best of its ability) the goal or mission, then science is essential for advocating which path(s) are best followed to achieve that goal or mission. The assistant director agreed, indicating the importance of placing science-based advocacy in context.

We need to be able to present this information unencumbered by organizational gags, spins, threats, smear tactics, and misinformation (Nelson and Vucetich, 2009). In addition, as US citizens, regardless of employer, we have First Amendment rights of free speech as well as an obligation to advocate for outcomes that our knowledge and ethical codes direct. See Chapter 7, Scientific Integrity and Advocacy: Keeping the Government Honest, for whistleblower protections, Chapter 11, Essays from the Trenches of Science-based Activism, for speaking out as a government scientist, and Ruch (2017) for shortcomings when doing so. Again, this must be unencumbered by employers. Not speaking truth to power or the public essentially abrogates the free speech rights for which our predecessors sacrificed their lives or careers, and it is what most of the public thinks it is paying us to do (Nelson and Vucetich, 2009).

Key Takeaway: It is Your Constitutional Right (if not an essential obligation as a citizen) to Speak Out. "Congress shall make no law respecting an establishment of religion, or prohibiting the free exercise therefore; or abridging the freedom of speech, or the press; or the right of the people peaceably to assemble, and to petition the government for a redress of grievances" (The First Amendment of the US Constitution).

Nonetheless we must understand personal contradictions or disconnects in our science-based advocacy. For example, say you enjoy fly-fishing or catch-and-release fishing. Are you advocating for a fly-only or catch-and-release area because you want to fish there or because that particular method is the only way to offer a fishery in that area? Also, you must maintain the position that the needs of the resource are preeminent regardless of the management decision (i.e., to walk the talk). Decades ago, there was an intense internal discussion in an agency regarding a fishery; technical staff were uniformly opposed because the fishery would overharvest at least some of the stocks. The management decision was to open the fishery. Nonetheless, many of the vociferously opposed staff demonstrated their moral malleability (hypocrisy) and participated in that fishery.

Professional scientific societies like The Wildlife Society, Society for Freshwater Science, Society for Conservation Biology, Ecological Society of America, Consortium of Aquatic Scientific Societies, American Fisheries Society, American Institute of Fishery Research Biologists, and American Institute of Biological Sciences can visibly enhance their roles in encouraging ethical forms of advocacy by their members. One way of doing so is by explicitly stating in their mission statements that science-based advocacy is expected professional practice by members who are consultants and employees in natural resource agencies.

Given the vast differences in opinions and personal biases across institutions and expanding on the *Four Agreements* in Chapter 1, The Nuts and Bolts of Science-Based Advocacy, we propose a set principles that can be used by professionals seeking to advocate for science while retaining their jobs, credibility, and sanity.

1. *Do no harm to the resource or biodiversity.* As in the medical and judicial professions, natural resource professionals need to endorse an ethic that when push comes to shove, the perpetuation of the natural world is the underlining basis of engagement (Nelson and Vucetich, 2009). It simply is not just a job. Indeed, because minimally human-disturbed areas generally have higher biodiversity and ecological complexity, studying markedly human-altered, simplified ecosystems with reduced ecological interactions limits scientific understanding for effective resource management and conservation (Hughes and Noss, 1992; Paine, 1994; Hughes, 2019).
2. *Be true to the science.* This includes stating assumptions, uncertainties, possible alternative explanations, and limitations while advocating for the use of best available science, including logical arguments, definitive experiments, and scientifically appropriate interpretation, filtering, synthesis, and risk analysis (Vadas, 1994; Sullivan et al., 2006; Lackey, 2007; Nelson and Vucetich, 2009; Dasgupta, 2017; Hossain et al., 2018). Critiques of proposed or ongoing human activities are critical for assessing environmental impacts (Carroll et al., 2012), especially if they provide testable hypotheses subject to peer review before distribution (Nelson and Vucetich, 2009) to minimize scientific errors.
3. *Be aware of publishing and speaking restrictions.* Make sure that your funding source is open to your publishing or presenting results regardless of findings, including management implications. This is best accomplished by signing a legal agreement to this effect.
4. *Be willing to speak out.* The Whistleblower Protection Act of 1989 was enacted to protect the identities of federal employees who disclose

"Government illegality, waste, and corruption" from adverse consequences related to their employment, including demotions, pay cuts, or replacement (see Chapter 7, Scientific Integrity and Advocacy: Keeping the Government Honest and Ruch, 2017). State governments often have similar laws.

5. *Be mindful of confirmation biases or cognitive dissonance.* This means constantly challenging your own perspective by remaining open-minded to the potential for paradigm shifts as new information is produced (Vadas, 1994; Nelson and Vucetich, 2009; Chapter 2, When Scientists are Attacked: Strategies for Dissident Scientists and Whistleblowers). We can never have all the answers regarding the natural world and long-term engagement in a particular issue can lead to inherent biases that exclude competing explanations that do not fit the dominant paradigm.
6. *Avoid making personal attacks or snide remarks about or to others.* Scientists should attack the problem, not the person, and in all cases avoid character assassination (Vadas, 1994; Chapter 2, When Scientists are Attacked: Strategies for Dissident Scientists and Whistleblowers).
7. *Do not take it personally.* Speaking out means that you will be subject to the criticism of others. Always take the high road by advocating for the science and remember to have a thick skin because nothing in this profession comes easy, especially when speaking out (Lackey, 2007).
8. *Find allies.* Find and collaborate with others in your institution as well as members of the public and scientists in other institutions who are willing to support your positions in speaking and publications.
9. *Learn how to translate science for decision makers and the public.* Professionals need to be able to compress their findings to communicate with those who may be focused on markedly different concerns. This often means developing the communication skills to turn complex findings into simple constructs when providing summaries for the media and decision makers (Sullivan et al., 2006; Olson, 2020). But choose your words carefully to reduce the probability of being misquoted. Several guides are available to help scientists become effective communicators (e.g., AAAS (American Association for the Advancement of Science), 2020; Olson, 2020). Be available to the media and to the public, but keep your supervisor informed, in writing if necessary, of the conversation.
10. *Keep at it.* Conservation is a marathon not a sprint. Change takes time, although given the biodiversity, climate, and ecosystem services crises we now face, scientists need to increasingly engage in policies affecting them (Nelson and Vucetich, 2009).
11. *Develop the ability to convincingly argue against your preferred position.* Only by totally understanding the other side can one fully articulate their own.
12. *Remember to have a clear dividing line between your personal life and workplace, work computer, and work phone.* For example, if you visit fishing or hunting websites, do not use work resources to do so unless it is in direct

support of employer policy. Likewise, use your home computer for emails that are tangential to your work or that may appear controversial to managers. Demand that your peers respect this dichotomy.

> *Key Takeaway*: Supervisors must uphold high ethical standards and support whistleblowers.
>
> A federal employee worked on a project overseen by two different federal agencies. The federal managers at the worksite falsified work hours and data, drank alcohol excessively at lunch, and encouraged excessive alcohol consumption during field training. The employee formally complained to the offsite supervisor, but his testimony had little effect and he was targeted by his onsite managers as the whistleblower. He resigned for the sake of his ethics and sanity and took another job where written rules are enforced.

Speaking truth to power: closing thoughts

There is irrefutable scientific evidence that (1) the planet's life support systems are in rapid disrepair and (2) we have very little time remaining to reverse course (Crutzen and Stoermer, 2000; Hooke, 2000; Sullivan et al., 2006; Ripple et al., 2017; Diaz et al., 2019; Masson-Delmotte et al., 2019; AFS (American Fisheries Society) and 110 world aquatic scientific societies, 2020). Scientists of all types of employment (NGO, state, federal, consultants, etc.) need to work together to present conservation-based science to those in power and to society in general, to mitigate or solve the planetary crisis before it is too late. To do otherwise, is ethically deficient and avoids one's citizen responsibility by being complicit to the problems at hand (Nelson and Vucetich, 2009, Chapter 1, The Nuts and Bolts of Science-Based Advocacy). That complicity results in serious environmental, social (Ballash, 2016; Wilderson, 2020), esthetic, and scientific impacts to humanity (Paine, 1994). If educational, transparent advocacy (Nelson and Vucetich, 2009) cannot be done through one's employer, then scientists should consider undertaking it on their own or through a relevant environmental NGO. Fortunately, there are indications that use and respect of science will increase in the post-Covid world among many persons, because the health and economic results of ignoring science are increasingly evident (Holford and Morgan, 2020).

Acknowledgments

We thank Dan Dauwalter for allotting us a full day to discuss these issues at the 2019 AFS/TWS conference, and for presentations at that conference by Brian

Czech, Chris Frissell, Chad Hanson, Lee Miller, Sara O'Neal, Leanne Roulson, Mary Scurlock, Cleve Steward, and Tina Swanson. We also thank Bob Zeigler for access to newspaper articles on his scientific advocacy efforts and Clay Antieau (Washington Native Plant Society) for reviewing the draft manuscript. None of the authors have a conflict of interest, and the opinions expressed herein are those of the authors and do not necessarily represent those of their organizations.

References

AAAS (American Association for the Advancement of Science), Center for Engagement with Science & Technology. 2020. <https://www.aaas.org/programs/center-public-engagement-science-and-technology>.

AFS (American Fisheries Society) and 110 world aquatic scientific societies, Statement of world aquatic scientific societies on the need to take urgent action against human-caused climate change, based on scientific evidence. World Climate Statement. Bethesda, MD. 2020. <https://climate.fisheries.org/world-climate-statement>.

Alwan, N.A., et al., 2020. Scientific consensus on the COVID-19 pandemic: we need to act now. Lancet 396, e71–e72. <https://www.thelancet.com/journals/lancet/article/PIIS0140-6736(20)32153-X/fulltext>.

Baker, P., Trump infected: what we know and don't know. The New York Times. 2020. <https://www.nytimes.com/2020/10/02/us/politics/trump-infected-what-we-know.html>.

Ballash, H. (Ed.), 2016. Building Cities in the Rain: Watershed Prioritization for Stormwater Retrofits, 2016. Washington Department of Commerce Publication (6). <https://www.ezview.wa.gov/site/alias__1780/34828/default.aspx>.

Banerjee, N., L. Song, and D. Hasemyer, Exxon: the road not taken. Inside Climate News. 2015. <https://insideclimatenews.org/content/Exxon-The-Road-Not-Taken>.

Barrett, P.M., 2019. Disinformation and the 2020 Election: How the Social Media Industry Should Prepare. Center for Business and Human Rights, New York University. <https://issuu.com/nyusterncenterforbusinessandhumanri/docs/nyu_election_2020_report>.

Bella, D.A., Ethics and the credibility of applied science, in G. H. Reeves, D. L. Bottom, and M. H. Brookes (technical coordinators). Ethical questions for resource managers. U.S. Forest Service, General Technical Report PNW-GTR-288, Portland, OR. 1992, pp. 19–32.

Bella, D.A., 2006. Emergence and evil. ECO 8 (2), 102–115.

Bigford, T., Tom's top ten policy issues. American Fisheries Society Policy News. 2015. <https://fisheries.org/2015/07/toms-top-ten-policy-issues>.

Boreman, J., 2013. On behalf of the fish. Fisheries 38, 343.

Callisto, M., et al., 2019. A Humboldtian approach to mountain conservation and freshwater ecosystem services. Front. Environ. Sci. 7, 195. <https://www.frontiersin.org/articles/10.3389/fenvs.2019.00195/full>.

Carroll, C., Rohlf, D.J., Noon, B.R., Reed, J.M., 2012. Scientific integrity in recovery planning and risk assessment: comment on Wilhere. Conserv. Biol. 26, 743–745.

Cohn, J.P., 2005. After the divorce: improving science at federal wildlife agencies. BioScience 55, 10–14.

Crutzen, P.J., E.F. Stoermer, The "Anthropocene." Global Change Newsletter 41, 2000, pp. 17–18.

Dalal, N., 2016. The Space Shuttle Challenger Explosion and the O-Ring. Priceonomics, <https://priceonomics.com/the-space-shuttle-challenger-explosion-and-the-o>.

Dasgupta, S., Experience or evidence: how do big conservation NGOs make decisions? Mongabay Series: Conservation Effectiveness. 2017. <https://news.mongabay.com/2017/11/experience-or-evidence-how-do-big-conservation-ngos-make-decisions>.

DellaSala, D.A., Hanson, C.T., 2015. The Ecological Importance of Mixed-Severity Fires: Nature's Phoenix. Elsevier, New York, NY.
Diamond, J., 2019. Upheaval: Turning Points for Nations in Crisis. Little, Brown & Company, New York, NY.
Diaz, S., Settele, J., Brondizio, E., Ngo, 26 coauthors, Summary for policymakers of the global assessment report on biodiversity and ecosystem services of the Intergovernmental Science-Policy Platform on Biodiversity and Ecosystem Services. 2019. <https://www.ipbes.net/system/tdf/ipbes_7_10_add.1_en_1.pdf>.
Fairley, P., How a plan to save the power system disappeared. The Atlantic. 20 August 2020.
França, J.S., Solar, R.S., Hughes, R.M., Callisto, M., 2019. Student monitoring of the ecological quality of neotropical urban streams. Ambio 48, 867–878.
Grossman, Z., Parker, A. (Eds.), 2012. Asserting Native Resilience: Pacific Rim Indigenous Nations Face the Climate Crisis. Oregon State University Press, Corvallis, Oregon.
Hanson, C.T., Bond, M.L., Lee, D.E., 2018. Effects of post-fire logging on California spotted owl occupancy. Nat. Conserv. 24, 93–105.
Harari, Y.N., 2015. Sapiens: A Brief History of Humankind. HarperCollins, New York, New York.
Harris, K., Kimson, A., Schwedel, A., 2018. Labor. 2030: Collis. Demographics, Autom. Inequal. Bain & Company. <https://www.bain.com/insights/labor-2030-the-collision-of-demographics-automation-and-inequality>.
Holford, M., R. Morgan, 4 ways science should transform after COVID-19. World Economic Forum. 2020. <https://www.weforum.org/agenda/2020/06/4-ways-science-needs-to-change-after-covid-19-coronavirus>.
Hook, E.B. (Ed.), 2002. Prematurity in Scientific Discovery. On Resistence and Neglect. University of California Press, London, England.
Hooke, R.L., 2000. On the history of humans as geomorphic agents. Geology 28, 843–846.
Hossain, M.Y., Vadas Jr., R.L., Ruiz-Carus, R., Galib, S.M., 2018. Amazon sailfin catfish *Pterygoplichthys pardalis* (Loricariidae) establishment in Bangladesh: a critical review of its invasive threat to native and endemic aquatic species. Fishes [online] 3 (1), 14. <http://www.mdpi.com/2410-3888/3/1/14>.
Hughes, R.M., 1997. Do we need institutional change? In: Stouder, D.J., Bisson, P.A., Naiman, R.J. (Eds.), Pacific Salmon and Their Ecosystems. Chapman & Hall, New York, NY, pp. 559–568.
Hughes, R.M., 2014a. Iron triangles and fisheries. Fisheries 39, 147.
Hughes, R.M., 2014b. Fisheries ethics, or what do you want to do with your scientific knowledge in addition to earning a living. Fisheries 39 (5), 195.
Hughes, R.M., 2019. Ecological integrity: conceptual foundations and applications. In: Wohl, E. (Ed.), Oxford Bibliographies in Environmental Science. Oxford University Press, New York, NY. <http://www.oxfordbibliographies.com/view/document/obo-9780199363445/obo-9780199363445-0113.xml?rskey=Mfte5h&result=21>.
Hughes, R.M., Noss, R.F., 1992. Biological diversity and biological integrity: current concerns for lakes and streams. Fisheries 17 (3), 11–19.
Karr, J.R., 2006. When government ignores science, scientists should speak up. BioScience 56, 287–288.
Kelderman, K., E. Schaeffer, T. Pelton, A. Phillips, and C. Bernhardt, The thin green line: cuts in state pollution control agencies threaten public health. Environmental Integrity Project. 2019. <http://www.environmentalintegrity.org>.
Krugman, P., This land of denial and death: Covid-19 and the dark side of American exceptionalism. The New York Times. 2020. <https://www.nytimes.com/2020/03/30/opinion/republicans-science-coronavirus.html>.
Kuhn, T.S., 1962. The Structure of Scientific Revolutions. University of Chicago Press, Chicago, IL.
Lackey, R.T., 2007. Science, scientists, and policy advocacy. Conserv. Biol. 21, 12–17.
Lancet, 2020. Reviving the US CDC. Lancet 395, 1521. <https://www.thelancet.com/journals/lancet/article/PIIS0140-6736(20)31140-5/fulltext>.

Lazer, D.M.J., et al., 2018. The science of fake news. Science 359, 1094–1096.

Leopold, A., 1949. A Sand County Almanac and Sketches Here and There. Oxford University Press, Oxford, England.

Lepore, J., 2018. These Truths: A History of the United States. W.W. Norton & Company, New York, NY.

Lerner, S., As the west burns, the Trump administration races to demolish environmental protections. The Intercept. 2020. <https://theintercept.com/2020/09/19/wildfires-trump-election-epa-environment/>.

Lichatowich, J., Managing for sustainable fisheries: some social, economic, and ethical considerations. Pages 11–17 in G. H. Reeves, D. L. Bottom, and M. H. Brookes, technical coordinators. Ethical questions for resource managers. U.S. Forest Service, General Technical Report PNW-GTR-288, Portland, OR. 1992.

Maas-Hebner, K.G., Schreck, C.B., Hughes, R.M., Yeakley, J.A., Molina, N., 2016. Scientifically defensible fish conservation and recovery plans: addressing diffuse threats and developing rigorous adaptive management plans. Fisheries 41, 276–285.

MacLean, N., 2017. Democracy in Chains: The Deep History of the Radical Right's Stealth Plan for America. Penguin Books, New York, NY.

Masson-Delmotte, V., Zhai, P., Pörtner, H.-O., Roberts, D., Skea, J., Shukla, P.R. (Eds.), 2019. Global Warming of 1.5°C: An IPCC Special Report on the Impacts of Global Warming of 1.5°C Above Pre-industrial Levels and Related Global Greenhouse Gas Emission Pathways, in the Context of Strengthening the Global Response to the Threat of Climate Change, Sustainable Development, and Efforts to Eradicate Poverty. Intergovernmental Panel on Climate Change, United Nations, Geneva, Switzerland. <https://www.ipcc.ch/site/assets/uploads/sites/2/2019/06/SR15_Full_Report_High_Res.pdf>.

Mayer, J., 2016. Dark Money: The Hidden History of the Billionaires Behind the Rise of the Radical Right. Penguin Books, New York, NY.

Mckee, J.K., Sciulli, P.W., Fooce, C.D., Waite, T.H., 2004. Forecasting global biodiversity threats associated with human population growth. Biol. Conserv. 115, 161–164.

Michael Jr., J.H., 2017. Managing salmon for ecosystem needs in the Pacific Northwest: limiting science input in ecosystem management—silos r us. Fisheries 42, 373–376.

Michaels, D., 2020. The Triumph of Doubt: Dark Money and the Science of Deception. Oxford University Press, Oxford, England.

Mottram, B., State ecologist transferred after he objects to lake fill: Bob Zeigler became librarian of a non-existent library. Tacoma News Tribune, May 8 1983.

Nelson, M.P., Vucetich, J.A., 2009. On advocacy by environmental scientists: what, whether, why, and how. Conserv. Biol. 23, 1090–1101.

Olson, R., 2020. The Narrative Gym: Introducing the ABT Framework for Messaging and Communication. Randy Olson, Los Angeles, CA.

Oreskes, N., Conway, E.M., 2010. How a Handful of Scientists Obscured the Truth on Issues from Tobacco Smoke to Global Warming. Bloomsbury Press, New York, NY.

Paine, R.T., 1994. Marine rocky shores and community ecology: an experimentalist's perspective. Excell. Ecol. 4, 1–152.

Peery, M.Z., Jones, G.M., Gutiérrez, R.J., Redpath, S.M., Franklin, A.B., Simberloff, D., et al., 2019. The conundrum of agenda-driven science in conservation. Front. Ecol. Environ. 17, 80–82.

Peterson, B., 2017. Wolf Nation: The Life, Death and Return of Wild American Wolves. Da Capo Press, Boston MA.

Poff, N.L., Olden, J.D., Merritt, D.M., Pepin, D.M., 2007. Homogenization of regional river dynamics by dams and global biodiversity implications. Proc. Natl Acad. Sci. U. S. A. 104, 5732–5737.

Potok, M. (editor), Greenwash: nativists, environmentalism & the hypocrisy of hate. Southern Poverty Law Center Special Report. Montgomery, AL. 2010. <https://www.splcenter.org/20100630/greenwash-nativists-environmentalism-and-hypocrisy-hate>.

Reiser, D.W., 2017. Science and advocacy in the American Fisheries Society: seeking the same side of the fence. Fisheries 42, 361–365.

Ripple, W.J., Wolf, C., Newsome, T.M., Galetti, M., Alamgir, M., Crist, E., et al., 2017. World scientists' warning to humanity: a second notice. BioScience 67, 1026–1028.

Rivers, M., Boeing's boss won't resign over the 737 Max crashes: will passengers stand for it? The Guardian. 2019. <https://www.theguardian.com/commentisfree/2019/nov/06/boeing-boss-737-max-crashes-consumer-confidence>.

Ruch, J., 2017. Emerging law of scientific integrity—a bumpy birth. Fisheries 42, 353–356.

Sanger, D.E., 47 coauthors, U.S. now leads the world in confirmed cases. The New York Times. 2020. <https://www.nytimes.com/2020/03/26/world/coronavirus-news.html>.

Scheufele, D.A., Krause, N.M., 2019. Science audiences, misinformation, and fake news. Proc. Natl Acad. Sci. U S Am. 116, 7662–7669.

Schuyler, T., Fish kill, food sources argued: Ziegler labels OPPD plant study 'whitewash'. Benson Sun, Omaha, NB, August 7 1975.

Singal, J., Meet the psychologists who helped the CIA torture. The Cut. 2014. <https://www.thecut.com/2014/12/meet-the-shrinks-who-helped-the-cia-torture.html>.

Stack, S., *Why is suicide on the rise in the US – but falling in most of Europe? The Conversation.* 2018. <https://theconversation.com/why-is-suicide-on-the-rise-in-the-us-but-falling-in-most-of-europe-98366>.

Sullivan, P.J., et al., 2006. Defining and implementing best available science for fisheries and environmental science, policy, and management. Fisheries 31, 460–465.

Vadas Jr., R.L., 1994. The anatomy of an ecological controversy: honey bee searching behavior. Oikos 69, 158–166.

Vadas Jr., R.L., Weigmann, D.L., 1993. The concept of instream flow and its relevance to drought management in the James River basin. Va. Water Resour. Res. Cent. Bull. 178. <https://vtechworks.lib.vt.edu/handle/10919/46625>.

Vitousek, P.M., Mooney, H.A., Lubchenco, J., Mellilo, J.M., 1997. Human domination of earth's ecosystem. Science 277, 494–499.

Wilhere, G.F., 2012. Inadvertent advocacy. Conserv. Biol. 26, 39–46.

Wilkerson, I., 2020. Caste: The Origins of Our Discontents. Random House, New York, NY.

Wood, M.C., 2014. Nature's Trust: Environmental Law for a New Ecological Age. Cambridge University Press, New York, NY.

Zakaria, F., The pandemic upended the present. But it's given us a chance to remake the future. Washington Post. 2020. <https://www.washingtonpost.com/opinions/2020/10/06/fareed-zakaria-lessons-post-pandemic-world>.

Climate reality leadership

Bill Bradbury
Former Oregon Secretary of State, Bandon, OR, United States

Keep hope alive

Al Gore was elected to Congress from his home state of Tennessee in 1976. During his first term he held the first Congressional hearings on climate change. Later, he carried those concerns to the US Senate and the office of Vice-President. After leaving office in 2001, Gore lectured on the dangers of global warming that led to the 2006 publication of An Inconvenient Truth and a companion documentary film, which won an Academy Award (Fig. 9.1).

Figure 9.1 The author and Al Gore at Gore's home for first climate training in 2006.

On December 10, 2007, Gore accepted a Nobel Prize for his work on global warming. In accepting the prize, he urged the world's biggest carbon emitters, China and the US, to "make the boldest moves, or stand accountable before history for their failure to act" (Biography.com, 2020).

Conservation Science and Advocacy for a Planet in Peril.
DOI: https://doi.org/10.1016/B978-0-12-812988-3.00006-5

He donated his share of the $1.6 million award to a new nonprofit organization, now known as the Climate Reality Project, devoted to taking action on climate change.

After he had given numerous presentations and a Hollywood documentary (An Inconvenient Truth) had been released about his global efforts, Gore clearly thought that was NOT enough. Mr. Gore decided he would train others to spread the word and give a slide show on climate change.

I was one of the first 50 people trained by Al Gore and the Climate Reality Project in 2006 in Nashville, Tennessee. We trained in Al Gore's barn in Carthage, Tennessee. This chapter is about my early experiences as a Climate Reality trainee and later as a mentor, the challenges of being one of the few politicians willing to speak truth to power on climate change, and how my thinking has evolved from focusing mainly on the problem to now on the solutions.

The problem

During our training event, Mr. Gore ran through the many slides of the "An Inconvenient Truth" presentation to his 50 trainees from all over the country and from many different walks of life. There were very few elected officials in the group, and I was the highest ranking one as Oregon Secretary of State. So, we were the guinea pigs—could we learn the climate science and present it in both an entertaining and informative way? So many facts, so much to learn—but that was really only half of it. We had to be ready to present to large groups AND answer questions and stay light and confident. My years of "training" as a politician came in very handy in this work. I actually loved presenting to groups—I had done so for years as a TV newscaster. When elected to the state legislature in Oregon, I presented about salmon recovery, renewable energy, and forestry practices to name a few.

My first audience after Gore's climate training was to my family at our annual Thanksgiving get-together in Chicago. I went through all the slides and tried to remember as much as I could about what each slide was showing. Needless to say, the show was way too long and frankly kind of boring. My family was very kind to me and excited about all the information but I and they knew it had to be shorter.

My family showing started a 14-year editing process that continues today—picking and choosing Gore climate slides and adding lots of local slides and figures. There are many examples of Oregon-specific climate effects: ocean dead zones; salmon on the way to Idaho dying in the increasingly warm water of the Columbia River; extreme-fire weather events; and receding glaciers in the Cascades to name a few.

Scientific certainty about climate change cause and effect has grown significantly since 2006. When I was first presenting, scientists were not willing to attribute massive storms to global warming; rather they would say the storm was strengthened by warmer oceans. Now scientists point to the record number and strength of storms and say there is little question the massive storm activity is directly related to a warming planet.

"One of the changes we've seen" says Dr. Katherine Hayhoe of Texas Tech University "is that the average humidity of our planet has increased by 4%. Warmer air holds more water vapor and so, on average, our atmosphere is 4% more humid than it was 30–40 years ago. So, when storms come through, there is more water for them to pick up and dump."

Rain bombs

And dump they do. In parts of the United States, massive "rain bombs" dump more water more quickly than they did 50 years ago. Extreme downpours are furthering climate chaos happening 30% more often than 70 years ago. In the middle of the 20th century, large rainstorms averaged one every 12 months, now they average one every 9 months (Madsen and Wilcox, 2007).

More downpours mean more severe flooding. In the past two decades, the world's 10 worst floods have done more than $165 billion in damage and have driven more than one billion people from their homes (Anthens, 2018).

Hurricanes

Warmer oceans fuel larger, stronger, and much wetter hurricanes. Hurricane Sandy in 2012 gave us a preview of unusually strong storms to come. The official number of deaths from Hurricane Sandy was 147, occurring almost equally in the Caribbean region and the eastern United States. At its greatest extent, Sandy measured more than 900 miles (about 1450 km) in diameter. The storm caused more than $70 billion in damage and was among the costliest natural disasters in the US history.

An estimated 8.5 million people lost electrical service as a result of Sandy. Several cities and towns along the Atlantic coast of New Jersey and New York were devastated, and the storm surge was made worse by high tides amplified by the full moon that occurred on October 29. In New York City a storm surge measuring nearly 14 feet (about 4.3 m) combined with heavy rain caused the Hudson River, New York Harbor, and the East River to flood the streets and tunnels of Lower Manhattan (Britannica, 2018).

Parts of subway lines were inundated while flooding and power outages forced the closure of the New York Stock Exchange, the longest weather-related closure of the exchange since 1888.

During the last 30 years, an average hurricane season has 12 tropical storms that grow big enough to be named, six grow into hurricanes, and three of them become major (category 3 or above) hurricanes with winds exceeding 110 miles per hour (177 km/h).

Scientists realize the hurricane season is becoming increasingly destructive due in large part to climate change. The Gulf of Mexico in 2020 is 3 F (1.7°C) warmer than normal and Colorado State University's 2020 hurricane season forecast says "The tropical waters of the Atlantic currently are warmer than normal, and these warmer than normal water temperatures are anticipated to persist for the next several months. Hurricanes live off of warm ocean water, so warmer (ocean) water fuels hurricane formation and intensification" (Colorado State University, 2020).

CSU has been issuing an annual Atlantic basin hurricane forecast for almost 40 years. Its 2020 prediction is the most damaging to date, with eight tropical storms turning into hurricanes and half of them being category 3 or above. CSU forecasts that 2020 hurricane activity will be about 140% of the average season, as compared with 2019 hurricane activity, which was about 120% of the average season.

Hurricane Dorian

In 2019, the Bahamas were totally devasted by Hurricane Dorian and CSU estimates the 2019 season was 20% above average. Hurricane Dorian was an extremely powerful and devastating Category 5 Atlantic hurricane and is regarded as the worst natural disaster in the Bahama's history. Damage in the Bahamas was catastrophic due to the prolonged and intense storm conditions, including heavy rainfall, high winds, and storm surge. Thousands of homes were destroyed and at least 84 people were killed. Dorian is by far the costliest disaster in Bahamian history, estimated to have caused almost $5 billion in damages (Wikipedia contributors, 2020).

Hurricanes have been around for a long time—and scientists say climate change is making storms stronger and more devastating. As I write this at the beginning of the 2020 hurricane season, it is hard to imagine what 40% above normal looks like.

Climate deniers

In my early presentations (2007–09) all across the State of Oregon, I could announce a "townhall" as Secretary of State and often over 100 people would attend. Obviously, in such a large audience, there would be climate doubters and deniers.

I remember a presentation in LaGrande in northeast Oregon—very rural with the main activities being farming wheat and raising livestock. Several doubters showed up with very real questions and asked them in a considerate and straightforward manner. I did my best to answer their questions and we all went away feeling just fine.

Deniers are a different story. As a state elected official, I announced my climate townhalls on my website and through press releases. A group of the same five deniers came to virtually every townhall for a while and "asked" the same question that would often be a 5-minute statement with no question attached. I learned that questions need to be at the end of the presentation, not during (otherwise I would never finish the presentation). I would ask participants to hold their questions to the end, and actually have questions—that generally worked!

Deniers were mostly saying "it's just the weather" not a fundamental shift in the planet's climate. As storms grew and droughts deepened in more recent years, deniers did not dispute the climate changes, and they just deny any human impact on the climate. Afterall, they argue, humankind can not really affect such a large planet, and all the changes are explained by the earth's normal cycles.

Look up and you see a vast sky overheard—how could we affect that? You get a different perspective when you see our atmosphere from space—a very thin layer separating us from the vacuum of space. If the earth was the size of a basketball, the thickness of our atmosphere would be equivalent to 5 or 6 sheets of thin paper wrapped on the outside. It becomes easier to conceive of human impacts on the earth and our climate with that understanding.

Deniers now say, "yes the climate has changed, and humans had some small role, but the problem is so large, so global, we as humans can't do much about it." Of course, we can. Changes will not be easy, and it will take time and global commitment. If we act quickly and decisively, we can head off the worst impacts of climate change. But it requires immediate action, and our opportunity is diminishing quickly!

The solution

In the early climate presentations, I would spend most of my talk on the problem: rising emissions trapping more heat, leading to more storms and droughts; very little time was spent on renewable energy, fossil fuel-free transit options, energy conservation in homes and office buildings to name a few.

Now, I spend a majority of my presentation on solutions. The storms have become so big, the droughts so deep, and the hurricanes so overwhelming that the challenge now is convincing people not that there is a global climate problem but, rather, that we can affect it.

> *"Solutions to the climate crisis are within reach, but in order to capture them, we must take urgent action today across every level of society."*
>
> **-Al Gore**

My current presentation starts with three big climate events: the 2019 Australian bush fires, Hurricane Dorian in the Bahamas in 2019, and the US West Coast forest fires in 2017 and 2018 (aided by extreme-fire weather and logged over landscapes). The fires burned down the town of Paradise, California and scorched Brookings, Oregon; a burnt strip along almost 900 miles (1448 km) of the Pacific coastline. The Australian bush fires affected millions of hectares of wildlife habitat and destroyed thousands of buildings. And although wildfires have beneficial ecosystem effects, a perfect storm is in motion now fueled by climate change and the explosion of human development in fire-prone areas. To combat this, we need to move swiftly into solutions.

Green energy—wind, solar, and conservation

Electricity from solar and wind was cheaper than new coal and gas plants in approximately 1% of the world in 2014. By 2019, only 5 years later, solar and wind provided the cheapest sources of new electricity in two-thirds of the world. Within 5 more years, these sources are expected to provide the cheapest new electricity for the whole world (Reback, 2019).

Wind energy has grown rapidly: in 1997 1.5 GW was generated; in 2017 540 GW was generated; and a 2025 estimate is 1000 GW will be generated. (Bloomberg New Energy Finance, 2019) Globally wind could supply worldwide electricity consumption 40 times over (Lu et al., 2009).

Solar is affordable and has immense potential. Enough solar energy reaches the earth every hour to fill all of the world's energy needs for a year. The cost of solar generated electricity has dropped from over $79 per watt in 1976 to $0.25 per watt in 2019 (Harrington, 2015).

The only problem is the wind and sun energy are inconsistent day to day. So, the power system needs a dependable alternative source of energy that is easy to turn on and off like batteries to effectively use only wind and solar.

Batteries

Battery technology is developing quickly, and costs are dropping. Tesla has installed a massive lithium-ion battery to complement a wind farm in South Australia (Thornhill, 2019). The cost of lithium-ion batteries has dropped 85% from 2010 to 2018 (Bloomberg New Energy Finance, 2019).

Lower costs mean greater use. In 2018, global annual energy storage additions more than doubled, to 9 GW, and is estimated to have surged over 70% in 2019 (Herring, 2019). The world is clearly moving to renewables with battery back-up as our preferred source of electricity.

Building efficiency

Buildings account for over half of electricity demand globally, and the market for existing building energy efficiency upgrades is growing dramatically.

In 2011, $118 billion was spent globally making existing buildings more energy efficient. In 2018, that total more than doubled to $298 billion employing an estimated 3.1 million people in the United States alone (Jordan, 2018).

Transportation

The second largest source of greenhouse gas is transportation—cars, trucks, and buses. Electrification of the sector puts a significant dent in greenhouse gas pollution, and significant progress is being made around the world. From *The Times of India*: "India is aiming for an all-electric car fleet by 2030, with petrol and diesel to be tanked" (Times of India, 2017).

India is joined in their effort to phase-out fossil fuel vehicles by countries all over the world. Fifteen of those countries including Canada, France, and Taiwan have pledged to begin the phase-out no later than 2040, and many will do so sooner (Roberts, 2017).

Buses are another large source of fossil fuel pollution. Nonetheless, about half the world's buses (approximately 1.2 million) will be electric by 2025. Thirty-five cities around the world, including London, Los Angeles, Mexico City, Paris, Rio de Janeiro, and Tokyo, have all committed to buying zero emission (electric) buses starting in 2025 (C-40, 2020). Shanghai and Shenzhen, China have already committed to buying only electric buses now (Cleaner Air, 2018).

Lighting

More energy-efficient LED lighting can save individuals, businesses, and governments significant money by their longer lifespan and increased efficiency. LED lights also provide only the level of illumination needed at a given time, have a high degree of control over light direction, reduce light pollution, and mitigate unintended consequences to humans and wildlife (US Department of Energy, 2017).

It is expected that as LED lighting gains market share (market share is estimated at 95% of the lighting needs by 2025), costs will continue their rapid drop. LED light output is measured in kilolumens (klm), and the cost per LED klm in 2008 was $155. By 2016, the price had dropped to $9 (Krishnaswami, 2019; Kennedy, 2020).

The government of Japan began phasing out incandescent light bulbs in 2015 with a goal to completely eliminate them by 2020 (Lin, 2015).

The number of solutions in my climate presentations can become overwhelming due to information overload. My purpose, however, is to inspire hope that we can solve the climate crisis in time. Many are already feeling the harmful effects of climate change, but it will get much worse quickly if we do not act to implement solutions around the world now.

Climate Reality Project

The need for swift global action makes it clear why Mr. Gore did not stop at an Academy Award for "An Inconvenient Truth." He started training climate leaders. The first 50 in 2006 has grown to almost 21,000 trained leaders from 154 countries around the world in 2020. Through 2019, all climate leaders attended an intensive 3-day training session with Mr. Gore, climate scientists and community leaders (Mayors, Governors, City Councilors, County Commissioners). Trainings have been held all over the world, including in China, Europe, and numerous sites in the United States.

Climate mentor

I have served as a "mentor" at six trainings, ranging from Chicago to Beijing. Each mentor is assigned about 30 trainees. Mentors sit at the table with their trainees, running table exercises, and coordinating participation. Mentors help trainees absorb the overwhelming dose of information. Some trainees talk too much, and others sit silently. The mentor's job is to even that out so everyone participates. At the end of a training, the trainees become Climate Reality Leaders (Fig. 9.2).

Figure 9.2
A climate training in Tokyo, Japan, in 2019.

Due to Covid-19, the first "virtual" training was held online over a 9-day period in July 2020. Of note, 11,000 participants from around the world were trained. More than 500 Climate Reality Leadership Corps members served as mentors and worked with trainees to deepen the training and commitment to action.

A second virtual training was held in late August, 2020 for thousands of trainees from over 130 countries.

Climate Reality Leaders have delivered almost 33,000 climate presentations and carried out almost 90,000 "Acts of Leadership" of which include letters to the editor, communicating with elected officials or participating in a public demonstration. What follows is a summation of how this is taking shape in different chapters around the world.

International chapters

The Climate Reality Project has 10 international chapters spanning the globe, as follows:

Africa

Number of trained Climate Reality Leaders: 670 (representing 32 countries)

Based in South Africa and serving as the point of contact for the African continent, the African Climate Reality Project ("ACRP") is a hub for wide-ranging educational and advocacy projects led by Climate Reality Leaders. ACRP strengthens citizen engagement through awareness and capacity-building programs and trainings. Through these activities, ACRP builds and executes campaigns that are action focused.

As climate change increases the risks of dangerous heatwaves, drought, and storms to countries across Africa, more and more people are searching for practical solutions that can not only address this crisis, but also power sustainable growth. In response, ACRP leads the Action 24 project, aimed at strengthening environmental governance and civic participation, in order to advance decarbonized sustainable and inclusive development in South Africa.

ACRP also conducts research in understanding climate literacy and the gendered effects of the climate crisis. Their own Zero Emissions/Omissions program mobilizes young activists at universities and in communities to organize and run their own fossil fuel divestment campaigns, while continuing the pressure on the African Development Bank to divest.

Australia and the Pacific

Number of trained Climate Reality Leaders: 1540

Launched in 2006 the Climate Reality Project Australia works to turn the growing awareness of the danger of climate change into support for action across Australia and the Asia-Pacific region.

The chapter focuses on connecting climate to the lives of everyday Australians, offering Leaders wide-ranging workshops and trainings in communicating with a broad range of culturally and linguistically diverse communities. Thanks to these efforts, more than one in 40 Australians have seen a Climate Reality presentation, helping drive greater public awareness of solutions.

Along with its commitment to public outreach and education, the chapter works with state and local governments to help them make increasingly ambitious commitments to climate action.

The urgency of this work has never been clearer. Those in the Pacific region know all too well what it means to live with the impacts of climate change, with devastating cyclones and rising sea levels already affecting communities. In late 2019 and early 2020, extreme heatwaves and drought fueled urban-devastating wildfires that became headline news across the planet.

And while the country has made only a weak commitment to reducing emissions through the Paris Agreement, other governments in the region are stepping up, with the Pacific Islands Forum issuing the strongest ever collective statement on climate (the Kainaki II Declaration)—and New Zealand passing the world's first Zero Carbon Act in 2019.

Brazil

Number of trained Climate Reality Leaders: 591

As South America's largest economy and home to the world's largest carbon sink—the Amazon Rainforest—Brazil has a unique and critical role in the global climate fight. Opened in 2014 The Climate Reality Project Brazil has quickly become a key partner in this fight, working with local governments to help slash emissions and drive progress in states and communities across the country.

In addition to its policy work, the chapter strives to build broad awareness of how climate change affects Brazil. The chapter develops nonpartisan educational trainings for government representatives, helping local leaders from a diverse array of political parties understand the issue and incorporate pro-environmental agendas into their work.

Working with Climate Reality Leaders the chapter organizes dialog circles as a tool to increase climate literacy among youth and impacted communities to spread a message of hope and solutions across Brazil.

Canada

Number of trained Climate Reality Leaders: 1,020

The Climate Reality Project Canada works to increase Canada's commitment to climate action, starting from the ground up.

In 2017, the chapter launched a network of Community Climate Hubs across the country. Through these Hubs, Canadian Climate Reality Leaders work with their municipalities to develop plans and policies for reaching net-zero emissions by 2050.

In addition, the chapter has developed the National Climate League Standings initiative, where Community Climate Hubs collect climate and related data from 82 different municipalities to help local policymakers take a science-based approach to climate action.

China

Number of trained Climate Reality Leaders: 864

One of the single most important climate stories of the decade has been China's ongoing transition from dirty coal power to solar and wind. While China's government is driving this shift at the policy level, everyday people and the business community have strong roles to play in ensuring the world's biggest polluter becomes its greatest climate leader.

Climate Reality China works to build public support for the country's energy transition and help business leaders pursue new opportunities for growth in a clean energy economy.

Europe

Number of trained Climate Reality Leaders: 1,665

Europe has long been a leader on international climate action, but with temperatures rising and the window for effective action shrinking, European nations, and the EU must do more to cut emissions and accelerate the transition to clean energy.

In this critical time, Climate Reality Europe works with Climate Reality Leaders and a wide range of partners to build popular support for ambitious climate action at both national and local levels across the continent.

The chapter particularly focuses on promoting climate solutions in traditional coal economies, partnering with business and political leaders to ensure the transition to a clean energy economy is a just one for all affected communities.

India

Number of trained Climate Reality Leaders: 579

As a rapidly developing nation blessed with both a vibrant entrepreneurial community and incredible solar and wind resources, India has all the raw materials to become a world leader on climate. With increasingly deadly climate-related heatwaves and intense storms regularly sweeping through the country, the need for action could not be clearer.

The country's leaders face a choice. Develop with fossil fuels and accelerate climate change or pioneer a new path to sustainable development that grows the economy and raises standards of living without threatening the health of the nation or planet. Massive investments in solar are encouraging, but much more remains to be done.

In response, Climate Reality India works throughout India to educate communities about the crisis and empower them to implement practical solutions in a number of areas ranging from clean energy to water conservation to reducing waste.

Starting with awareness, the chapter's flagship Teacher's Training Program has trained over 4000 teachers at more than 500 schools to share the truth about climate change and renewable energy sources in their own communities.

The highly regarded Green Campus programs works with schools to increase their climate change education, as well as conserve energy and natural resources on their property.

It also works to plant thousands of trees to sequester carbon from the atmosphere and help India meet its Paris Agreement commitment.

Indonesia

Number of trained Climate Reality Leaders: 296

The climate crisis is already transforming Indonesia, leading to coastal ecosystem destruction, decrease in water availability, changes in crop productivity, and the loss of biodiversity. Climate change also has an impact on health, food security, and economic prosperity, both at the local and national level.

Climate Reality Indonesia seeks to inform young people and decision makers on the causes and the risks of the global climate crisis and to empower them to take actions and implement solutions.

Recognizing that younger generations have the most at stake in a future shaped by climate change, The Climate Reality Project Indonesia works to educate young Indonesians about the crisis and how we solve it through its flagship Youth Leadership Camp for Climate Change ("YLCCC"). Already, the program has trained over 2000 young people.

In addition, the chapter organizes activists at the yearly Indonesia Climate Change Forum and Expo and the Indonesia Pavilion at UN climate conferences—CRI works with textile and apparel groups to promote the use of sustainable, local materials while maintaining traditional weaving techniques; hosts flood reduction workshops for communities; and supports YLCCC alumni in piloting their own projects throughout Indonesia.

Latin America

Number of trained Climate Reality Leaders: 1310

Although Climate Reality Latin America is headquartered in Mexico City, Latin America is a large region with many different cultures and political climates. Strong engagement in Colombia and Mexico is allowing for relationships to cross borders and foster climate action throughout the entire region.

The Climate Reality Project Latin America office works side by side with government officials, journalists, and business leaders to influence climate policy and train the next generation of climate negotiators.

The chapter is working to broaden public understanding of the issue and build strong support for government action. Climate Reality Latin America brings together broad coalitions of stakeholders from business leaders to local officials to partner on practical solutions on the ground.

Key to these efforts is a focus on creative outreach, with the chapter using environmental art, films, and cutting-edge technology to reach and inspire a wide range of audiences with a message of urgency, hope, and solutions.

Alongside its own outreach, the chapter also works with media outlets to increase the accuracy and reporting of climate news throughout the continent.

Philippines

Number of trained Climate Reality Leaders: 592

After Typhoon Haiyan/Yolanda devastated Tacloban and other communities across the Philippines in 2013, the nation has emerged as one of the world's moral voices for climate action, confronting the crisis at home while calling on developed nations to slash emissions.

The Climate Reality Project Philippines is an instrumental player in this process, working closely with the Filipino government, industries, and civil society to push for a stronger Paris Agreement commitment and establish the policies to achieve it.

As part of this effort, the chapter has successfully advocated for increasing the country's carbon tax and phasing down use of dangerous hydrofluorocarbons.

The chapter also works on expanding green finance opportunities trains young people as international negotiators, fights to reverse pro-coal resolutions at the local level, and trains activists through its Climate Connectors program (CRP, 2020).

United States

Number of trained Climate Reality Leaders: ~12,000

The Climate Reality Project has 101 chapters in the United States, each keeping Climate Reality Leaders in touch with one another and coordinating group climate action.

College campuses are a center of climate action, and The Climate Reality Project USA has formed 26 chapters of the Campus Corps spanning the country. These chapters distribute current climate information and help students find ways to take action.

Global change can happen

It is a strong global effort to educate about climate change and push governments, businesses and citizens to act to address the changing climate. We know climate science-based education is valuable and always worth it, as is the case for all the chapters of this book. What we do not know is whether education will lead to action soon enough to avoid the worst impacts of climate change.

It seems a long shot, but we must keep working. In the course of human events, especially recently, technologies change very quickly. Two examples tell the story: the transition of telephone landlines to cell phones; and the global adoption of renewable energy.

The explosion of cell phones

In 2005, the number of cell phones in the developing world exceeded the number of cell phones in the developed world for the first time. Since then, cell phones have increased a small amount in the developed world (to about 1.5 billion) and skyrocketed in the developing world (to about 6 billion users).

According to a 2015 Pew Research Center study on Communications Technology in Emerging and Developing Nations, a median of 84% of people in emerging and developing nations owned some type of cell phone.

The study also found that while cell phone ownership has increased drastically over the past decade, particularly in Africa, landline connections have remained relatively low—likely due to the lack of infrastructure required for reliable connections. Instead of waiting for landline access, many in emerging and developing nations have bypassed fixed phone lines in favor of mobile technology (Pew Research Center, 2019).

Renewable energy

The same rapid growth scenario is playing out with renewable energy installations around the world.

Chile has announced it will close eight coal-fired power stations by 2025. The country plans to switch entirely to renewable electricity by 2040. The Chilean solar market has grown dramatically in recent years from 1.6 GW at the end of 2016 to over 17.75 GW either approved or under construction now (Comisión Nacional de Energía de Chile, 2017).

India's energy policy has recognized the obvious, the sun shines for free and the wind blows at no charge. By 2030, India plans 450 GW of renewable electricity capacity, and it is projected to cost at least 20% less than the average cost of coal-fired electricity in India.

"You'd have to be quite courageous to invest in coal (in India) at this point," said Navroz Dubash of New Delhi's Center for Policy Research in December 2018. The largest Indian coal power producers are looking to renewable projects to build new capacity. Adani Power has invested more than $600 million in a solar plant in the southern Indian state of Tamil Nadu (Mundy, 2019).

Speaking truth to power: thoughts and solutions

There was a time when implementing solutions to the climate crisis appeared unaffordable. Many were saying that climate change was not costing humanity anywhere near as much as the proposed solutions.

I will never forget a frustrating experience in 2014 when I was serving on the Northwest Power and Conservation Council. Every five years the Council develops a 20-year plan for affordable and available electricity in the four-state region of Oregon, Washington, Idaho, and Montana.

Columbia River hydro is the backbone of the Northwest's electricity system, but as the region grows additional sources of electricity are needed. Conservation was our largest "resource" to meet new electrical demand, but it was estimated that power tradeoff would leave us 15% short of our 20-year need. Members of the council leaned toward renewable sources (solar and wind) to fill that remaining 15%, but renewables were more expensive at that time than a natural gas peaker plant. The law required us to plan for the most affordable solution to our expected needs so that gas peaker plant stayed on our list. That would not be likely today given the continued reduction in the cost of renewables.

When Al Gore talks about solutions at the end of his slide presentation he focuses on renewable energy and conservation. Wind, solar and conservation are being developed world-wide because they are now the most affordable choice for anyone needing new electrical capacity.

So clearly, things can change very quickly. When it is not just "good public policy" but "less expensive good public policy" energy systems and transit will change faster than we ever thought possible.

As an advocate for rapid change to address the climate crisis, I am obviously pleased at this turn of events. I have always been frustrated by society's slow acceptance of the warming problem and slow implementation of solutions. There is still a very big question: how much damage will be done by the greenhouse gases we have already released into the atmosphere?

Hurricanes, fires, droughts, and rain bombs are becoming more problematic every year. We have unleashed a climate catastrophe on ourselves. Science made it clear that the problem would get worse, and acting now would be less expensive than acting later.

But the politics of climate change are very divided and government action is often thwarted. Polling shows clearly that Democrats and Republicans are on opposite sides of the fence when it comes to government actions aimed at climate change.

According to a Pew Research poll of over 10,000 Americans taken on April 29 to May 5, 2020, a majority of Americans (63%) say that climate change is affecting their local community some or a great deal. Fewer (37%) say climate change is impacting their own community either not at all or not too much.

Partisanship is a large factor in views of the local impact of climate change. A large majority of Democrats (83%) say climate change is affecting their local community some or a great deal. By contrast, far fewer Republicans (37%) believe climate change is affecting their local community at least some; most Republicans (62%) say climate change is impacting their local community not too much or not at all (Kennedy, 2020).

Fossil fuel interests (gas and oil dealers, truckers etc.) shut down an Oregon effort to address climate change (see Chapter 13, To Zero Emissions, and Beyond? Oregon Stumbles Forward). Similar stop climate efforts are taking place nationwide as our planet heats up. Governments are likely to slow the transition to renewables because of fossil fuel's huge lobbying power, but they are unlikely to stop it. Lower cost will eventually win. The only question: will it be soon enough?

In order to keep hope alive, scientists and politicians need to work together by speaking the truth of climate science.

References

Anthens, E., 2018. A floating house to resist the floods of climate change. The New Yorker. January 3, 2018. Available from: https://www.newyorker.com/tech/elements/a-floating-house-to-resist-the-floods-of-climate-change

Biography.com, Editors 2020. Al Gore Biography, May 12, 2020. Available from: https://www.biography.com/political-figure/al-gore

Bloomberg New Energy Finance, 2019. New energy outlook, 2019. Available from: https://bnef.turtl.co/story/neo2019/?teaser = true

Britannica E., Superstorm Sandy, 2018. Available from: https://www.britannica.com/event/Superstorm-Sandy

Cleaner Air, et al., Electric buses in cities: driving towards cleaner air and lower CO2 bloomberg new energy finance, 2018. April 10, 2018. Available from: https://about.bnef.com/blog/electric-buses-cities-driving-towards-cleaner-air-lower-co2

Colorado State University, Department of Atmospheric Science, 2020. CSU extended range forecast of Atlantic seasonal hurricane activity, 2019. http://tropical.colostate.edu

Comisión Nacional de Energía de Chile, Reporte Mensual ERNC CNE—Volumen No. 13" (September 2017). 2017. Available from: https://www.cne.cl/wp-content/uploads/2015/06/RMensual_ERNC_v201709.pdf

CRP, 2020. Climate Reality Project International Branches, Available from: https://www.climaterealityproject.org/internationalbranches

Harrington, R., 2015. This incredible fact should get you psyched about solar power. Business Insider, September 29, 2015. Available from: https://www.businessinsider.com/this-is-the-potential-of-solar-power-2015-9

Herring, G., 2019. Amid global battery boom, 2019 marks new era for energy storage. S&P Global Market Intelligence, January 11, 2019. Available from: https://www.spglobal.com/marketintelligence/en/newsinsights/trending/9GIYsd7qF8tNpiopwH7KSg2

Jordan, P., 2018. Energy efficiency: America's job-creation powerhouse. Environmental and Energy Study Institute, October 25, 2018. Available from: https://www.eesi.org/briefings/view/102518efficiency

Kennedy B., Pew Research Center, 2020. Most Americans say climate change affects their local community, including 70% living near coast Factank, June 29, 2020. Available from: https://www.pewresearch.org/fact-tank/2020/06/29/most-americans-say-climate-change-impacts-their-community-but-effects-vary-by-region-2/

Lin J., 2015. Japan to phase-out incandescent and fluorescent lights by 2020, LEDinsider, November 27, 2015. Available from: https://www.ledinside.com/news/2015/11/japan_to_phase_out_incandescent_and_fluorescent_lights_by_2020

Lu X., M.B. McElroy, J. Kiviluoma, 2009. Global potential for wind-generated electricity, Proceedings of the National Academy of Science. https://www.pnas.org/content/106/27/10933.short

Krishnaswami, A., 2019. Revolution now-cost reductions, Tableau Public. Available from: https://public.tableau.com/profile/arjun.krishnaswami#!/vizhome/RevolutionNow-CostReductions-Jan2019/LEDsWorksheet

Madsen, T., Wilcox, N., 2007. When it rains, it pours: global warming and the increase in extreme precipitation from 1948 to 2011. Environ. Am. Res. Policy Cent. 2012. Available from: https://environmentamerica.org/reports/ame/when-it-rains-it-pours-global-warming-and-rising-frequency-extreme-precipitation-united

Mundy, S., 2019. India's renewable rush puts coal on the back burner. Financial times. January 1, 2019. Available from: https://www.ft.com/content/b8d24c94-fde7-11e8-aebf-99e208d3e521

Pew Research Center, 2019. Internet seen as positive influence on education but negative on morality in emerging and developing nations, March 19, 2015. Available from: http://www.pewglobal.org/2015/03/19/1-communications-technology-in-emerging-and-developing-nations/

Reback, S., 2019. Solar, wind provide cheapest power for two-thirds of globe. Bloomberg, August 27, 2019. Available from: https://www.bloomberg.com/news/articles/2019-08-27/solar-wind-provide-cheapest-power-for-two-thirds-of-globe-map

Roberts, D., 2017. The world's largest car market just announced an imminent end to gas and diesel cars. Vox, September 13, 2017. Available from: https://www.vox.com/energy-and-environment/2017/9/13/16293258/ev-revolution

Thornhill, J., 2019. Tesla set to bulk up the world's largest lithium-ion battery. Bloomberg, November 18, 2019. Available from: https://www.bloomberg.com/news/articles/2019-11-19/tesla-set-to-bulk-up-the-world-s-largest-lithium-ion-battery

Times of India, 2017. India aiming for all-electric car fleet by 2030, petrol and diesel to be tanked, April 30, 2017. Available from: https://timesofindia.indiatimes.com/auto/miscellaneous/india-aiming-for-all-electric-car-fleet-by-2030-petrol-and-diesel-to-be-tanked/articleshow/58441171.cms

US Department of Energy, Office of Energy Efficiency, 2017. 5 common myths about LED street lighting, June 6, 2017. Available from: https://www.energy.gov/eere/articles/5-common-myths-about-led-street-lighting

Wikipedia contributors, 2020. Effects of Hurricane Dorian in The Bahamas (May 12, 2020) Wikipedia. Available from: https://en.wikipedia.org/w/index.php?title = Effects_of_Hurricane_Dorian_in_The_Bahamas&oldid = 956247372

III

The Politics of Science in Decision Making

Out of the ivory tower: campaign-based science messaging for the public

Richard McIntyre[1,2]

[1]The Community Governance Partnership (CGP), Sacramento, CA, United States [2]Cannabis Removal on Public Lands Project (CROP Project), Oakland, CA, United States

Does science matter?

The challenge of translating science into an equally compelling public story has always represented an obstacle to the scientific community. In some ways the skills required to do so mean stepping away from the linear thinking that defines the field. It is not enough to be scientifically correct, because in today's world, that is, for many people, entirely subjective. It is not enough to provide hard, peer-reviewed data and presume that will convince the public or decision makers, because the data itself are considered suspect by fringe groups and political figures with a clear antiscience bias. We can rail at the injustice of it all, but science presented purely as such is now less important to decision makers than any time in recent United States history. For instance, Dr. Anthony Fauci, the United States top infectious disease expert and Director of the National Institute of Allergy and Infectious Diseases, has been the target of Donald Trump's incessant antiscience diatribes. The attacks are not furtive or disguised; they are straight up assaults on science by people whose agendas are threatened by medical and scientific fact.

Communications has never been the strong suit of scientists. With the rare exception of great storytellers like Carl Sagan or Bill Nye, science is seen as droll or exceedingly technical by much of the American public (Chapter 1, The Nuts and Bolts of Science-Based Advocacy). That same disconnect is evident in the positions and votes of their political representatives. It is critical that scientists find understandable ways of presenting information that connects science more directly to communities and working-class people and puts a human face on the data. Utilizing earned media, securing respected public voices, understanding and playing the political dynamics, and allowing others to tell the story is something the scientific community should learn.

Conservation Science and Advocacy for a Planet in Peril.
DOI: https://doi.org/10.1016/B978-0-12-812988-3.00012-0

In response to the main theme of this book—speaking truth to power—in this chapter I will set a historical context for the dismissal and denigration of science, and examine how a shift in communication strategies and campaign development can garner the public support needed to turn scientific findings and imperatives into policy. I will provide a case study of such an approach, and outline the key steps required for a winning, science-based campaign.

Snowball climate denial

The scientific community understandably bemoans the extent to which the public dismisses, ignores, or distorts the results of credible scientific research, sometimes to the point of disdain. Frequently, results are painted with a broad partisan brush that allows elected officials to wrap their ideology around the findings, which are then bent to their liking. The US example of Senator James Inhoffe of Oklahoma holding aloft a snowball (Fig. 10.1) scooped from the steps of the Capital to dismiss climate change was ludicrous to many, but the reality is that it was stunningly effective as a communications tool, rebroadcast on every network for days, and used in conservative media as a cudgel on science. Big on imagery, short on words, it worked—an important lesson to be learned. The national press tends to amplify moments such as these, because it makes for good television and print copy. In doing so, they make themselves complicit in the outcomes, leaving it to a confused public to do their own due diligence on the underlying facts. This follow-up, as we well know, often never occurs, or is subject to ideological spin.

While the players have changed, denigrating science is not a new process. Science has been the subject of disdain and distrust for centuries, and those

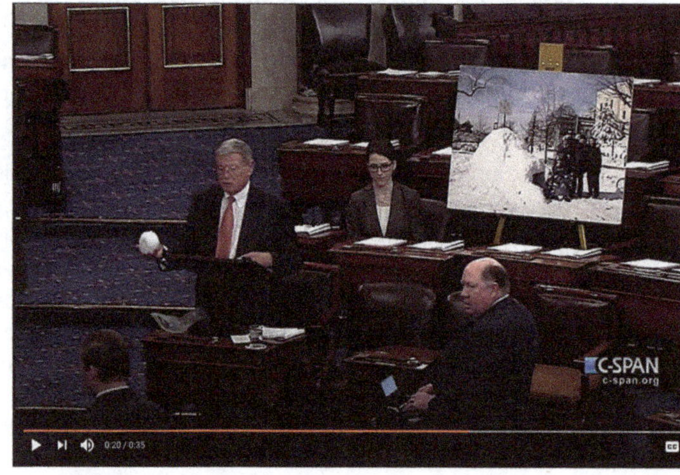

Figure 10.1 Jim Inhofe, Chair of the US Senate Committee on Environment and Public Works, holds a snowball at a climate change hearing. *Courtesy CSPAN, public domain.*

seeking to block its advance have always found willing allies. Galileo may not have had to deal with the spin of conservative media in his time, but the Catholic Church played a very similar role in his Inquisition trial in 1683 (see Preface). Galileo's embrace of science and humanism crossed swords with scholasticism and the absolutism that dominated the Catholic Church at the end of the Renaissance. Astronomy and physics were the fake news of his time.

In a letter he wrote to Johannes Kepler during his trial, he stated, "My dear Kepler, what would you say of the learned here, who, replete with the pertinacity of the asp, have steadfastly refused to cast a glance through the telescope? What shall we make of this? Shall we laugh, or shall we cry?" These rhetorical questions are ones which Dr. Anthony Fauci could reasonably ask over three centuries later.

The politicization of science

In our century the political polarization of American society has created our own equivalency. The US controversy over face masks is a germane example. The science on the efficacy of face masks to reduce the risk of exposure to the coronavirus is clear, but disinformation from conservative political leadership, enhanced by foreign actors, has challenged both the science and the need. They, too, have little interest in looking through the proverbial telescope.

Science is intrinsically political because, by itself, it does not generally allow for a convenient interpretation that matches political agendas. Science is fact based, but in an era where "facts" are viewed by much of a politicized public as highly subjective, the challenge to move findings to public policy cannot be understated.

> *"Political values are unavoidably wrapped up with scientific research, because science tells us what's possible. Science is inherently controversial because nobody wants to hear that their options are limited"* (Suhay and Druckman, 2015).

Politics and science exist in the same space, but the relationship between them is troubled at best. The world of fact-based science is, by nature, at odds with the ideological political world, and struggles to find solid footing in public policy debates. Scientific findings are selectively discounted or subject to "spinning" to match specific agendas, as science struggles to maintain credible autonomy.

One's scientific training and expertise provides insight and knowledge beyond that of the general public. Your training points you toward fact-based solutions, filled with the terminology of equations, statistics, absolutes, and probabilities.

In the public forum, with policy change as a goal, those elements, in the absence of a broader, relatable context, are a recipe for failure.

Science as a stand-alone is an abstraction. Making it germane to decision makers and the public is central to seeing credible, issue-based science protected from interpretation and spin. In the final analysis, this is a political communication issue. Persuasion and politics have never been the strong suit of the scientific community. That needs to change.

Scientists as storytellers

Public policy outcomes often hinge on the ability of those messaging scientific facts to use common language storytelling as part of a broader strategy. The results of that effort historically have been uneven, in part because the scientific community has not recognized or embraced the need for a multidisciplinary, campaign-based approach for their core message to be understood and to resonate with the public.

Recent social science clearly indicates that a germane story increases the effectiveness of communicating science and proposed policy. What is required is compelling storytelling, that is, understandable, timely, and interesting. Paul Cairney, in his 2017 treatise "Telling Stories that Shape Public Policy," Jones and Crow (2017) stated,

> *"The primary means by which human beings render complexity understandable and reduce ambiguity is through the telling of stories. We "fit" the world around us and the myriad of objects and people therein, into story patterns. We are by nature storytelling creatures. And if it is true of us as individuals, then we can also safely assume that storytelling matters for public policy where complexity and ambiguity abound"* (Jones and Crow, 2017).

A good story that results in changes to public policy requires careful planning and execution. It is not linear—some of it is intuitive, and as such represents a challenge for scientists. It requires a campaign approach that involves organizing, solid scientific data, strategic use of the press, identifying common ground between diverse interests, public support, and having the right messengers. It involves judgment calls, emotion, and political pragmatism, which is somewhat antithetical to science-based approaches.

The intersection of science, politics, and public policy now has a component that Galileo could never have imagined. It is estimated that the national media, from network news and radio to on-line outlets, reaches some 9 out of every 10 Americans on a daily basis. Understanding that intersection, and embracing it, is key to seeing science as central to emerging public policy.

What, then, are the elements of a publicly compelling, science-based story? The best stories—of any kind—have drama, protagonists, antagonists, and a resonating message or lesson. The best of them has a common enemy. They are understandable (e.g., do not require an advanced education to comprehend), utilize science to buttress the core story, are central to public policy that is solution based, and use the press to tell the story and gain the attention of decision makers at the state and national level. There is a current, germane example of campaigning with science and storytelling.

The Cannabis Removal on Public Lands story: a case study in science-based campaigns

The CROP (Cannabis Removal on Public Lands) Project (www.cropproject.org) was formally established in 2018 by two California nonprofits, the Community Governance Partnership (CGP) and the California Wilderness Coalition (CalWild), funded by a pair of foundations who immediately saw the political and environmental opportunity the project represented. We will use this project to outline the steps taken to elevate science as a central element of a campaign.

CROP was created to address a profound environmental issue largely unknown to the public and unaddressed on the federal lands of California. Thousands of trespass cannabis grows on public lands, primarily run by cartels, exist on nearly every national forest in the state. They are responsible for the poisoning of wildlife on a landscape level, including imperiled terrestrial and aquatic species, the dewatering and poisoning of streams and public water supplies, and for large wildfires.

Trespass grows steal up to 9 billion gallons (34 billion liters) of water annually, in a state where drought is a way of life. Their grow sites contain quantities of banned pesticides and rodenticides; a quarter teaspoon of Carbofuran (a US-banned pesticide) will kill a 600-pound (270 kg) bear and has been weaponized to target law enforcement. The growers are heavily armed, and unsuspecting hikers and other recreationalists sometimes find themselves staring down the barrel of an AK-47. Law enforcement is frequently outgunned, and on federal lands, grossly understaffed, with one law-enforcement officer for every 250,000 mountainous acres (100,000 ha). Large areas of the National Forests have become "no-go zones" for everyone. Up to 50% of all black-market cannabis is grown on federal lands in California, poisoning wildlife and users alike, while depressing the legal market.

Out of sight and largely relegated to remote areas, the issue had remained off the political and public radar screen for decades. Here were the raw ingredients for a public campaign, based on science, to create new public policy and solve a seemingly intractable problem.

Science takes center stage—The science behind the issue, while not widely known or disseminated, was far from ambiguous. Extensive research by the Integral Ecology Research Center (IERC.org) in California had documented the effects of US-banned or strictly controlled pesticides and their lethal effects on wildlife of Sarin-based Malathion, Brodifacoum, Carbofuran, Methamidophos, and Cholecalciferol. With years of sampling in multiple locations, they found profound impacts to numerous species, including the federally listed Northern Spotted Owl (*Strix occidentalis caurina*) (70% with high levels of rodenticide), Pacific fisher (*Martes pennanti*; candidate for listing with 85% exposure to five rodenticides), as well as major impacts to the endemic Humboldt marten (*Martes americana humboldtensis*); the first animal to receive endangered species status listed due to impacts from chemical exposure at trespass grows. Further, IERC documented extensive mortality among raptors, Black bears (*Ursus americanaus*), and other wildlife up and down the food web. Levels of banned pesticides were documented in game animals such as mule deer (*Odocoileus hemionus*), placing hunters and tribal subsistence gatherers at great risk. Simplified, the entire food web was compromised by the toxics found at trespass grows.

Importantly, Dr. Mourad Gabriel of IERC conducted research that demonstrated that the same toxic chemicals killing wildlife on a landscape scale were leeching into public water supplies in Trinity and Mendocino County.

The missing ingredient—How is it possible this issue was not better known? Why was it not on the radar screens of political leadership and the national press? Why was an outraged public not demanding action against the international drug trafficking organizations responsible for the slaughter of wildlife and the poisoning of the communities? The answer is both simple and complex, but for our purposes, it has one answer: the science had not been presented in a manner that would cause anything to change. There was no compelling story, no narrative for science to take center stage. There was no strategy on how to fundamentally address this seemingly intractable problem, no enabling tactics to deploy. Presentations were made at conferences, and papers were published—and it made no difference on the ground. Trespass grows continued to explode across the landscape, and reclamation funding was miniscule.

What was needed was a well-designed, science-based campaign to raise public awareness, elevate the issue, and generate outrage that would demand the attention of political decision makers. That campaign would use science as an anchor but speak in language that the press, the public, and politicians could understand. It would also incorporate two elements rarely found in science-based work: storytelling and emotion. It begins with an understanding of what and who it will take to tell the story, a comprehensive, staged strategy and defining what victory looks like.

The initial message delivered to the public must command attention, be informative without being boring, and touch an emotional chord. It needs to frame the story in understandable language and impacts people and the things they value. Like a good actor or musician, you want to leave the audience moved and wanting more. For the CROP Project, it was a 174-word core message that framed the issue.

"Trespass cannabis grows on public lands in California have flown under the public radar for decades. Thousands of grow sites exist in nearly every National Forest in the state. They are controlled primarily by foreign drug trafficking organizations (cartels) who use banned, imported pesticides that indiscriminately slaughter wildlife on a landscape scale, destroys critical habitat, and poisons public water supplies. Rodenticide use (aka "rat poison") is widespread and is now documented in 90% of California's Mountain lions, 70% of ESA-listed California Spotted Owls (*Strix occidentalis occidentalis*), and 80% of Pacific fishers (*Pekania pennanti*). The Humboldt marten has become the first animal to be ESA-listed strictly as a result of trespass grow–related activity. A one-fourth teaspoon of the pesticide Carbofuran, often found on-site by the liter, is enough to kill a 600-pound black bear, and is being weaponized by the cartels to target law enforcement. These growers, heavily armed, confront hikers, hunters, and other forest users, making entire areas "no-go zones." They produce up to 50% of all black-market cannabis in the state, depressing the legal market and killing local tax revenues."

As one may surmise, this is not a science memo. It contains statements that admittedly scream for supporting data. It uses language that some may find overly simplistic ("slaughter"), terms that dance on the edge of sensationalism ("heavily armed"), and political correctness ("cartels"). It purposefully conjures up unpleasant images and clearly identifies the common enemy. It is usually accompanied by photos of charismatic fauna in the advance stages of poisoning—or already deceased. It goes directly for the gut—and it works.

With the core message crafted, and well before it is formally unveiled, a campaign strategy needs to be developed. Naming a complex problem and making it understandable to the public is pointless without a well thought out strategy with staged tactics to carry it out.

At the heart of a campaign of any kind is organizing, which, properly planned and implemented, provides project proponents with allies, visibility, political support, and capacity. It involves identifying power (and ways to influence it), building relationships, coalition development, and political acumen. It is based on pragmatism (which the scientific community understands) and presenting facts in an understandable way (which it often struggles with).

The Midwest Academy (www.midwestacademy.com), the foremost organizing program in the United States, has trained thousands of activists in community organizing (including the author), using a complex strategy chart that embodies many of these concepts. Many of these elements are in the steps that follow.

Step one in a campaign: identify your goals and define success

In 2017, community and government outreach started in NW California to better understand and quantify the problem. Those early meetings confirmed the extent of the problem, a strong interest among local communities and governments in solving this seemingly intractable problem—and a vacuum of leadership to address the problem

First, we identified our short-, intermediate-, and long-term goals and defined what victory would look like. In this instance, our long-term goals matched our strategies which were to: (1) secure federal and state funding to reclaim thousands of trespass cannabis grows on public lands; (2) increase US Forest Service law-enforcement officers on federal lands to interdict and discourage growers; and (3) increase federal penalties for bringing toxics onto federal lands. We also established short-term goals to establish bipartisan credibility as key steps in carrying out our strategy.

Step two: campaign considerations and resources

Campaign and organizational resources must be identified well in advance of the launch of the campaign. Those resources include available funding, a rough campaign budget, identifying internal staff capabilities, social media, research capacity, ability to secure press, political experience, and policy implementation experience.

We examined what it would take to initially develop the campaign and strengthen the campaign organization. In this case, it included:

- securing initial financial support based on realistic budgets to cover staff time and project costs;
- expanding leadership of the effort to bring expertise and credibility to the campaign, as well as economic and political power;
- developing the basic campaign messaging, and utilizing storytelling to express it in a way that was relatable and easily translated by the press;
- securing respected science voices to support the effort and explain the environmental outcomes; and
- securing local and regional political voices to support the effort and act as project endorsers.

Step three: identify the allies needed to succeed

Any issue-based science campaign seeking allies must answer a series of questions. Who cares enough about the issue to help us? Whose problem is it? What do they gain if we win? What risks are they taking? Most importantly, what power do they have over the targets we need to influence?

As important—and in the case of science, more so—is "who are our opponents?" What will victory costs them, and what will they do to oppose you? How strong are they, and what power do they have over your targets?

In the case of the CROP Project the answers were simpler than is often the case in science-based campaigns, where deep corporate pockets are often at the heart of the problem (think fossil fuels and timber production). The issue of public land trespass grows was well known among regional county governments, the state, federal land management agencies, and tribes whose subsistence rights on National Forests were compromised by cartel grows. Underfunded, outgunned, and unable to gain meaningful funding to address the problem, they would endorse the project concept and work with us to forward the project. Once educated on the issue through press engagement, an outraged public would voice their support, as would a legal cannabis industry whose production was undercut by the black market. We were fortunate (so far) that the opponents of the CROP Project were foreign drug trafficking organizations with little influence over our public and political targets. Their influence, should it come, usually comes in the form of threats and violence which, fortunately, has yet to develop.

CROP identified who was needed as allies to carry out the initial strategy, which was to establish credibility for the project at the local, regional, statewide and federal levels. In this instance it was: (1) scientists to explain the environmental and toxicant damage (IERC and the US Forest Service); (2) local and statewide political leadership to establish political viability (from county Boards of Supervisors to the Governor's Office); (3) federal land agencies to provide credibility and forward the work at the federal level, specifically the US Forest Service and Bureau of Land Management; (4) the legal cannabis industry; (5) local tribes; and (6) statewide and national environmental organizations.

From those allies, we carefully chose our Advisory Board, both to help frame the project and provide bullet-proof credibility for carrying out the strategy. That included key county Supervisors from the northern part of the state, Forest Service National Forest Supervisors and USFS law enforcement, Dr. Gabriel from IERC, the director of the California Cannabis Industry Association, Natural Resource Director for the Karuk Tribe, the Director of the North Coast Environmental Center, as well as the Executive Directors of CGP and CalWild. Our coalition was intentionally diverse, politically and socially. It was bipartisan, narrowly focused on public lands, and avoided national issues such as legalization, or state issues such as licensing. Individuals were vetted before being invited onto the Advisory Board to insure they were collaborative and concurred with the project principles, objectives, and approach. That approach would be validated throughout the project, with remarkable cooperation among divergent interests who leveraged their contacts and expertise at key times.

Step three note: Be careful and selective when choosing allies. Consensus on goals, defining success, and willing collaboration are key to avoiding conflict or mixed messaging. A winning campaign does not need everyone—it needs the people and organizations aligned with your strategy and outcomes, who bring power to the table to influence decision makers.

Step four: emotion is your friend

Never underestimate the power of emotion to influence public opinion. Images and pained testimonials are tangible to the public and often are at the heart of their opinions. They put a human face on scientific fact, and make it relatable to people who may otherwise be unmoved by facts alone.

In presentations to member of Congress, the data on the CROP Project were compelling. Nine billion gallons of water are stolen annually to grow black-market cannabis. Pesticide-laden streams are going into public water supplies. Streams dried up, trees cut down, riparian areas destroyed. What they could not stop looking at, however, were not the data or the credentials behind the science.

Instead, it was the video of a Pacific fisher pathetically flopping around suffering from Carbofuran poisoning. It was the photo of the dead litter of Humboldt martens, and the dead, lactating black bear next to a grow site. It was the image of martens and fishers hanging from baited hooks, and of pristine wilderness turned into massive dumps of garbage, propane tanks, and toxic pesticide containers [The CROP Project (Cannabis Removal on Public Lands), 2017].

That, too, became a central theme of the extensive national press the CROP Project received, emotional anchors to tell a story anchored in scientific fact. The Associated Press article, *Illegal Pot Farms on Public Land Create Environmental Hazard*, coverage would end up in 200+ papers across the country. Atlantic, in covering the site visit in an article entitles, "The Environmental Catasrophe in Your Joint" (Lowrey, 2019).

> *"When deputies raided the remote clearing in the woods Sept. 9, they found hundreds of pounds of harvested marijuana, thousands of pounds of trash and more than 3 miles (4.8 kilometers) of plastic irrigation piping, according to the Trinity County Sheriff's Office. They also discovered bottles of carbofuran, a banned neurotoxicant used to kill rodents that also has been linked to the deaths of spotted owls, fish and mountain lions. A quarter-teaspoon can kill a 600-pound bear"* (Lowrey, 2019).

Experts say illegal sites like the one found in the Shasta-Trinity National Forest, about 100 miles (160 km) south of the Oregon line, siphon valuable water, pollute legal downstream grows and funnel potentially tainted cannabis onto the streets (Figs. 10.2 and 10.3).

> *These places are toxic garbage dumps. Food containers attract wildlife, and the chemicals kill the animals long after the sites are abandoned," said Rich McIntyre, director of the Cannabis Removal on Public Lands (CROP) Project, which is dedicated to restoring criminal grow sites on state and federal property in California. "We think there's a public health time bomb ticking.*

Figure 10.2 Irrigation infrastucture removal during decommissioning of trespass grow site, Klamath National Forest. *Source: CROP/Vuh.*

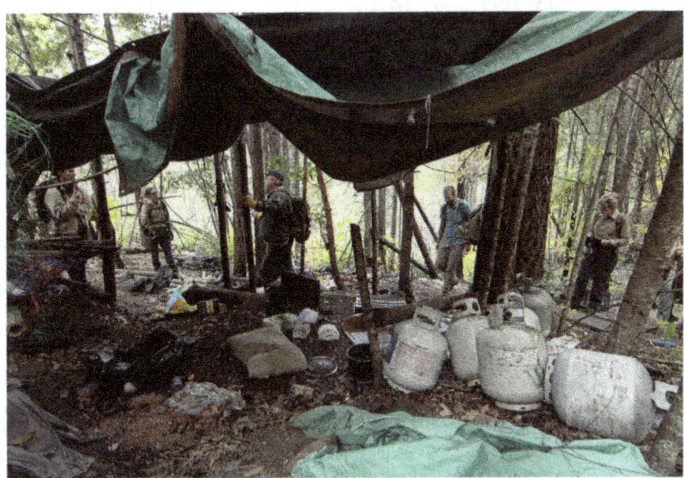

Figure 10.3 Trespass grow camp, Shasta-Trinity National Forest. *Source: CROP/Vuh.*

Step five: identify your targets (decision makers)

One of the great misconceptions and classic blunders of a campaign is identifying institutions or elected bodies as your targets for influence. The primary target is always a person, a person who has the power to give you what you want. In the instance of the CROP Project, for example, we did not need the entire Board of Supervisors of a specific county to endorse our project. We needed an influential board member to represent the county on our Advisory Board. We did not need the endorsement of every conservation organization or cannabis grower group—we needed the engagement and support leadership of the most prominent. Finally, we did not need to sell CROP to the entire United States Congress—we needed a key Republican and key Democrat to carry the message and sponsor enabling legislation (in this case the Protecting Public Lands Against Narcotic Trafficking (PLANT) Act, being introduced in the 117th Congress).

Secondary targets—people who can influence the decision maker—can be very useful in these situations. It may be a campaign donor, a local politician, the local newspaper editor, or the target's minister. Whoever it is, they have a personal relationship that can be leveraged.

Step six: design and deploy your tactics

Once key targets are identified, tactics finally come into play. Key allies and supporters will have different targets, and tactics must be individually designed for each of those targets to achieve the intermediate and short-term goals previously established.

The messenger is often not the primary project proponent, rather it is an ally recruited to the project, often for this specific moment. The objective is pressure, respectfully but forcefully applied. In the context of the strategy, what power does a science-based argument have over your target?

Always including at least one face-to-face meeting, your tactic must be backed up by a specific form of power. In the instance of political bodies, it all about money, votes, and press.

> *Congressman, my name is _______, and I am the Executive Director of _____. We have 312 voting members in your district. We want your support for ______, which will address this long-standing problem.*

- Tactics may include public hearings.
- Letters to the Editor and Op-eds (Opinion editorials).
- Phone, email, petitions.
- Town Halls.
- Media.

Step seven: use the media

The national press understands the power of a compelling story; it is, after all, the manner in which they try to make the news of the day interesting to as broad an audience as possible. To provide them with the basis for a new story—one generally unknown, but with a serious hook—requires strategy and perseverance. The best reporters are inundated with story pitches and have a good ear for those that stand out. The ones that break through share a compelling story with a strong human element that grabs the attention of the reader, listener, or viewer. They are, most importantly, pitched the right way: tight, compelling, and confidently. A professional communications firm can be invaluable in securing the attention of the national press. The services of such a firm usually start in the 20,000 USD range, for which consideration they leverage their past successful work with press outlets to pitch your story. They assist in editing outreach materials, fact sheets, and photos used in the pitch effort. They can often secure attention quickly, saving time and effort. Additionally, press begets press: a groundbreaking story attracts other media attention. A single quality article by the Associated Press on the CROP Project was reprinted in over 230 newspapers, with connected coverage in the international press. The National Public Radio (NPR) Morning Edition story on the CROP effort, "Illegal Pot Operations in Public Forests are Poisoning Wildlife and Water," was heard on hundreds of NPR stations across the country (Westervelt, 2019). The response of an outraged public helped change the narrative around the problem.

If making the approach on your own, brevity and quality descriptive photos are critical. Plan the pitch, have your back-up data and photos, have your compelling story ready to tell (quickly). Be ready to move quickly if the journalist is interested, and understand they may see a different angle than you do. If you get someone on the phone, you have ten seconds to get their attention. For example, "Cartel marijuana growing on National Forests is slaughtering wildlife, poisoning public water supplies with banned pesticides, outgunning law enforcement—and no one knows about it. I'd like to tell you that story." If you can get a meeting, use the same approach, but have compelling photos to drop as you are speaking.

If it is by email, realize that follow-up will be critical. The average reporter is hit with scores of emails a day. It has to stand out, be brief, and grab them. The subject line is the only thing they may read, and it has to be perfect. "Cartels growing illegal cannabis are taking over our National Forests" may sound dramatic, but it has a high likelihood of causing the reader to go on. The keywords need to be there (as italicized above), less than 45 characters, including spaces. Avoid niceties—get to the point. The pitch must be short and descriptive—less than 200 words, preferably in bullet points. Forget superlatives and exclamation points, and never fall back on clichés. Do not attach anything like press releases or data; the only exception is a compelling photo that underlines your story and will make them think about it. Your job is to convince them of the "why now?," "why will people care?," "would I want to read this myself?," and "why would I want to cover this?"

If you are lucky enough to get that in-person meeting, a lifetime of doing this has taught an important lesson: know your audience. You will again stick to facts, compelling phrases, human impacts, and storytelling. You will, again, avoid pleasantries like the plague. That stated, journalists like recognition, to be known for their work. Research what they have written, and review some of it. "Before we get started, I just wanted to say how interesting I found your piece on honey production. Fascinating—I had no idea." Proverbial ice, broken.

Site visits, well-orchestrated with expert voices, are a powerful tool in getting your story out. Reporters want to see and feel what they are reporting on. When the Atlantic covered a CROP site visit, writer Annie Lowry wrote this compelling opening to a major article on the trespass grow issue:

> *On a cold morning this fall, I was clinging to a paracord line, descending through heavy brush on what felt like a near-vertical hill in Shasta-Trinity National Forest, in front of me a team of influential ecologists, behind me a team of affable, heavily armed law-enforcement agents. We were deep enough into the forest—bushwhacking into vegetation, miles and miles down a fire road—that anything human-made would have looked out of place. Then I saw a suitcase. A child's coat. Pots and pans. A pair of jeans. Chapstick. Ramen packages. It looked a little like the site of a plane crash, except for the thousands and thousands of dead marijuana plants.*
>
> *We'd arrived at a "trespass grow," one of hundreds in the state of California. If you bought weed on the black market recently, there's a good chance it came from a grow like this one. There's a good chance, then, that your joint was contributing to a spiraling environmental catastrophe* (Lowrey, 2019).

The use of press in educating and influencing decision makers cannot be overstated. When the CROP team went to DC to meet with key Congressional offices, we were pleased to learn many had already seen the press—and knew exactly who we were. Those who did not were suitably impressed. Decision makers, especially at the Congressional level, still get their information from traditional public news sources: newspapers, television, radio. Positive press can open doors during initial contacts, providing immediate credibility.

Speaking truth to power: closing thoughts

As was stated at the outset, science has never been accomplished at self-promotion and the mechanics of influencing public opinion. To the scientific mind, facts should stand on their own, and not require what is effectively a public relations campaign to change public opinion, and by association, public policy. That opinion is understandable, but no more valid today than in Galileo's time.

Science needs people in its ranks capable of storytelling, people that embrace the need for a multidisciplinary approach to policy change. Communications, strategy, and campaign skills are critical to humanizing and personalizing science, making it accessible and understandable to the voters and decision makers that will ultimately determine whether science is translated into effective policy, or, if like Galileo, the proverbial telescope is placed back in the closet.

References

Jones, M.D., Crow, D.A., 2017. Telling Stories That Shape Public Policy. Paul Cairney: Politics & Public Policy.

Lowrey, A. (2019). 'The environmental catastrophe in your joint', *The Atlantic*, 9 December.

Suhay, E., Druckman, J.N., 2015. The politics of science: political values and the production, communication, and reception of scientific knowledge. Am. Acad. Pol. Soc. Sci.

The CROP Project (Cannabis Removal on Public Lands), 2017. Available at: www.cropproject.org, accessed November 2020.

Westervelt, E. (2019). 'Illegal pot operations in public forests are poisoning wildlife and water', *National Public Radio*, 12 November.

11

Essays from the trenches of science-based activism

Joel Clement[1,*], Randi Spivak[2] and Jennifer Mamola[3]
[1]Harvard Belfer Center for Science and International Affairs, Cambridge, MA, United States
[2]Public Lands Program, Center for Biological Diversity, Tucson, AZ, United States [3]John Muir Project of the Earth Island Institute, Washington, DC, United States

How the Trump Administration tried to cancel the Interior Department

Joel Clement, Senior Fellow, Harvard Belfer Center for Science and International Affairs

As a scientist and senior executive at the U.S. Department of the Interior (DOI), I had a front-row seat for climate and public lands policy making in two administrations. Not 6 months into the new Trump Administration, it became clear that this new group of political appointees was ignoring science, climate change, and the environment while marginalizing scientists and experts. This was no mere political stunt, their willful disregard for science was compromising the safety of Alaska Natives faced with the dramatic impacts of climate change in a rapidly warming Arctic. I had no choice but to blow the whistle, and did so on July 20, 2017, becoming the first public whistleblower of the Trump Administration.

Politics trumps climate assistance

It began with an email on the evening of June 10, 2017. After being tipped off by a friend that I might want to check my inbox, I found a message from the acting Deputy Secretary telling me that I'd been reassigned. This was not shocking, as members of the Senior Executive Service can easily be reassigned. The bizarre thing about it was the explanation for the reassignment:

**Editor's note: For his work as the first Trump Administration whistleblower, Joel received the prestigious LaRoe III Memorial Award from the Society for Conservation Biology in 2018 given to employees of government agencies that demonstrate leadership in conservation biology.*

Conservation Science and Advocacy for a Planet in Peril. DOI: https://doi.org/10.1016/B978-0-12-812988-3.00005-3

As the Director of the Office of Policy Analysis, it said, I supervised economists, and therefore I must be familiar with math and numbers. Shaking my head at the preschool tone thus far, I read on to learn that I'd been reassigned to the Office of Natural Resources Revenue, the office responsible for collecting royalty checks from fossil fuel companies operating on public lands.

My area of expertise was climate change, an issue President Trump and his minions were profoundly ignorant of, and I was focused on climate impacts to American Indians and Alaska Natives, who this administration had little interest in helping. So, I did not expect support from the new administration at the time given their record on climate denial. Sending me to the office that collects royalty checks, however, was a transparent attempt to get me to quit, and I was surprised by this retaliatory display of (unlawful) adolescent behavior.

When I learned that I was one of dozens of senior executives reassigned that night in an unprecedented purge of career civil servants, I realized something that I should have seen coming: These guys (they were all men) were ham-fisted amateurs, and everything I learned from that point forward confirmed this to be true.

The week after the reassignment letters went out, DOI Secretary Ryan Zinke testified to Congress that he intended to use reassignments to trim the DOI workforce. It didn't seem to occur to him that reassignments don't trim the workforce, so he was essentially admitting that his intention was to use reassignments to get people to quit—which is unlawful. More importantly, he was intentionally taking me off the job of coordinating the federal response to help Alaska Natives faced with dangerous climate impacts, thereby increasing risk to these American citizens.

Department of the Interior in the crosshairs

As Director of DOI's Office of Policy Analysis, it was my job to understand the most recent scientific and analytical information regarding matters that affected the mission of the agency and to communicate that information to agency leadership.

I never assumed that agency leadership would make their decisions based entirely upon that information, but I did assume that they would take it into consideration. That proved true for 6 years while I was in that position prior to the Trump Administration.

That all ended with the Trump political team at DOI, which aggressively sidelined scientists and experts, flattened the morale of the career staff, and quickly got to work hollowing out the agency. They and their industry sponsors were eager to hobble the agency responsible for ensuring that public lands are managed for all Americans, not just the fossil fuel industry.

This created a conundrum for career civil servants, who are well aware that an incoming Republican Administration will always focus on resource extraction over conservation. Regardless of who is in the White House, the career staff have pledged to support and defend the Constitution and advance the mission of the agency.

But what if their leaders are trying to break down the agency? What if their directives run counter to the agency mission as dictated by Congress? What if political appointees are intentionally suppressing science that indicates they are doing harm, and in the process putting Americans and the American economy at risk?

During the Trump Administration, career staff had to ask themselves these questions nearly every day, or at least decide where their red line is. For me, the Trump appointees crossed it by putting American health and safety at risk and wasting taxpayer dollars.

Ignoring imminent danger to Alaska Natives

Rapid climate change is impacting every single aspect of the agency mission, and it was my job to evaluate and explain these threats. For example, as the federal trustee for American Indians and Alaska Natives, DOI is partially responsible for their well-being. With over 30 Alaska Native villages listed by the Government Accountability Office as acutely threatened by the impacts of climate change, and a handful of villages in imminent danger, it should be a top priority for DOI to help get these Americans out of harm's way as soon as possible.

"Imminent danger" means some villages are one storm away from being wiped off the map. It's a one-two punch for coastal villages—the permafrost beneath their feet is thawing while the sea ice barrier that would normally protect them from waves and storms no longer sets up until well into the winter. The long, narrow spits and barrier islands that these villages sit on are small to begin with, and now there are waves chewing away meters of shoreline at a time. For villages along rivers, the combination of permafrost thaw and river flooding is eroding the land right out from under their homes.

For 6 years, I worked with an interagency team to address this and related Arctic issues and spoke publicly about the need for DOI to protect Americans and address climate impacts in the region. When the Trump Administration came in, however, this was frowned upon. That June 10 email came 1 week after I spoke at the United Nations on the importance of building resilience to climate change.

So, it was pretty clear to me and my colleagues that the reassignment was retaliation for my work highlighting DOI's responsibility to address climate change and protect American citizens, and that as a result of this retaliation people could suffer. Recognizing this, I submitted a whistleblower complaint to the Office of Special Counsel, penned an Op Ed for the Washington Post, and became the Trump Administration's first public whistleblower.

The ensuing Inspector General investigation of the Senior Executive purge revealed that in their eagerness the Trump team had broken every single one of the Office of Personnel Management guidelines for reassigning senior executives and left no paper trail to justify their actions. They checked every box for management failure, including discrimination, as over a third of the reassigned executives were American Indian.

More importantly, in my view, the personnel action also sent a signal that scientific information and the needs of Americans in danger were no longer a priority. In the years that followed, there were abundant examples to demonstrate their assault on science.

Among them, former DOI Secretary Bernhardt ignored and failed to disclose over a dozen internal memos expressing concern about the impacts of oil and gas exploration in the Arctic National Wildlife Refuge; former DOI Secretary Zinke canceled a National Academy study on the health impacts of coal mining right before lifting a moratorium on coal leasing; and Zinke instituted a political review of science grants, led by an old football buddy, that bottlenecked research and canceled studies.

Not only did this group ignore science and expertise, they crossed the line by actively suppressing it—at the expense of American health and safety, our public lands, and the economy.

They intentionally left their best player on the bench, with the intent to throw the match.

Speaking truth to power: betraying the public trust

From that point until the end of the Trump Administration, the morale bottomed out in the agency, with career staffers looking over their shoulders and trying to keep their heads down. Political appointees showed no hesitation to reassign, relocate, or otherwise make life difficult for career employees—particularly the scientists and experts that they considered a threat.

Agency scientists self-censored their reports and deleted the term climate change to avoid being targeted by political appointees. They were barred from speaking to reporters without advance permission from the agency, faced new barriers to attending the professional conferences that are part of the job, and their work was incompletely communicated to the public, if shared at all.

As I told two congressional committees in the summer of 2019, these conditions do not reflect a culture of scientific integrity, but one of fear, censorship, and suppression that was keeping incredibly capable federal scientists from sharing important information with the public or participating as professionals in their field.

This is no accident. As empowered by Congress, an effective DOI with high-functioning bureaus and offices must operate on behalf of Americans to ensure the conservation or sustainable use of our natural resources into the future, must look out for American Indians and Alaska Natives, and it should prevent private industries from laying waste to public lands.

If, however, the agency is being led by representatives from those very same industries, it is in their interest to hobble the agency so that even when they are no longer in the driver's seat, the agency will struggle to enforce regulations and stand against them. An added bonus to hobbling the agency and its scientific enterprise is that it also compromises the public's trust in the agency, furthering an industry-first agenda.

It goes without saying that this is a betrayal of the public trust. It is bad faith. It is disingenuous. This collapse of ethics and integrity norms at the agency is truly shameful.

To rebound from this assault on science and democracy, we must do the following:

- Support and strengthen laws preventing the politicization of science.
- Pass new laws that provide real accountability in the case of ethics violations.
- Hire a new legion of scientists and experts to provide support and backup to the weary civil servants who have been through this assault.
- And climate-proof agency operations so public lands can be part of the solution to climate change, not the other way around.

The amateurs did their best to hobble the agency. Like parents cleaning up a child's room, the Biden Administration must put time and effort into repairing the damage. However, this re-boot does provide the opportunity to imagine what it would take to elevate the agency to a high-functioning steward of our nation's public lands, an effective trustee of American Indians and Alaska Natives, and a leader in addressing the urgent demands of the climate crisis. Some other ways to clean up DOI are as follows:

- Repurpose funding and staff for the overinflated fossil fuel leasing and permitting programs to address the chronic restoration and maintenance needs on public lands.
- Invest in the staff and grant programs necessary to help American Indians and Alaska Natives thrive in the face of the climate impacts that their frontline communities face.
- Set a carbon budget and actively begin to draw down the proportion of US greenhouse gas emissions that come from federal lands.
- Once again hire and invest in the science enterprise that is so essential to effective policy making.

At a time when America is simultaneously facing climate, biodiversity, social justice, and public health crises, it is essential that federal agencies like DOI evolve into a new form of public service that is not beholden to industry, but that is driven by clean and renewable energy, cares for its people, and values natural landscapes for the many life-giving ecosystem services that made the United States such a global power.

Why scientists should talk to elected officials

Randi Spivak, Public Lands Program Director, Center for Biological Diversity

Public policy should be firmly grounded in the best available science to ensure that the crucial goals of biodiversity conservation and the climate are achieved. While this may sound like an observation of the obvious, it is anything but ensured when it comes to laws, regulations, and policies at the federal, state, and local levels. Public policy reflects societal values and is often a product of political compromise. Unfortunately, there's no strong correlation between the quality of our science and the policy we derive from it: good policy does not come automatically from good science.

I have worked to protect public lands, wildlife and biodiversity for over 25 years as a policy expert and lobbyist on Capitol Hill. As director of the public lands program at the Center for Biological Diversity, I now work to ensure that the proper use of the best available science is the foundation of public policy. Central to that success is bringing—and translating—science to decision makers.

The challenge is that too often the best science poses an "inconvenient truth" that conflicts with an elected official's worldview, the financial interests of major political donors, or the perceived ability to get re-elected. For example, a member of Congress who represents a district where the fossil fuel industry plays a dominant role may either downplay the role fossil fuel burning plays in climate disruption or dismiss climate change outright. A senator from a state where the timber industry is prominent may push back on science that shows industrial logging is harmful to a range of forest-dependent species and watershed integrity.

The good news is that not every elected official needs to be convinced. In the case of the House of Representatives, only a majority is needed. In the Senate—taking into account the filibuster rule—it takes a three-fifths vote to win.

Below I discuss what works and what doesn't for scientists to engage with elected officials and other policymakers (appointed officials, staff, etc.). I have brought scientists to educate lawmakers in Washington, DC, as well as helped them prepare congressional testimony and briefings and reach out to media.

Scientists operate in a rational world, where hypotheses are accepted or rejected based on rigorous, replicable methodologies and empirical data (see Chapter 1, The Nuts and Bolts of Science-Based Advocacy). The operating environment of politicians is a different, though still rational, landscape in which re-election or advancement to higher office is the central motivator. "Political science" might better be called "political art:" The goal of peer-reviewed science is to advance truth through method, where the goal of policymakers is to maintain popular support—a goal that may or may not align with truth-telling.

Scientists and policymakers have different goals, attitudes toward information, and languages, and can also have different perceptions of time.

Natural-resource decisions are often controversial, meaning the interests of special interests are threatened—still another reason why scientist voices are critical for ensuring policies reflect the best science. While policy experts like me rely on scientific papers to underpin policy recommendations, scientists talking directly to congressional staff and agency decision makers can have a significant positive impact. Scientists bring increased credibility and context to conversations, giving decision makers a deeper understanding on the likely outcomes of varying policy approaches.

What to say and how to say it

Most often, as a scientist, you will be meeting with congressional staff, rather than the representative or senator. This is not a slight; it is the way business gets done. The staff you'll meet with will be responsible for environmental issues for that elected official, and such staffers' level of knowledge varies greatly. For example, staff for the committees of jurisdiction for House Natural Resources or Senate Energy and Natural Resources are fairly knowledgeable compared to those whose member is on the committee of jurisdiction but who handles many other issues. In almost all cases, staff will know more than their bosses—even overworked and stretched thin, staff have more time than elected officials to learn the issues.

Plan on a half-hour meeting, though you may get more or less time. Staffers are very busy and have many other meetings lined up for the day. So be respectful of their time, make sure you arrive a few minutes early, and try not to schedule meetings back-to-back. You'll need flexibility if one meeting starts late or runs long.

Here a few tips on how to present your material.

1. *Rehearse:* If your meeting group includes more than just you, make sure you rehearse with your group before the meeting. Decide which person will open the meeting and state the purpose for the meeting, what the order of speakers will be, who will cover what issues or points, and who will close the presentation. Rehearse out loud. You may be tempted to skip this step, but *please don't*: Your meeting will be far more effective if your presentation is succinct and well organized and your "ask" is clearly articulated.

2. *Set up the meeting objectives:* State your purpose up front and restate the reason you asked for the meeting, for example: "Thanks for taking time to meet with me/us. I know the senator is considering signing onto the 'trillion trees' bill that would mandate tree planting on national forests. I'd like to talk to you about my research on forest carbon and discuss the potential unforeseen consequences the trillion trees bill would have on climate change and forest resiliency. If your goal is to increase forest carbon sequestration, I am going to suggest better ways to achieve that."
3. *Establish your credentials:* Be brief but establish your expertise. For example, "My field of study is forest carbon and disturbance processes. I'm a professor of biological science at Faber College. I have done field research for the past 15 years and published 25 papers on this topic."
4. *Be succinct and policy relevant:* Policymakers often want a "bottom line" and may not have time for a deep dive into nuance. When presenting the science, summarize the findings up front (see Chapter 1, The Nuts and Bolts of Science-Based Advocacy). Policymakers need a simple one-line answer to what are often, especially to scientists, complex issues. Nuance is not well embraced by a process that in the end comes down to a yes-or-no vote, so presenting your results to public policymakers is *not the same* as presenting your results to colleagues. Failing to state caveats to your colleagues is deadly to careers, but excessive caveats to policymakers can leave them uncertain of what to believe. Never misrepresent scientific findings or degree of uncertainties; rather try to simplify what is known and unknown in plain language, as though talking to a nonscientist friend. Importantly, you can help shed light on the consequences of specific legislation or policy based on your knowledge of the relevant body of science.
5. *Don't use jargon:* Every area of expertise has special language. The congressional staffer will likely not be familiar with many technical terms and acronyms, so simplify highly technical concepts or findings. Don't say "C," say "carbon." Instead of "Our research found that there was a significant deviation from the mean, rejecting our null hypothesis," say, "Our research showed that intact habitat was the most accurate predictor of species reproductive success."
6. *Be sure to follow up:* A short email or letter (via email) thanking the staffer for making time for the meeting is always appreciated. Be prompt and include any follow-up material you committed to provide; briefly summarize your position, scientific findings, and/or the recommendations you made in the meeting.

Best practices for congressional testimony

If you are asked by a congressional committee to speak at a congressional hearing, the following are a few tips on best practices. A congressional hearing is the principal formal method by which congressional committees of jurisdiction collect and analyze information and opinions in the early stages of legislative policy making. Hearings usually include oral testimony from witnesses and questioning of the witnesses by members of Congress. There are several good guides online for writing and preparing to speak at a congressional hearing. The staff person who extended the invitation will work with you to help you prepare.

1. *What to include in testimony*. There are two parts to congressional testimony: (1) your oral presentation and (2) your written statement. Even if you are not invited to present testimony to the committee, any member of the public can submit written testimony for the record. Your oral testimony should be short—no longer than 5 minutes—and cover the highlights of your written statement. As a technical expert, you should appear agnostic on the politics and focus on delivering the scientific facts and your professional opinion on the consequences of the proposed legislation. You should state what legislation or issue area you are testifying about, briefly establish your scientific credentials, and thank the committee chair and members of the committee. Open your testimony with a summary of your main points. The rest of your testimony should support your points, including, if appropriate, dispelling a popular but scientifically unsupported finding. If you work for a science organization or academic institution you should consider having your written testimony reviewed in advance by your peers. It can also help to have your testimony be reviewed by a nontechnical expert to make sure it will be understood by an intelligent, but inexpert, audience.
2. *Know your audience.* Before you testify you should familiarize yourself with members of the Committee and their known policy positions on the topic of the legislation or subject of the hearing. You can research what statements members have made and/or bills they have introduced to give you a sense of who they are. With this knowledge, you can craft your testimony to address for example, scientifically unproven claims as well as anticipate questions you may get from certain members of the panel. Try, as well, to be up to date on current events related to the hearing topic so you are aware of relevant context for the hearing. Knowing who else is testifying can also inform your oral testimony. For example, you may want to coordinate with other witnesses who may be testifying and share your perspective to prevent repeating information and use the limited hearing time more effectively.
3. *Be ready for questions.* Congressional hearings include time for members of the Committee to ask questions of the witnesses. You may get both friendly and hostile questions. Some members will use their time to make a long-winded statement, then demand a yes-or-no answer. Your preparation should include being ready to answer questions from both sides of the debate. Make sure you respond directly to the question with a very succinct answer; time will likely not allow for elaboration. Get to the point right away, offer support for your answer, and if time allows, summarize it again. For example, a member of the committee might cite the amount of carbon emissions from wildfires and carbon-sequestered wood products to support policies to suspend environmental laws to increase logging. You might answer that while wildfires do emit

carbon and wood products do store some carbon, a full life-cycle analysis shows that only a fraction of the carbon is stored in wood products and logging emits more carbon than wildfires. Always be respectful, but don't feel you can't disagree with the viewpoint of a member of Congress. If you get hit with a question for which you don't have a ready answer, it is very perfectly acceptable to say, "Respectfully, I'm not prepared to answer your question at this time, but I will submit an answer for the hearing record." Hearing records usually close 10 business days after the hearing.

Speaking truth to power: is it worth it?

Meeting with members of Congress and testifying before Congress takes time and resources. It is fair to ask whether the returns justify taking time away from doing your research.

It's not only worth it but vital. Susan Solomon, an atmospheric chemist who coled the Intergovernmental Panel on Climate Change (IPCC) that won the 2007 Nobel Peace Prize, said "science does have a duty, when called upon, to provide information that's important to society." In the context of the IPCC, she added, the goal was not to make policy-prescriptive statements but policy-relevant statements.

As noted earlier, natural-resource policy debates are often contentious and socially divisive. Bringing unbiased scientific information to the fore is critically important as decision makers will weigh other elements in making their complex political deliberations.

A classic policymakers' joke goes: "I've made up my mind, so don't confuse me with the facts." Scientist engagement, bringing high-quality, relevant science to the policy debate can have an important impact on policy decisions. None of these actions crosses the line between science and advocacy if you are careful to discuss the policy implications of your research without stipulating a preference for a particular policy decision. It's not about your personal preference but rather what science, uniquely, can reveal about the consequences of policy options. Congress is a critically important power source with direct influence on the most pressing natural-resource and climate change challenges of our time and scientist input into those policies, is vitally important.

Lobbying 101: tips to effective legislative advocacy online

Jennifer Mamola, John Muir Project

Lobbying decision makers is one of the most effective ways to influence public policy. Effective policy making requires citizen participation. Any citizen can and should lobby decision makers on proposals and policies that impact their community, livelihood, and profession. Constituents play a critical role in educating legislators, especially scientists that have specialized knowledge that can inform policies. Scientists interested in lobbying can also team with conservation groups that know their way around DC. There are several—some examples include—Center for Biological Diversity, John Muir Project, and Union of Concerned Scientists. Each of these groups, and many others, maintain lobbying staff that you can team with to get around the Hill, on or off-line.

Don't reside around the DMV (District-Maryland-Virginia) area and have no future plans to travel here? No worries! With the 2020 pandemic, meetings have shifted online or over the phone, which, in my personal experience, has been for the better. Don't get me wrong, I miss roaming the halls of Congress terribly (never thought I'd say that), but I've noticed that my meetings last a solid 30 minutes at minimum and I get to bring beloved colleagues along for the ride. Whereas precovid, when I was able to play Congressperson bingo and guess what political faces were rushing by me, I was lucky to get 15 minutes with a very distracted staffer or fellow.

Depending on the topic at hand, I usually aim to chat with relevant committee members or what districts might be most impacted by our work. For example, with the 117th Congress just starting in January 2021, I began by requesting meetings from Freshman and Western state members and those that sit on Natural Resources and Energy committees. Occasionally, I would throw in Appropriations members when funding talks begin in March usually. However, this will likely vary for you.

Once you've determined what offices you'd like to meet with, you'll want to obtain the staffers email address as it's the best avenue to request a meeting. You can try looking up the person on the internet, but I find it much easier to just call the DC office, which you should be able to find online. Depending on the issue you're wanting to discuss, that will determine the staffer you're wanting to get a meeting with. Typically, I chat with Energy &/or Environment committee staffers and occasionally they're the same person. The staffer answering the phone should be able to assist you in figuring out which staffer is best aligned with the topic you're wanting to cover, so have a brief summary of what you're wanting to share with the office. When dealing with a House member the staffer email is: firstname.lastname@mail.house.gov, whilst the Senate is: firstname_lastname@membername.senate.gov.

Keep the email request brief, for instance, here is a template to consider using:

> *Dear Staffer,*
> *My name is _____ and I'm the ____ expert with the _____ organization. I'd greatly appreciate the opportunity to sit down and discuss forest related policy and issues, and the inherent link between forests and mitigating climate change. Please find a brief fact sheet attached.*

Might you be available the first week of _____ to sit down and chat about preserving our forest here in the United States as they're pertinent to solving our climate crisis? Thank you for your time, and I hope to meet with you next week.

I like to email about 20 staffers at a time, depending on how busy I'd like to keep the next week or so as one can never be certain how many will respond. And don't let lack of response deter you, as it might take a follow-up email or two to get the staffers attention. Just remain polite and firmly short in your friendly reminder emails. Once I have a date and time confirmed I like to inquire if they favor whichever video conferencing platform you have at your disposal or old-school phone call. Personally, I like to try and encourage video as I prefer seeing the person and try to gage their attention and interest. Albeit they don't always log-in on video, but alas it's worth a try! Lastly, I try to inquire how much time they have. It tends to be 30 minutes, but occasionally can go longer, but it's nice to be mindful of their time.

If I have colleagues and/or scientists attending any of my virtual meetings, whether via video or phone, I tend to send an email around the day before or morning of with the structure of how the meeting will go and a brief background of the staffer(s) if I can locate them on Linkedin. Whomever requested the meeting should open and thank the staffer for their time and give a brief background, allow others to introduce themselves and then follow with the suggested layout above from Randi Spivak. It's the hope that the office will be engaged and inquisitive, but don't fret if they don't seem so as they merely might be distracted with other issues at hand as they're juggling many issues and portfolios.

Now that you have their attention, it's helpful to keep them informed. I try to send relevant articles or studies that I believe might be useful in highlighting and keeping our work at the forefront. Depending on any upcoming hearings that might be pertinent, try to keep them on a monthly basis. Unless you find your topic making headlines and want to keep misinformation at bay. An office once told me that Capitol Hill tends to be 20 years behind the science, so don't shy away from sharing your expertise and knowledge.

Speaking truth to power: preparing for DC lobbying (what to bring, wear, get around)

Now that you are prepared for the congressional meeting or testimony, here's what to expect if you choose to skip the virtual and trek the long corridors and hallways of the US Senate and House of Representatives. It's always best to dress in business attire, even on those sticky summer days. But don't fret, the air conditioner is usually on full blast in the summer, and the heater is cranked up in the winter. Definitely wear your most comfortable shoes as miles can be clocked on any day strolling the halls of Congress.

While there are cafeterias and vending machines in the basements, it is recommended that one bring their reusable water bottle and/or coffee mug. There are water coolers in every office and usually every hallway. Some offices have local beverages and snacks that are made in their district or state, but always best to come prepared with your own, especially if you have allergies as peanuts can be found in many offices.

The Capitol building is flanked by the House offices to the South and the Senate and to the North. Cannon, Longworth and Rayburn are where you'll find your 435 Representatives from the 50 states and 6 nonvoting Delegates that represent the District of Columbia and the US territories; while Hart, Dirksen, and Russell house the 100 US Senators. Be prepared for security at each entrance, but not to worry as there are underground tunnels that connect the three buildings on either side of Capitol Hill. However, you'll need special access to utilize the tram that connects the Capitol to either side. You can usually obtain this by contacting your local Representative or one of your Senators to request a tour at least 3 months in advance.

The District is very accessible by metro and bus, you can purchase a metro pass at any metro station. Otherwise, there are usually plenty of cabs about town, especially around Capitol Hill, or can utilize your preferred ride sharing app.

Last but not least—you are in DC—enjoy the culture as a side benefit. The city has many outstanding museums, botanical gardens, art galleries, pubs and restaurants all within walking or metro distance of the Capitol. And the majority of the sites are free of charge!

Shifting the burden of proof to minimize impacts during the science-policy process

Kara A. Whittaker and Peter Goldman
Washington Forest Law Center

Who should carry the burden of proof during an adaptive management process?

Renowned astrophysicist Dr. Martin Rees once wrote that "the absence of evidence is not evidence of absence." While Dr. Rees may have been referring to cosmology and the purported absence of life elsewhere in the universe, his profound point applies equally to how government agencies sometimes make scientifically-risky permit and regulatory decisions that could harm people or the environment because evidence of harm does not yet exist. The precautionary principle (Box 12.1) arose in part to address this problem *and* because "...the pace of efforts to combat problems such as climate change, ecosystem degradation, and resource depletion is too slow and that environmental and health problems continue to grow more rapidly than society's ability to identify and correct them" (Kriebel et al., 2001).

Box 12.1 Definition and application of the precautionary principle Raffensperger and Tickner (1999).

The definition of precautionary principle that we use herein originated from a consensus statement of the Wingspread Conference convened by the Science and Environmental Health Network (Raffensperger and Tickner, 1999):

(*Continued*)

Conservation Science and Advocacy for a Planet in Peril.
DOI: https://doi.org/10.1016/B978-0-12-812988-3.00016-8

Box 12.1 (Continued)

When an activity raises threats of harm to human health or the environment, precautionary measures should be taken even if some cause and effect relationships are not fully established scientifically. This idea, known as the "Precautionary Principle," is seen by environmentalists and public health experts as the key to protecting ecological and human health. The precautionary principle has four central components:

1. taking preventive action in the face of uncertainty
2. shifting the burden of proof to the proponents of an activity
3. exploring a wide range of alternatives to possibly harmful actions, and
4. increasing public participation in decision making

We are environmental professionals, a scientist and a lawyer, who regularly observe instances where important long-term environmental policy decisions are made based on vigorously negotiated scientific assumptions. When regulators disregard the precautionary principle (also see Chapter 1, The Nuts and Bolts of Science-Based Advocacy) and habitat-degrading activities continue during the ensuing period of uncertainty, the burden often falls on threatened or endangered species or an adjacent community that can ill afford such a burden. The problem is exacerbated when regulatory agencies assume that their regulatory prescriptions governing habitat-degrading activities will not have an appreciable adverse environmental impact while prescriptions are studied, and that any regulatory shortcomings will be addressed after years of monitoring, study, and negotiation. The defect with this assumption is that it assumes the potential impact *will* be studied, that there *will* be a *timely* post-monitoring/study policy response, and that irreversible harm to natural resources or communities *will not occur* during the often-lengthy regulatory response period. Although ongoing environmental degradation has helped fuel a paradigm shift in natural resource management from a traditional multiple-use, sustained yield approach towards a more holistic, interdisciplinary, ecosystem management approach (Yaffee 1999), we have witnessed the pitfalls of this approach in practice.

In this chapter, we provide two case studies where, in our view, federal and state regulators have disregarded the precautionary principle and where, as a result, species and adjacent communities have been forced to bear the cost of scientific uncertainty. The first involves a 50-year federally-issued permit for the Washington Department of Natural Resources (DNR) to conduct logging in mature forests otherwise needed for nesting and recovery by the Marbled Murrelet (*Brachyramphus marmoratus*), a federally threatened bird species listed under the Endangered Species Act. The second involves the State of Washington's delayed regulatory response to strengthen the rules governing forest practices (Rules) conducted on and around steep and unstable slopes susceptible to logging-induced landslides. Both case studies stem from federally permitted Habitat Conservation Plans (HCPs) with adaptive management components. HCPs are formal agreements between federal agencies and nonfederal parties (such as state agencies or private corporations) that are a requirement for receiving an incidental take permit under section 10(a)(1)(B) of the Endangered Species Act. HCPs describe the anticipated effects of the proposed taking (direct or indirect harm to the species), how those impacts will be minimized and mitigated, and how the HCP is to be funded (U.S. Fish and Wildlife Service and National Marine Fisheries Service, 2016).

These two examples demonstrate that federal and state natural resource agencies—often under considerable political pressure from regulated industries, such as the timber industry in Washington—sometimes delay environmental policy decisions, and these decisions often rely heavily on monitoring and the commitment to revise policy prescriptions as the science develops. The absence of evidence (that a new prescription is needed to prevent environmental harm) is wrongly taken as the evidence of absence (that the existing prescription works because it has not yet been disproven). This practice unfairly puts the burden of proof on those people, wildlife, and ecosystems *subject* to the impacts of concern as long as the application of science in a policy setting is delayed. To minimize the potentially harmful impacts of emphasizing and prolonging the absence of evidence, we urge a more precautionary approach, one that shifts the burden of proof to those conducting on-the-ground activities and which better forces and accelerates the science-policy process. This approach is consistent with the Hippocratic Oath to 'do-no-harm' to the 'patient' and appropriately places the burden of proof on the proponent of an action to demonstrate no harm will occur (Kriebel et al., 2001; Chapter 1, The Nuts and Bolts of Science-Based Advocacy).

The DNR HCP: Marbled Murrelet Long-Term Conservation Strategy

The DNR manages roughly 809,000 hectares (ha) of state forests across Washington. In 1997, under section 10 of the Endangered Species Act, the U.S. Fish and Wildlife Service (USFWS) approved a 70-year "incidental take permit" under the State Trust Lands HCP. This permit allows DNR to incidentally "take" (harm) federally listed threatened and endangered species in the course of the State's timber sales program, which ranges from about 944,000 to 1,180,000 m^3 (400 to 500 million board feet) per year.

The Marbled Murrelet (murrelet), a unique and iconic seabird that nests in mature coniferous forests along the Pacific coast, is one of the DNR HCP-covered species (Fig. 12.1). The murrelet's population decline in

Figure 12.1 Marbled Murrelet on its mossy nest platform. *Photo courtesy of Brett Lovelace and Oregon State University.*

Washington has been linked to loss and fragmentation of nesting habitat, as well as to threats in the marine environment (U.S. Fish and Wildlife Service, 1997). It was federally listed as threatened in 1992 (U.S. Fish and Wildlife Service, 1992) and uplisted to state endangered in 2017 (Desimone 2016). In its HCP, DNR committed to "...[h]elp meet the recovery objectives of the U.S. Fish and Wildlife Service, contribute to the conservation efforts of the President's Northwest Forest Plan, and make a significant contribution to maintaining and protecting marbled murrelet populations in western Washington over the life of the HCP" (Washington Department of Natural Resources, 1997, p. IV.44). DNR manages roughly 567,000 hectares of forest within the murrelet's range (within 88 km of marine waters; Fig. 12.2), and as of 2019, there were roughly 84,000 hectares of murrelet habitat on DNR-managed lands (Washington Department of Natural Resources, 2019). Compared with other land ownerships, DNR-managed forests are disproportionately important to murrelet conservation in both the short- and long-term—spatially, they are relatively productive, well-distributed, low elevation forests, and temporally, existing habitat on DNR-managed lands is needed to serve as a "bridge" to support the murrelet population while it is most vulnerable to extirpation over the next 30–50 years (Raphael et al., 2016; Peery and Jones 2018).

At the time the DNR HCP was first implemented, the USFWS and DNR lacked a solid understanding of the murrelet's habitat needs, and the agencies agreed more time was needed to conduct the science to properly inform a policy agreement over such large spatial and temporal scales. Accordingly, as a condition of its HCP, DNR agreed to identify, model, and survey murrelet nesting habitat for "occupancy" to avoid take under Section 9 of the Endangered Species Act. In adaptive management fashion, these steps constituted the

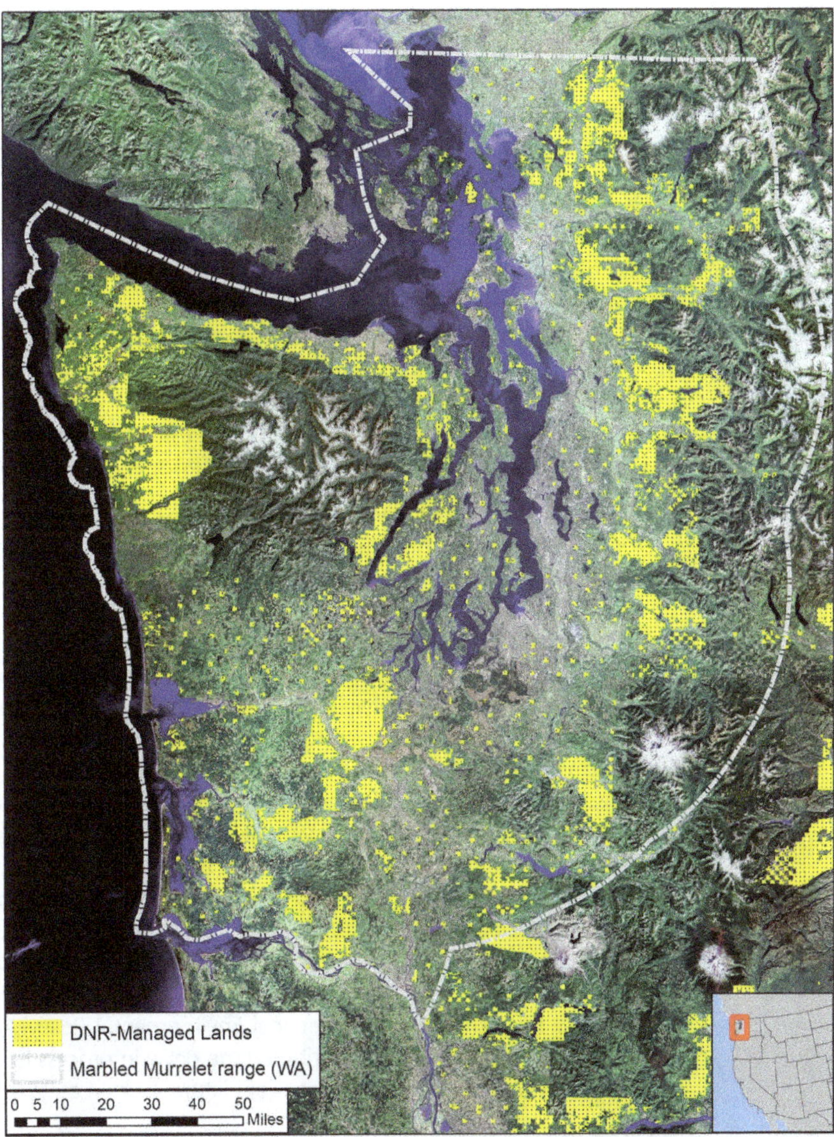

Figure 12.2
The range of the Marbled Murrelet in Washington and DNR-managed lands covered by the DNR HCP. *DNR*, Department of Natural Resources; *HCP*, Habitat Conservation Plan.

Marbled Murrelet Interim Conservation Strategy, followed by a Long-Term Conservation Strategy (LTCS) to be implemented for the remainder of the term of the HCP (until 2067; Washington Department of Natural Resources, 1997). The DNR's interim habitat relationship studies and other new murrelet research was designed to enable a more rigorous understanding of the murrelet's habitat characteristics and threats to its reproductive success, both necessary components of effective conservation efforts.

To synthesize the best available science, in 2004 DNR convened a panel of Marbled Murrelet experts whose mission was to develop a detailed set of recommendations and supporting analysis of conservation opportunities for the LTCS (*aka* the Science Team Report; Raphael et al., 2008). The purpose of the Science Team Report was to "provide the foundation for a credible, science-based LTCS that meets DNR's HCP requirements" consistent with recovery objectives for the species across its listed range (U.S. Fish and Wildlife Service, 1997). Sufficient information was available at the time to generate conservation recommendations for the murrelet in four of six DNR HCP planning units. The Science Team's modeling demonstrated that, if their recommendations were adopted as the LTCS, then the biological goals for murrelet population size, population stability, geographic distribution, and resilience could be achieved across all landownerships statewide. The Science Team Report was considered "a thoughtful and innovative piece of scientific work unlike that done by any other landowner for the conservation of this federally threatened species" (Raphael et al., 2008).

After a decade of extensive data collection, analysis, and synthesis, the stage was set for DNR to finalize a Marbled Murrelet LTCS. During the Science Team process, in 2006, DNR conducted the first concurrent State and National Environmental Policy Acts public scoping processes for the LTCS, and the Science Team anticipated the LTCS would be adopted by *2010* after completion of these public processes, adoption of the LTCS by the Board of Natural Resources, and approval of the HCP amendment by the USFWS (Raphael et al., 2008). Unfortunately, rather than implementing the Science Team recommendations, DNR essentially *ignored* them and continued to auction timber sales in areas the Science Team had mapped as important to murrelet conservation and recovery (Marbled Murrelet Management Areas). When forest bird conservationists, such as us, became aware this was occurring, we began to legally appeal the approval of these timber sales to prevent DNR from logging areas of forest that could preclude recovery of the species. With a coalition of conservation organizations, we publicly advocated for the implementation of the Science Team Report as the long-awaited LTCS. Under increasing pressure from us and the USFWS, DNR withdrew its plans to continue logging in Marbled Murrelet Management Areas and resumed the joint State and National Environmental Policy Acts' public process for the LTCS in 2012. DNR later attributed the delay in process to the national economic downturn (Washington Department of Natural Resources, 2019).

DNR spent the next seven years working on the LTCS in a public process. They expanded upon a complex habitat model first proposed by the Science Team, developed an analytical framework to estimate the amounts of habitat take (log) and mitigation (deferral), and contracted a Population Viability Analysis (Peery and Jones 2018). DNR also engaged heavily with USFWS behind closed doors to negotiate various aspects of the LTCS.

In 2016, DNR presented a set of LTCS alternatives in a Draft Environmental Impact Statement, then a 2018 Revised Draft Environmental Impact Statement and draft HCP amendment, and lastly a 2019 Final Environmental Impact Statement and final HCP amendment. Our coalition commented and advocated extensively on behalf of the murrelet at every opportunity, proposing our own "Conservation Alternative" that exceeded the conservation proposed by any of the alternatives analyzed, because none of the proposed alternatives sufficiently mitigated or minimized the take of Marbled Murrelets or made the "significant contribution" to murrelet conservation committed to in the HCP.

Among the alternatives analyzed, DNR concluded that the alternative patterned on the Science Team-recommended approach (Alternative F) was most likely to help achieve murrelet population biological goals, but DNR rejected this alternative as too costly to the trust beneficiaries. Under the direction of the Board of Natural Resources (the ultimate decision-makers for state trust land policies), DNR sought a different approach. DNR sought to balance, on a 1:1 basis, the amount of incidental take against the amount of mitigation the LTCS would provide, and this became Alternative H. Although the federal Biological Opinion of the LTCS concluded that Alternative H would not 'jeopardize' the existence of the species across its *listed* range, it nevertheless found that DNR's mitigation may not be sufficient in light of future natural disturbances, and determined an 'environmentally preferred alternative' much more aligned with the Science Team approach (Alternative G; U.S. Fish and Wildlife Service, 2019). Despite the scientifically-questionable 1:1 take-to-mitigation ratio, the USFWS approved DNR's amended permit to take murrelets for the next fifty years.

The Washington murrelet population has continued to decline by four percent per year on average since monitoring began in 2001, and it has yet to stabilize (McIver et al., 2020, Fig. 12.3). This statistically significant negative trend represents a loss of approximately 4902 murrelets (47% of the population size) over the past two decades, plus previous population losses. In endangered species management, time is of the essence to avoid passing the point of no return because small populations are especially susceptible to stochastic extinction vortices (Gilpin and Soulé 1986; Carden 2006). Functional extirpation of the Marbled Murrelet from Washington State remains a real possibility, which could impact the population dynamics of the species across its listed range (Desimone 2016). While numerous factors contribute to the murrelet's population trend, we believe the species' chance of stabilizing and recovering would be greater had the LTCS erred on the side of caution and been completed sooner. This is a crystal-clear example of how policy negotiators have disregarded the precautionary principle in favor of permitting extensive logging of habitat (15,000 hectares, Washington Department of Natural Resources, 2019). In the case of the LTCS, a true win-win situation for the murrelet *and* the trust beneficiaries may not be possible without major system reform, and as a result, the murrelet is forced to compete with economic interests and carry the burden of inadequate conservation on DNR-managed forests.

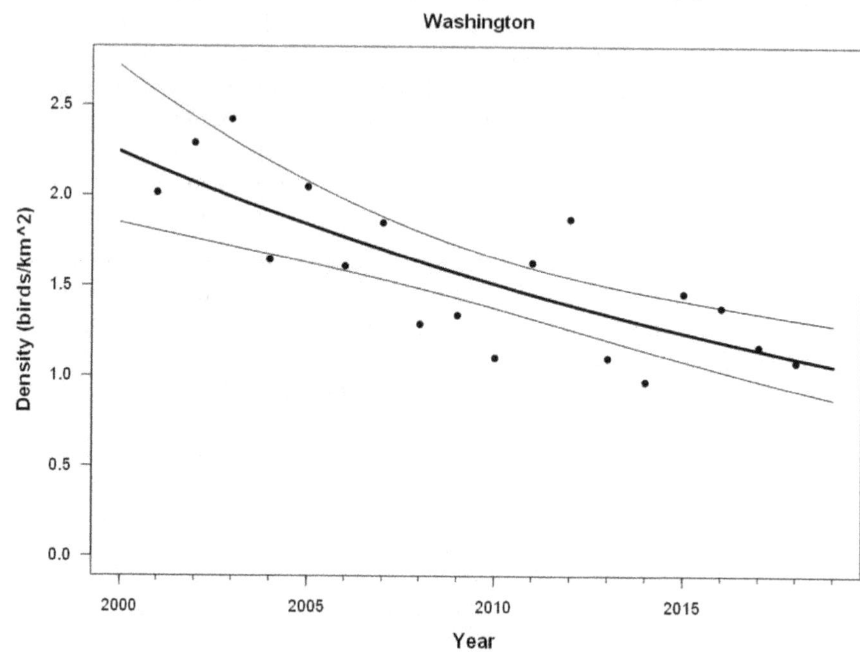

Figure 12.3 Long-term population decline of the Marbled Murrelet in Washington State derived from at-sea surveys (McIver et al., 2020, public domain). Washington exhibited a significant declining trend between 2001 and 2018 (−3.9% per year; 95% CI: −5.6% to −2.1%) where the most recent the population estimate is approximately 5600 murrelets (CI = 2800–8300).

Forest Practices HCP: unstable slopes and landslide hazards

A second case study demonstrating delays or stakeholder obstruction of policy-driven science and failure to apply the precautionary principle involves the forest practices adaptive management process in Washington State. This process is, in theory, designed to adapt the Rules over time, including those governing logging on potentially unstable slopes, in response to new scientific information. Although Washington experienced massive landslide events in 2007 and 2014, resulting in tragic loss of life and property, and the State invested significant resources in its adaptive science-policy process, this process has yielded only minor refinements to the Rules, and for the most part, 'business as usual' continues. We describe this chain of events in detail below.

Since 2005, Washington has a federal Forest Practices HCP that covers activities impacting threatened and endangered aquatic and riparian-dependent species on more than 3.6 million ha of private and state-owned forestlands in Washington State (Washington Department of Natural Resources, 2005). The Rules associated with the Forest Practices HCP regulate forest management activities and includes a multi-stakeholder Adaptive Management Program intended to conduct and then apply scientific research to improve the Rules over time and better protect public resources (water, fish, wildlife, and capital improvements of the state; Washington Administrative Code (WAC) 222-16-010). In order to meet the goals and objectives for water quality and fish habitat, the purpose of the Adaptive Management Program is to:

> *... make adjustments* as quickly as possible *to forest practices that are not achieving the resource objectives... (and) shall incorporate the best available science and information, include protocols and standards, regular monitoring, a scientific and peer review process, and provide recommendations to the board on proposed changes to forest practices rules to meet timber industry viability and salmon recovery.* (RCW 76.09.370(7), emphasis added)

The Adaptive Management Program includes four entities: Timber, Fish and Wildlife Policy Committee (Policy); Cooperative Monitoring, Evaluation, and Research Committee; Adaptive Management Program Administrator; and Forest Practices Board (Board; Fig. 12.4). DNR operationally implements the

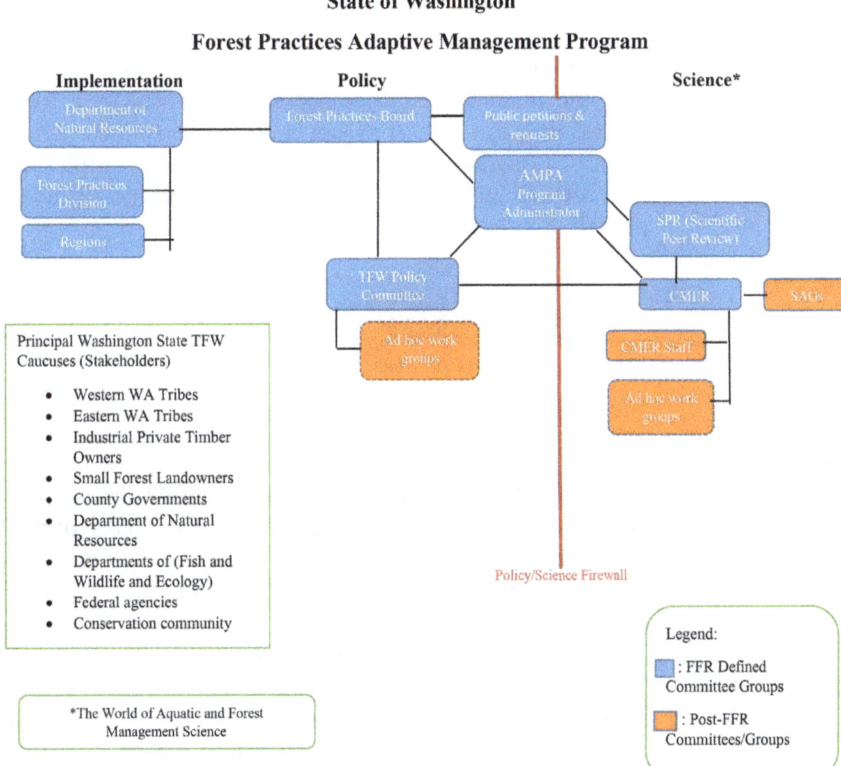

Figure 12.4 Schematic of the structure of the Washington State Adaptive Management Program. FFR refers to the Forests and Fish Report (U.S. Fish and Wildlife Service, 1999, public domain).

program. CMER Scientific Advisory Groups review existing science and conduct original research to directly inform the adaptive management process. Policy consists of members of nine caucuses that make recommendations to the Board for majority-based decisions on Rule changes and guidance (Fig. 12.4). The "firewall" between CMER and Policy is intended for the science process to produce unbiased technical information uninfluenced by non-science interests.

CMER and Policy were designed to be collaborative consensus-based forums with specified time periods during which decision makers must "close the adaptive management loop." This process breaks down, though, when consensus does not occur, specifically

> *. . .in an arena where aquatic resource protection necessitates some level of restriction of forest practices activities and where changes to established rules could have a significant economic impact on forest owners or pose a significant risk to the aquatic resources, disputes can arise at many decision junctures. Left unresolved, disputes could slow or stop the adaptive management process by delaying recommendation or preventing them from reaching the Board altogether* (Washington Forest Practices Board, 2013, p. 18)

Because consensus-based decision-making has the inherent potential to stall progress, the Adaptive Management Program includes a formal dispute resolution process that, in theory, provides a time-driven structure ranging from two to five months. If the dispute resolution process is unsuccessful, then the Board makes the final determination on the matter by vote, with politics and sometimes the media playing strong roles.

The science and policy surrounding the improvement of rules governing logging on potentially unstable slopes

The Forest Practices HCP, Rules, and the 1999 Forest and Fish Report (U.S. Fish and Wildlife Service, 1999, which served as the basis for the Forest Practices HCP) include goals for the management of potentially unstable slopes on non-federal (i.e., state and private) forest lands. The agreed intent of the Rules is to avoid accelerating rates and magnitudes of landslides that could deliver sediment or debris to a public resource or that have the potential to threaten public safety (WAC 222-16-050 (1)(d)). This is because landslides can degrade aquatic habitat contributing to the decline of salmonid, amphibian, and other native aquatic species that are unable to adapt to the altered disturbance regime (Cederholm and Lestelle 1974; Welsh and Ollivier 1998; Montgomery 2004). Regulating forest practices on steep and unstable slopes also "provide[s] clean water and substrate and maintain channel forming processes by minimizing to the maximum extent practicable, the delivery of management induced coarse and fine sediment to streams (including timing and quantity) by protecting stream bank integrity, providing vegetative filtering, protecting unstable slopes, and preventing the routing of sediment to streams" (U.S. Fish and Wildlife Service, 1999).

To identify those forested areas that have the greatest potential to deliver sediment to public resources and/or threaten public safety, the Rules include a list of "Rule-Identified Landforms" (RILs) observed to be susceptible to shallow and/or deep-seated landslides, especially when subjected to forest practices (predominantly logging and road construction). Because logging on or near a RIL requires a site-specific geologic report and review under SEPA, forest landowners typically avoid RILs by buffering or "bounding them out" of their proposed Forest Practices Applications (FPAs) to avoid triggering the more onerous SEPA review process.

An important area of research within CMER concerns the relationship between forest practices and potentially unstable slopes on non-federal forest lands. While the scientific understanding of *shallow* landslide initiation and downstream runout in response to forest practices is relatively well-established (Rood 1984; Swanson et al., 1987; Montgomery et al., 2000), the same cannot be said for *deep-seated* landslides (Miller 2016, 2017). "Unlike shallow (translational) landslides and debris flows that may occur repeatedly and are better understood, secondary failures within compound landslides are less common and present an *unrecognized hazard potential*" (Washington Forest Practices Board, 2016, emphasis added). In order to meet the intent of the Rules to protect public resources and public safety, CMER's Upslope Processes Scientific Advisory Group conducts research focused on the susceptibility of unstable landforms to logging and road construction (Cooperative Monitoring, Evaluation and Research Committee, 2019).

In 2007 and 2014, Western Washington State experienced two large landslide events in forests under the jurisdiction of the Forest Practices Rules and Adaptive Management Program. These two landslide events raised significant concerns about the effectiveness of the Rules and provided the foundation for our thesis in this chapter: that too much logging is taking place under Rules for which there is little science and that Policy negotiators seem to be intractably stuck in their ability to improve upon the 'status quo.'

The December 2007 storm

In December 2007, a series of large storms passed through southwestern Washington and northwestern Oregon. Over the course of three days, the storms brought heavy precipitation (up to 48 cm), rapid snowmelt, and hurricane-force winds (Mote et al., 2007). This storm event triggered at least

Figure 12.5 Shallow landslides on private forest lands following the December 2007 storm. *With permission from David Perry.*

2503 landslides in southwestern Washington, primarily in the form of debris avalanches, flows, and torrents (shallow landslides; Sarikhan et al., 2008; Turner et al., 2010). Landslides blocked and damaged roads and bridges (isolating communities needing emergency response), significant flooding took place on numerous rivers (inundating numerous homes and a major interstate highway with as much as ten feet of water), and at least one person died as a consequence of the storm. The large pulses of sediment and woody debris induced by the storm also severely impacted riparian habitats. The vast majority of landslides recorded in association with the December 2007 storm initiated on private managed forest lands (Sarikhan et al., 2008, Fig. 12.5).

Post-Storm Assessments

Anticipating a large storm event such as this, CMER was well-poised with a study design in hand, to initiate the Mass Wasting Effectiveness Monitoring

Project (*aka* "Post-Mortem Study;" Stewart et al., 2012). The primary objective of the Post-Mortem Study was to evaluate the effectiveness of the Rules (RIL buffers) at reducing landslide density after a major storm event. In the 236 km^2 study area, the Upslope Processes Scientific Advisory Group found that most landslides initiated on hillslopes (as opposed to roads) and delivered sediment and woody debris to stream networks. A sizable proportion of the delivering landslides originated from hillslopes that did *not* fit a RIL definition and did not appear to be correlated with either precipitation intensity or geology. A lower density and volume of landslides was observed where clearcut logging on unstable slopes was avoided. The results of the Post-Mortem Study confirmed that logging without RIL buffers increases impacts to public resources, and when combined to the magnitude observed in December 2007, threatens public safety as well.

Due to the apparent inconsistency between the Post-Mortem Study findings and the Rules, DNR subsequently conducted the Southern Willapa Hills Retrospective study ("Willapa Hills Study;" Washington Department of Natural Resources, 2013). DNR reviewed a subset of 103 landslides and 37 associated FPAs from the same geographic areas examined in the Post-Mortem Study, and also found the majority of landslides sampled initiated from non-RILs. The Willapa Hills Study also examined why a large extent of RIL were logged and found that all but one of these landslides were associated with FPA-approved geotechnical reports, Watershed Analyses, or both. At that time, FPAs in areas with approved Watershed Analysis prescriptions for unstable slopes were exempted from the SEPA review process, creating what turned out to be a destructive loophole in the Rules. In 2011, the Board strengthened the Watershed Analysis Rules, representing an appropriate adaptive policy response to the December 2007 storm.

Slow Policy Responses

To further help achieve the intent of the Rules, in 2013 the Conservation Caucus proposed to Policy a preliminary set of Rule changes, improvements in Rule implementation, and further analysis of RIL, non-RIL, and FPA. To our knowledge, only one of our recommendations is being implemented, the CMER Unstable Slopes Criteria Project (Cooperative Monitoring, Evaluation and Research Committee, 2019). The purpose of this project is to determine the landslide susceptibility of different slopes/landforms to enable evaluation of the current RIL and to identify and characterize additional potentially unstable landforms. The Unstable Slopes Criteria Project has been implemented in 2020.

One of the Rule amendments the Conservation Caucus strongly recommended to further reduce the impacts of landslides from managed forest lands was simple, yet disregarded within the Adaptive Management Program. We argued that a more precautionary approach to buffering potentially unstable slopes was warranted because of the extreme nature of the impacts to public safety and aquatic ecosystems witnessed in the December 2007 storm. For example, a more conservative regulatory definition of RIL with a lower slope criterion (<70% gradient) could have been a simple yet effective way to reduce landslide rates closer to background/natural levels. Using data from the Post-Mortem Study, we found that if the RIL slope criterion was lowered from 70% to 60%, then many more (90% of RIL landslides and 55% of non-RIL landslides) may have fit this more conservative RIL criterion and required evaluation by a qualified expert (Fig. 12.6). To fairly shift the burden of proof, we recommended this straightforward, science-based Rule change to Policy pending further analysis demonstrating the Rule change is *unnecessary*.

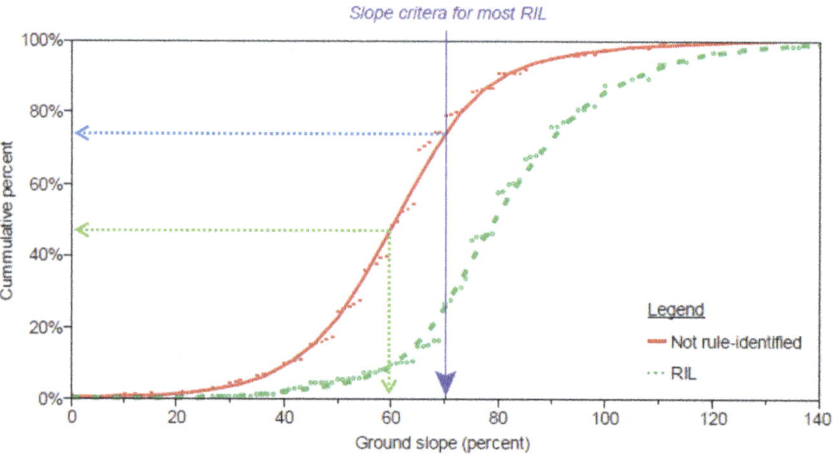

Figure 12.6 Cumulative percent of delivering hillslope landslides on rule-identified landform (RIL) and non-RIL by slope gradient under the current RIL criterion (70% gradient) and our proposed RIL criterion (60% gradient; Stewart et al., 2012 (public domain); dotted arrows added for emphasis).

To date, none of the studies or policy recommendations following the December 2007 storm have resulted in other Rule changes or research other than those two described above. The potential for meaningful adaptive management was minimized by resistance to change and non-consensus at Policy, both driven by economic interests.

The 2014 Oso landslide

In March 2014, a massive deep-seated landslide devastated the community of Oso, Washington. Rather than following a particular large storm event, the Oso landslide occurred after a three-week period of unusually high rainfall locally, though winter precipitation in the region was not atypical for the Pacific Northwest. The ~200 m high hillslope failed catastrophically and traveled more than a kilometer across the valley floor, rapidly inundating

"Steelhead Haven," a neighborhood of approximately 35 single-family residences that was established in the 1960s. After four months of search and rescue operations in treacherous conditions, it was determined the Oso landslide took the lives of 43 people, the deadliest landslide event in U.S. history, ten people survived, several with serious injuries. Approximately 600 m of highway was buried under up to 6 m of debris, closing this major state transportation route for over two months (the slide is also referred to as the "SR530 Landslide"). Washington State officials estimated economic losses due to the landslide exceeded $50 million (Keaton et al., 2014) (Fig. 12.7).

Figure 12.7 Aerial photo of the catastrophic 2014 Oso deep-seated landslide (upper right corner) in northwest Washington State including the full extent of its initiation and runout. *Photo credit Mark Reid, USGS (public domain).*

The 2014 Oso landslide is the site of an ancient landslide and has also failed several times since the 1930s, most recently in 2006. It was previously known as the "Hazel Landslide," and in 2006 traveled over 100 m and blocked the North Fork Stillaguamish River, but stopped short of the Steelhead Haven neighborhood. Although this deep-seated landslide had displayed signs of activity periodically over past decades, Washington DNR nevertheless permitted logging in the Hazel landslide's groundwater recharge area. Groundwater recharge areas to glacial deep-seated landslides are a RIL (WAC 222-16-050 (1)(d)(C)) because logging reduces evapotranspiration, increases groundwater inputs, potentially decreasing slope stability for roughly 20 years after logging.

Shortly after the tragic Oso landslide occurred, a Geotechnical Extreme Events Reconnaissance interdisciplinary team of scientists documented conditions at the landslide, collected potentially perishable field data, and proposed preliminary hypotheses regarding landslide initiation, mobilization, and run-out behavior (Keaton et al., 2014). The team concluded:

1. the accumulation of a large amount of water from both direct precipitation and groundwater was the primary catalyst of the landslide,
2. changes to groundwater recharge resulting from timber harvest above the slide possibly contributed to the slide but further site-specific investigation is needed to quantify this contribution, and
3. the groundwater recharge area delineation approved by DNR was undersized to the extent it excluded portions of the Whitman Bench [the terrace upslope of the Oso landslide], which also feeds neighboring glacial deep-seated landslides.

It became readily apparent to the Conservation Caucus (as well as to many elected officials, citizens, and non-timber industry geologists) that regulatory, policy, and enforcement changes for forest practices conducted on or near glacial deep-seated landslides were required to better protect public safety and public resources. This should not have been new news—Benda et al. (1988) and Miller and Sias (1998) documented *numerous* correlations between landslide activity following logging on the Whitman Bench between 1940 and 1998. Although a revised groundwater recharge area map was available when the Hazel Watershed Analysis was completed (Miller and Sias 1998), DNR applied an outdated groundwater recharge area map (Benda et al., 1988) to the approval of logging on the Whitman Bench in 2004, 2009, and 2011. DNR allowed such logging even though, contrary to the Rules, no one performed a five-year review of the watershed analysis under which the logging took place. Consequently, up to 96 ha might have fallen within a larger, more accurate groundwater recharge area of the glacial deep-seated landslide below the Whitman Bench and not permitted to be logged. Although it's difficult to quantify the impacts of these forest practices on the volume of groundwater feeding the Oso/Hazel landslide, this omission of information demonstrates that we *must* do better to incorporate the best available scientific information into the Rules, Board Manual guidance, and implementation in a timely manner.

Board manual section 16 revision

Another example of regulatory inaction despite known risk is the "Board Manual" that the DNR uses to implement the Rules governing logging on steep and unstable slopes. After the Oso landslide, per Board direction, DNR initiated a review of the adequacy of the Board Manual. Beginning in 2014, our Caucus participated on a multi-stakeholder technical committee tasked with the revision of Board Manual Section 16 to update Rule guidance to reflect the lessons learned from the major landslide events of 2007 and 2014. Our Caucus' policy position has been that, although only a "guidance-only" document, the Board

Manual functionally implements the Rules and therefore must provide sufficient up-to-date information to enable qualified expert geologists and general practitioners to do so aptly and consistently. Unfortunately, Policy member influence on work products essentially obstructed the ability of the technical committee members to revise the manual consistent with the best available science and a precautionary approach, even though a greater protection of public safety and public resources should result. Despite many hours of discussion, revisions by the technical members, and negotiation, the stakeholder group was unable to reach agreement on a number of key sections, and the Board Manual still lacks important updates (Washington Forest Practices Board, 2016).

In 2015, the Conservation Caucus again advocated for the Board to adopt a precautionary approach and risk management decision-making consistent with the intent of the Rules and to apply a moratorium to logging on and above deep-seated landslides. Based on the professional opinions of three highly-qualified geologists with experience in the forest sector and our own experience in the Board Manual revision process, we presented to the Board seven different ways the revised Board Manual was inadequate guidance for Rule implementation. We argued that when it comes to potential threats to public safety, it is both appropriate and responsible to apply the precautionary principle to landslide hazard assessment, because in the face of scientific uncertainty, land management decisions that err on the side of caution will best protect the environment and public well-being (Kriebel et al., 2001). A declaration to the Board by a leading geologist in the region stated:

> *The SR 530 [Oso] Landslide highlights the need to incorporate landslide hazard, risk, and vulnerability assessments into land-use planning and to expand and refine geologic and geo-hazard mapping throughout the State. In the absence of policies or programs conducting such assessments, a precautionary approach is prudent for protecting public resources and public safety on and around forest practices.*

Landslide activity level and reactivation potential

One of our Caucus' primary concerns with the Board Manual is its failure to adequately address the principle of deep-seated landslide reactivation. The additional groundwater generated by forest practices may have the potential to reactivate Board Manual-defined "relict," "dormant-distinct," or "dormant-indistinct" landslides that have not shown signs of movement for decades or even centuries (according to the available evidence). Based on an apparent 'absence of evidence,' it is generally assumed that landslides of these "activity levels" will *not* reactivate due to logging, and as a result, *not* all deep-seated landslides are treated equally in practice/ permitting. This significant assumption is based on professional judgement alone without empirical science to substantiate it. We *routinely* see approval of FPAs to log trees within "inactive" deep-seated landslides and their groundwater recharge areas because the Board Manual guidance is used to interpret them *not* to be RILs without any further evaluation or analysis required. The unsubstantiated assumption that "inactive" deep-seated landslides cannot be reactivated by logging is challenged by qualified experts, especially where increased risks to public safety is a concern:

> *The Board Manual's implicit assumption that relict and dormant/indistinct deep-seated landslides are not by default susceptible to hydrologic changes associated with forest practices is a potentially dangerous one, especially in light of the higher annual precipitation (and thus groundwater) forecasted for our region* (Montgomery 2015).

To address these potentially dangerous assumptions with a more precautionary approach, we recommended that where the threat to public safety is moderate, high, or uncertain for a deep-seated landslide of any "activity level," the proposed FPA should include a quantitative assessment of whether a potential increase in groundwater recharge from timber harvest will affect the stability of the landslide. To date, this approach has not been adopted in the Rules or Board Manual, and unless and until greater precaution is exercised in FPA permitting decisions, the burden of proof will remain on the people and public resources subject to the impacts of potentially risky forest practices.

Bedrock deep-seated landslides

Another example of how logging continues under the Board Manual's unsubstantiated scientific foundation is the Manuals' guidance for non-glacial (or bedrock) deep-seated landslides (BDSL). For no justifiable geologic reason(s) and despite the potential threat to public safety, the Board Manual arbitrarily requires less geologic review and protection for BDSLs. Consequently, we *routinely* see approval of FPAs to log timber within BDSLs and their groundwater recharge areas because the Board Manual guidance is used to interpret them *not* to be RILs without any further evaluation or analysis required. Again, an apparent 'absence of evidence' that BDSLs will reactivate due to forest practices leads to their essential dismissal in practice/permitting without empirical science to substantiate this broad assumption.

An especially concerning example of forest practices proposed on BDSLs is the 2015 "North Zender" timber sale on DNR-managed lands in northwestern Washington. Two logging units (~40 ha total) and a new road segment (~0.7 km) were proposed *directly* on two large BDSLs (Fig. 12.8). An elementary school, at least 15 residences, access roads, a domestic water supply intake, and a fish hatchery exist within the potential runout paths of these

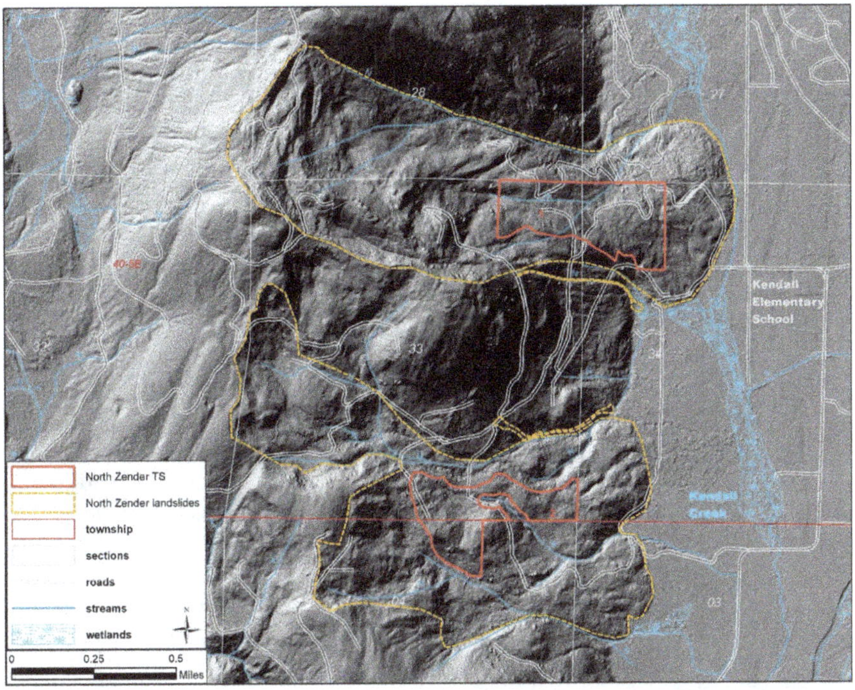

Figure 12.8 Light detection and ranging bare earth image (Puget Sound LiDAR Consortium 2010) and the proposed North Zender timber sale units on two bedrock deep-seated landslides (BDSLs) immediately above the town of Kendall, Washington.

two large deep-seated landslides. There are also a significant number of salmon runs (including Coho, Bull Trout, Pink, Chinook, Steelhead, and Cutthroat) and some wetlands with valuable wildlife habitat directly downstream of the two BDSLs. Geologic field review, office review, and local knowledge provided strong evidence of the potential for further slope movement. Two highly qualified geologists independently concluded that the proposed North Zender timber sale would increase the risk of reactivation of the underlying BDSLs, and that the proposed logging directly over these landslides was imprudent, if not reckless. In circumstances such as these, the Conservation Caucus *strongly* argues that the confidence in the slope stability assessment supporting the proposed FPA must be high because the potential for harm to the public and public resources if one or both landslides were to destabilize is also high.

On behalf of highly concerned local residents, we legally appealed the FPA approval, and DNR withdrew the North Zender FPA before logging commenced, but DNR has never made a formal declaration that the FPA will not be refiled. Our clients have a reasonable fear that should the FPA ever be refiled and approved, the resultant deforestation above their home could cause devastating harm to their property, their families, their recreational interests, and their community.

Unstable slopes proposal initiation

In 2016, the Board responded to the list of concerns with the Board Manual we brought forward on behalf of the Conservation Caucus. Those concerns "not near resolution" (needing either more science or Rule-making) were translated into a Proposal Initiation as defined in the Adaptive Management Program (Timber, Fish and Wildlife Policy Committee, 2016). The Adaptive Management Program Administrator proposed two items for Policy:

- Development track(s) and timeline(s) for completion of the review of the unstable slopes elements of concern;
- When developed, these products will be included in the Policy recommendations to the Board for
  - o Unstable slopes Rule making and associated Board Manual guidance, and
  - o Potential identification of additional research.

With respect to our Board Manual concerns (including landslide reactivation potential and BDSL dismissal), the Proposal Initiation included a list of explicit questions from the Board and directed the Adaptive Management Program to determine how the questions could be addressed and to bring these recommendations back to the Board. The questions were placed on either a "science track" to CMER or a "policy track." Two of our seven concerns were partially addressed in another Board Manual revision, and other questions were later incorporated into a Deep-Seated Landslide Research Strategy (Cooperative Monitoring, Evaluation and Research Committee, 2018). Unfortunately, it will take *many* more years before the results of these interrelated research projects will be considered for Rule changes to better protect public safety and public resources. In the meantime, 'business as usual' proceeds with the burden of proof on the public and aquatic ecosystems waiting for the generation of 'evidence' to close the adaptive management loop.

An urgent call for precautionary forest climate policy

Many of the concerns we described above regarding endangered species habitat and landslide hazards will likely be further exacerbated by climate change in future decades. For example, the best available science predicts greater threats to murrelet nesting habitat due to increases in drought-related fire, tree mortality, extreme flooding, landslides, and windthrow events and decreases in forest growth rates in the near- and long-term (van Mantgem et al., 2009; Littell et al., 2010; Snover et al., 2013). Precautionary management of forest practices on unstable slopes will also become increasingly important to minimize landslide hazards as the Pacific Northwest is expected to experience more frequent and intense storms (Dale et al., 2001; Christensen et al., 2007;Karl et al., 2009; Warner et al., 2015).

The Pacific Northwest has long been recognized for having some of most productive forests on Earth. More recently, scientists have also become aware of the potential for less intensive forest management in our region to contribute *significantly* to climate change mitigation and the preservation of biodiversity (Millennium Ecosystem Assessment, 2005). Preserving high-carbon-priority forests, such as those found in western Washington and Oregon, will be especially important to partially offset fossil fuel emissions and to promote climate resilience (Brandt et al., 2014; Krankina et al., 2014; Buotte et al., 2020). As noted by Kriebel et al. (2001), "The great complexity, uncertainty, and potential for catastrophe from global climate change are among the strongest motivators for those urging precaution in environmental policy." Science-based forest climate policy is urgently needed, because these older forests are essentially irreplaceable on the timescale needed for this function (carbon storage and sequestration) if they are lost in the near-term.

Speaking truth to power: who should carry the burden of proof during an adaptive management process?

The two complex, long-term science-policy case studies we described in this chapter demonstrate the importance and urgency of applying the precautionary principle by shifting the burden of proof to the proponents of logging in order to minimize the negative impacts of forest management and 'do-no-harm' to natural and human systems. The planetary Hippocratic Oath does not permit the absence of evidence to be taken as the evidence of absence (Chapter 1, The Nuts and Bolts of Science-Based Advocacy). Science continues to show us the need to act conservatively and manage adaptively in order to deal with the inherent complexity and uncertainty within ecosystems and to enable us to change course in response to new information (Holling, 1978; Lee, 1993; Walters, 1986; Yaffee, 1999). By shifting the onus to those who are creating potentially adverse impacts, we, and many before us, see the precautionary principle as the key to better protecting ecological and human health (Raffensperger and Tickner, 1999; Kriebel et al., 2001).

The DNR HCP science-policy process for the LTCS was an attempt at ecosystem management over large spatial and temporal scales with multiple stakeholders involved. Unfortunately, balancing conservation and economic interests within the LTCS under the current state forest management paradigm is untenable. When a species like the Marbled Murrelet is on a trajectory of extirpation at the state scale, management decisions must be made quickly, because time is a luxury imperiled species cannot afford (Brunner and Clark 1997). Given the United States' commitment to species protection under the Endangered Species Act, "if scientists' best judgment indicates that a species is at risk, *the burden should not be on proponents of protection to ensure that the species receives the protection it deserves*; the scientists' determination should hold sway" (Carden 2006, p. 228, emphasis added). Ecosystem management as an emerging paradigm has the potential to create true win-win situations in theory, but the combination of factors contributing to the LTCS have made this goal elusive in practice, including attempts of political interference on the behalf of the timber industry.

The Forest Practices HCP science-policy process is an attempt at ecosystem management intended to minimize the impacts of forest practices on landslide hazards, public resources, and public safety. The long-term survival of many endangered fish species will also depend on a management paradigm that emphasizes the restoration of habitat integrity and landslide rates closer to natural background levels (Reeves et al., 1995; Montgomery 2004). Although the Adaptive Management Program was designed with good intentions for timely adaptive management of aquatic ecosystems (Washington Forest Practices Board, 2013), the slow pace it has exhibited across a large spatial scale in an interminable search for consensus among stakeholders essentially keeps it in the same gridlocked position of decades past (Haynes and Weigand 1997). Perpetually discrediting science or waiting for more science in a policy setting results in a form of "paralysis by analysis" (Doremus 1997). When impacts of natural resource extraction are allowed to continue during artificially or intentionally lengthy scientific inquiries driven by excessive questioning or challenging of policy-driven science, consensus among stakeholders and an adaptive response to forest management are precluded. As long as ecosystem management goals are tempered by political opposition, "[t]raditional forest policy mechanisms will continue to be employed because they are conservative and reflect only modest changes in the status quo" (Cubbage 1997, p. 354), and the impacts of 'business as usual' will also continue.

Despite originating from an ecosystem management paradigm, we believe the adaptive management processes we have documented in this chapter are associated with a prolonged decline of endangered species populations and risk to the public. Robust and timely ecosystem-based and precautionary forest management approaches are needed to avoid or minimize "the real environmental and human problems caused by past management paradigms...because past approaches have not succeeded" (Yaffee 1999, p. 715). To create a more balanced, equitable forest management system, we must collectively accelerate the science-policy process, shift the burden of proof, and overcome the resistance to changing the status quo for the greater good.

References

Benda, L., Thorsen, G.W., Bernath, S., 1988. Report of the I.D. team investigation of the Hazel Landslide on the North Fork of the Stillaguamish River, Report F.P.A. 19-09420. Report to the Washington Department of Natural Resources, Northwest Region, Sedro Woolley, Washington. 13 pp.

Brandt, P., Abson, D., DellaSala, D., Feller, R., von Wehrden, H., 2014. Multifunctionality and biodiversity: ecosystem services in temperate rainforests of the Pacific Northwest, USA. Biol. Conserv. 169, 362–371.

Brunner, R.D., Clark, T.W., 1997. A practice-based approach to ecosystem management. Conserv. Biol. 11, 48–58.

Buotte, P.C., Law, B.E., Ripple, W.J., Berner, L.T., 2020. Carbon sequestration and biodiversity co-benefits of preserving forests in the western United States. Ecol. Appl. 30 (2), e02039.

Carden, K., 2006. Bridging the divide: the role of science in species conservation law. Harv. Environ. Law Rev. 30, 165–259.

Cederholm, C.J., Lestelle, L.C., 1974. Observations on the Effects of Landslide Siltation on Salmon and Trout Resources of the Clearwater River, Jefferson County, Washington, 1972–73. Final Report, Part I. Fisheries Research Institute. College of Fisheries, University of Washington, Seattle.

Christensen, J.H., Hewitson, B., Busuioc, A., Chen, A., Gao, X., Held, I., et al., 2007. Regional climate projections. In: Solomon, S., Qin, D., Manning, M., Chen, Z., Marquis, M., Averyt, K.B., Tignor, M., Miller, H.L. (Eds.), Contribution of Working Group I to the Fourth Assessment Report of the Intergovernmental Panel on Climate Change. Cambridge University Press, Cambridge, UK, pp. 847–940.

Cooperative Monitoring, Evaluation and Research Committee. (Jan.) 2019. 2019–2021 Biennium CMER Work Plan. 146 pp.

Cooperative Monitoring, Evaluation and Research Committee. 2018. Deep-Seated Landslide Research Strategy. Upslope Processes Scientific Advisory Group. 40 pp.

Cubbage, F.W., 1997. The public interest in private forests: developing regulations and incentives. In: Kohm, K.A., Franklin, J.F. (Eds.), Creating a Forestry for the 21st Century: the Science of Ecosystem Management. Island Press, Washington, D.C, pp. 337–356.

Dale, V.H., Joyce, L.A., McNulty, S., Neilson, R.P., Ayres, M.P., Flannigan, M.D., et al., 2001. Climate change and forest disturbances. Bioscience 51 (9), 723–734.

Desimone, S.M., 2016. Periodic Status Review for the Marbled Murrelet in Washington. Washington Department of Fish and Wildlife, Olympia, Washington, p. 28, + iii pp.

Doremus, H., 1997. Listing decisions under the Endangered Species Act: why better science isn't always better policy. Wash. Univ. Law Rev. 75 (3), 1029–1153.

Gilpin, M.E., Soulé, M.E., 1986. Minimum viable populations: processes of species extinction. In: Soulé, M.E. (Ed.), Conservation Biology: The Science of Scarcity and Diversity. Sinauer, Sunderland, Mass, pp. 19–34.

Haynes, R.W., Weigand, J.F., 1997. The context for forest economics in the 21st century. In: Kohm, K.A., Franklin, J.F. (Eds.), Creating a Forestry for the 21st Century: the Science of Ecosystem Management. Island Press, Washington, D.C, pp. 285–301.

Holling, C.S., 1978. Adaptive environmental assessment and management. Wiley, New York.

Karl, T.R., Melillo, J.M., Peterson, T.C., 2009. Global Climate Change Impacts in the United States. Cambridge University Press, Cambridge, UK.

Keaton, J.R., Anderson, S., Benoit, J., deLaChapelle, J., Gilbert, R., Montgomery, D.R., 2014. The 22 March 2014 Oso Landslide, Snohomish County, Washington. Geotechnical Extreme Events Reconnaissance Association Report No. GEER – 036. 186 pp.

Krankina, O., DellaSala, D.A., Leonard, J., Yatskov, M., 2014. High biomass forests of the Pacific Northwest: who manages them and how much is protected? Environ. Manag. 54, 112–121.

Kriebel, D., Tickner, J., Epstein, P., Lemons, J., Levins, R., Loechler, E.L., et al., 2001. The precautionary principle in environmental science. Environ. Health Perspect. 109 (9), 871–876.

Lee, K., 1993. Compass and gyroscope: integrating science and politics for the environment. Island Press, Washington, D.C.

Littell, J.S., Oneil, E.E., McKenzie, D., Hicke, J.A., Lutz, J.A., Norheim, R.A., et al., 2010. Forest ecosystems, disturbance, and climatic change in Washington State, USA. Clim. Change 102, 129–158.

McIver, W., Baldwin, J., Lance, M.M., Pearson, S.F., Strong, C., Lynch, D., et al., 2020. Marbled Murrelet Effectiveness Monitoring. At-Sea Monitoring - 2019 Summary Report, Northwest Forest Plan, p. 23.

Millennium Ecosystem Assessment, 2005. Ecosystems and human well-being: biodiversity synthesis. World Resources Institute, Washington, D.C., USA.

Miller, D. 2016. Literature synthesis of the effects of forest practices on glacial deep-seated landslides and groundwater recharge. Prepared for the Upslope Processes Scientific Advisory Group, Cooperative Monitoring, Evaluation and Research Committee. Washington State Adaptive Management Program, Olympia, Washington. 139 pp.

Miller, D. 2017. Literature synthesis of the effects of forest practices on non-glacial deep-sated landslides and groundwater recharge. Prepared for the Upslope Processes Scientific Advisory Group, Cooperative Monitoring, Evaluation and Research Committee. Washington State Adaptive Management Program, Olympia, Washington. 105 pp.

Miller, D.J., Sias, J., 1998. Deciphering large landslides: linking hydrological, groundwater and slope stability models through GIS. Hydrol. Process. 12, 923–941.

Montgomery, D.R., 2004. Geology, geomorphology, and the restoration ecology of salmon. GSA Today 14, 4–12.

Montgomery, D.R., Schmidt, K.M., Greenberg, H., Dietrich, W.E., 2000. Forest clearing and regional landsliding. Geology 28, 311–314.

Mote, P., Mault, J., Duliere, V., 2007. The Chehalis river flood of December 3–4, 2007. Office of Washington State Climatologist. University of Washington, Seattle, WA, p. 4.

Peery, M.Z., Jones, G.M., 2018. Using Population Viability Analyses to Assess the Potential Effects of Washington DNR Forest Management Alternatives on Marbled Murrelets, version 2. Unpublished report submitted to WDNR and USFWS. 79 pp.

Raffensperger, C., Tickner, J., 1999. Protecting Public Health and the Environment: Implementing the Precautionary Principle. Island Press, Washington, D.C.

Raphael, M.G., Shirk, A.J., Falxa, G.A., Lynch, D., Nelson, S.K., Pearson, S.F., et al., 2016. Factors Influencing Status and Trend of Marbled Murrelet Populations: An Integrated Perspective. Chapter 3 In: Falxa and Raphael (Eds.), Northwest Forest Plan—The First 20 Years (1994–2013): Status and Trend of Marbled Murrelet Populations and Nesting Habitat. Gen. Tech. Rep. PNW-GTR-933. Portland, OR. U.S. Department of Agriculture, Forest Service, Pacific Northwest Research Station.

Raphael, M.G., Nelson, S.K., Swedeen, P., Ostwald, M., Flotlin, K., Desimone, S., et al., 2008. Recommendations Supporting Analysis of Conservation Opportunities for the Marbled Murrelet Long-Term Conservation Strategy. Washington State Department of Natural Resources, Olympia, WA, p. 337.

Reeves, G.H., Benda, L.E., Burnett, K.M., Bisson, P.A., Sedell, J.R., 1995. A disturbance-based ecosystem approach to maintaining and restoring freshwater habitats of evolutionarily significant units of anadromous salmonids in the Pacific Northwest. American Fisheries Society Symposium 17, 334–349.

Rood, K.M. 1984. An aerial photograph inventory of the frequency and yield of mass wasting on the Queen Charlotte Islands, British Columbia. British Columbia Ministry of Forests, Land Management Report 34, Victoria, BC.

Sarikhan, I.Y., Stanton, K.D., Contreras, T.A., Polenz, M., Powell, J., Walsh, T.J., et al., 2008. Landslide Reconnaissance Following the Storm Event of December 1–3, 2007, in Western Washington. Open File Report 2008–5, Washington Department of Natural Resources, Division of Geology and Earth Resources, Olympia, WA. 22 pp.

Snover, A.K., Mauger, G.S., Whitely Binder, L.C., Krosby, M., Tohver, I., 2013. Climate Change Impacts and Adaptation in Washington State: Technical Summaries for Decision Makers. State of Knowledge Report prepared for the Washington State Department of Ecology. Climate Impacts Group, University of Washington, Seattle.

Stewart, G., Dieu, J., Phillips, J., O'Connor, M., Veldhuisen, C., 2012. The Mass Wasting Effectiveness Monitoring Project: An examination of the landslide response to the December 2007 storm in Southwestern Washington; Cooperative Monitoring, Evaluation and Research Report CMER 08-802; Washington Department of Natural Resources, Olympia, WA. 138 pp.

Swanson, F.J., Benda, L.E., Duncan, S.H., Grant, G.E., Megahan, W.F., Reid, L.M., et al., 1987. Mass failures and other processes of sediment production in Pacific Northwest forest landscapes. In: Salo, E.O., Cundy, T.W. (Eds.), Streamside Management: Forestry and Fishery Interactions: Seattle, WA. Institute of Forest Resources, University of Washington, pp. 9–38.

Timber, Fish & Wildlife Policy Committee. 2016. Unstable slopes proposal initiation; recommendations from TFW Policy Committee to the Forest Practices Board. August 4, 2016. Olympia, WA.

Turner, T.R., Duke, S.D., Fransen, B.R., Reiter, M.L., Kroll, A.J., Ward, J.W., et al., 2010. Landslide densities associated with rainfall, stand age, and topography on forested landscapes, southwestern Washington, USA. For. Ecol. Manag. 259, 2233–2247.

U.S. Fish and Wildlife Service. 1997. Recovery Plan for the Threatened Marbled Murrelet (*Brachyramphus marmoratus*) in Washington, Oregon, and California. Portland, OR. 203 pp.

U.S. Fish and Wildlife Service. 2019. Biological Opinion of the Washington State Department of Natural Resources Marbled Murrelet Long-term Conservation Strategy Amendment to the 1997 Habitat Conservation Plan. Lacey, Washington. 250 pp.

U.S. Fish and Wildlife Service and National Marine Fisheries Service. 2016. Habitat Conservation Planning and Incidental Take Permit Processing Handbook. 402 pp.

U.S. Fish and Wildlife Service, National Marine Fisheries Service, U.S. Environmental Protection Agency, Washington Governor's Office, Washington State Department of Natural Resources, Washington State Department of Fish and Wildlife, Washington State Department of Ecology, Colville Confederated Tribes, Northwest Indian Fisheries Commission, Washington State Association of Counties, Washington Forest Protection Association, and the Washington Farm Forestry Association. 1999. Forests and Fish Report: Washington Department of Natural Resources. 182 pp.

U.S. Fish and Wildlife Service, 1992. Determination of threatened status for the Washington, Oregon and California population of the marbled murrelet. Federal Register 57, 45328–45337.

van Mantgem, P.J., Stephenson, N.L., Byrne, J.C., Daniels, L.D., Franklin, J.F., Fule, P.Z., et al., 2009. Widespread increase of tree mortality rates in the western United States. Science 323 (21), 521–524.

Walters, C.J., 1986. Adaptive management of renewable resources. McGraw-Hill, New York.

Warner, M.D., Mass, C.F., Salathé Jr., E.P., 2015. Changes in winter atmospheric rivers along the North American West Coast in CMIP5 climate models. J. Hydrometeorol. 16, 118–128.

Washington Department of Natural Resources. 1997. Final State Trust Lands Habitat Conservation Plan. Olympia, WA. 223 pp.

Washington Department of Natural Resources. 2005. Final Washington State Forest Practices Habitat Conservation Plan. Olympia, WA.

Washington Department of Natural Resources. 2013. Southern Willapa Hills Retrospective Study. Olympia, WA. 43 pp.

Washington Department of Natural Resources. 2019. Final Environmental Impact Statement (FEIS) for a Long-term Conservation Strategy for the Marbled Murrelet. Olympia, WA.

Washington Forest Practices Board. 2013. Board Manual Section 22, Guidelines for Adaptive Management Program. Olympia, WA. 33 pp.

Washington Forest Practices Board. 2016. Board Manual Section 16, Guidelines for Evaluating Potentially Unstable Slopes and Landforms. Olympia, WA. 93 pp.

Welsh Jr., H.H., Ollivier, L.M., 1998. Stream amphibians as indicators of ecosystem stress: a case study from California's redwoods. Ecol. Appl. 8, 1118–1132.

Yaffee, S.L., 1999. Three faces of ecosystem management. Conserv. Biol. 13 (4), 713–725.

To zero emissions, and beyond? Oregon Stumbles forward

Angus Duncan[1,2]

[1]Natural Resources Defense Council—Consultant, Former Chair, Oregon Global Warming Commission, Salem, OR, United States [2]Former Chair, Northwest Power Planning Council, Portland, OR, United States

"Owing to past neglect, in the face of the plainest warnings, we have now entered upon a period of danger...The era of procrastination, of half-measures, of soothing and baffling expedients, of delays, is coming to its close. In its place we are entering a period of consequences... We cannot avoid this period; we are in it now."

Winston Churchill, in the House of Commons, November 1936

"If it was really that serious, someone would be doing something about it."

"Wouldn't they?"

Focus group response to the question how seriously we should be taking climate change?

Introduction

At the onset of the 21st century, Oregon and many other states and countries bethought themselves to rise to their climate and greenhouse gas (GHG) reduction responsibilities. Oregon's state agencies and serious legislators have taken past actions that identified emissions reduction goals and compiled repeated lists of aspirational actions, but until 2016 the State had largely deferred

Conservation Science and Advocacy for a Planet in Peril.
DOI: https://doi.org/10.1016/B978-0-12-812988-3.00008-9

enforceable commitments. In that year, Oregon finally adopted an obligatory utility coal closeout (by 2030). But the big test, the climate policy grail, was an economy-wide carbon cap similar to California's 2006 adoption of the Global Warming Solution Act (AB 32). In two tries, 2019 and 2020, the Oregon Legislature failed to carry this bucket of water up the hill. Ours is the story, repeated in many jurisdictions elsewhere, of fitful and equivocal policy success, notwithstanding a general appreciation among citizens that the findings of climate science needed to be taken seriously. As a coda, this chapter will parse the reasons why, not that in the end the atmosphere will care in the least for our excuses.

Prologue: a tale of three sessions

2010: the blooming of bipartisanship

Oregon's November 2010 legislative elections produced an unusual if not unprecedented outcome: an evenly split House of 30 Republican and 30 Democratic Members (hereafter, R's and D's). Since the state Senate and Governor's position were controlled by D's, an observer from 2020 might have expected House R's would opt for gridlock. They could have done so effectively by shutting down bill action in the House on anything meaningful, anything that would divide along partisan lines, much as the R's controlling Congress locked down on the Obama Administration after 2010's national elections.

Instead the Oregon House reinvented itself with cospeakers, one each from the moderate side of each caucus. They agreed that each Committee would have cochairs. There were rules adopted that enabled Members from each party to either move or block a bill and others that enabled a blockage to be overridden by bipartisan action.

The session that might have melted down instead was deemed productive by both sides of the aisle. Senator Michael Dembrow (D-Portland) reported to his constituents that the session "...managed to eliminate or shrink ... tax credits and loopholes" that had been hanging on past their pull date and "draining scarce resources." The two parties agreed on "a compromise legislative and congressional redistricting plan" that would have customarily left blood in the streets; and on higher education reforms, post-Great Recession budget outcomes, and more.

Oregon's political history had contained both productive and gridlock legislatures, and eras when each party was in the electoral ascendancy. Extreme outcomes have included pioneering Black Exclusion laws in the 1840's and 50's, labor union suppression at the turn of the 19th century, and laws favoring investor-owned electric utilities over the consumer-owned models that dominated neighboring Washington in the 1930's and 40's. But from the mid-20th century on, moderation by both parties was rewarded by voters. Control of the legislature and the Governor's office moved back and forth while the state forged ahead with a bipartisan sharing of postwar prosperity among public services and private business.

Environmental values were early on expressed as common ownership of Oregon's beaches to the high-water mark (1967). Some of the nation's first statewide land-use laws were adopted in the 1970's, sold as a way to protect farmland and forests for farming and forestry with a side of environmental values, but mostly aimed at keeping commercial and residential development in check. Governor Tom McCall, a Republican, led his party in support of these kinds of actions, whereas energy conservation and Jane Jacobs-style urban environmental values were championed by Portland Mayor Neil Goldschmidt, a Democrat. Senator Mark Hatfield (R) and Congressman Bob Duncan (D) collaborated to bring federal light rail funding to Portland in the 1970's, notwithstanding that they had been on opposite sides in Salem a few years earlier as Governor and House Speaker, respectively.

Oregon was a national leader on environmental issues during this era, a reputation that by the end of the decade had become a little threadbare.

2019: the turn to partisan divisiveness

Fast-forward to the 2019 and 2020 legislative sessions where "carbon cap and trade" (C&T) language was offered. Fifteen years earlier a Governor's Advisory Group on Global Warming recommended Oregon adopt such a measure, similar to the earlier, famously successful federal models that controlled other pollutants (SOx and NOx emissions from mid-western power plants) at a fraction of the cost of prescriptive regulation. Shortly after the report to Governor Kulongoski (D), the three west coast states—California, Oregon, and Washington—agreed to collaborate on a western carbon trading mechanism. California proceeded, but efforts in Oregon and Washington were waylaid by the 2008 recession. It took a decade to get back on track.

Getting the Oregon train back on its track did not get it to the station on time. Instead, in two succeeding sessions and notwithstanding a clear majority of (Democratic) votes in support, the carbon cap train went over the cliff of partisan divisiveness. As the end of the decade approached, there was not a single R in either chamber that supported moving this core carbon bill forward, notwithstanding (1) the demonstrated success of California's AB 32 setting up its carbon market, (2) Washington's Clean Energy Transformation Act (2019), (3) mounting evidence of climate impacts in Oregon (drought; wildfire; falloff in ocean fisheries; etc.) and projections by Oregon academics of more to come, (4) corollary evidence of falling costs of low-carbon energy alternatives (wind; solar; electric vehicles), and (5) compliance concessions in the bill for industry, rural areas, trucks, forestry, and others who showed up at the door complaining of impending catastrophic injury.

"... in two succeeding sessions and notwithstanding a clear majority of (Democratic) votes in support, the carbon cap train went over the cliff of partisan divisiveness."

By the end of the decade, as in the Congress, a history of advocacy and compromise had given way to scorched earth resistance to environmental proposals or any solutions that involved bigger government; bigger, by definition, equating to more oppressive.

The absence of R votes should not have been a problem, as D's had substantial majorities in both chambers. But Oregon's Constitution includes a peculiar quorum rule that advantaged the minority by requiring, not a simple majority, but two-thirds of Members to be on the floor; and in 2019 the absence of a quorum prevented the Senate, and thus the Legislature, from moving a carbon bill or transacting any other business. As further discussed below, the D leadership capitulated, the R's returned, business resumed but the carbon bill was dead.

2020: 2019 redux

In Oregon's 2020 "short" session (a 30-day session held in even-numbered years), legislative committees brought back the carbon bill with still more concessions, and with a clear majority of votes for. None of it—not substantive arguments, not compromises, not the majority of voters who sent a clear majority of legislators to Salem to act on climate—mattered. This time R's in both chambers walked out, locking up the democratic process.

Oregon's changing climate

In the last decades of the 20th century, climate scientists were seeing isolated instances of the terrestrial effects of GHG accumulations in the atmosphere. The advent of the Intergovernmental Panel on Climate Change (IPCC) reports in the 1990's pulled together and elevated the visibility of these predicted effects and linked them clearly and primarily to emissions from fossil fuel combustion, challenging both the energy scaffolding on which the global economy was built and the political power of the fossil fuel corporations. The resistance would be not only economic and political but cultural as well. Cars and trucks are icons of advanced economies, as are homes that are heated and cooled by burning fossil fuels. Manufacturing consumes very large quantities of energy to supply consumer goods; almost all that energy comes from burning coal, oil, and gas. These activities have historically supplied good jobs and wages without demanding of workers the burden of acquiring an advanced education.

In Oregon as elsewhere, early predictions of effects coming from climate scientists, but still absent tangible evidence, were having little visible impact on business-as-usual (BAU). Aspirational goals were adopted but then honored in the breach. The policy and political apparatus had more immediate options and concerns.

In 2007, the Legislature set up an Oregon Climate Change Research Institute (OCCRI) and charged it with reporting on effects as they might emerge, and with projecting the future course of climate change in the state (OCCRI also participates in and contributes findings to the National Climate Assessment mandated by the Global Change Research Act of 1990). The first OCCRI Oregon Climate Assessment for Oregon offered these sobering projections:

> *"Temperatures will continue to increase in Oregon through the 21st Century . . . Climate models point to hotter, drier summers . . . evidence that Oregon's marine environment is changing . . . likely substantial increases in ocean temperatures . . . mountain snowpacks will diminish and summer water supply will likely decline . . . water quality likely to be impacted (affecting) stream ecology and salmon habitat . . . wildfire will likely increase . . . climate change poses economic risks to the state . . . agriculture . . . is highly sensitive to climate. . . ."*
>
> **First Climate Assessment, Executive Summary" (Oregon Climate Change Research Institute, 2010)**

Only 8 years later, the companion Assessment could array physical evidence of the impacts arriving in the state, affecting communities and ecosystems.

> *"Oregon is already experiencing statewide impacts of a changing climate . . . some of the worst air quality on the planet (in 2018) owing to smoke from wildfires . . . lack of water from a low winter snowpack and a hot and dry summer . . . (increases in) extreme precipitation . . . sea level rise . . . human health risk . . . disadvantaged communities are the most vulnerable."*
>
> **"Fourth Climate Assessment, Executive Summary" (Oregon Climate Change Research Institute, 2019)**

The Oregon Global Warming Commission (the Commission, or OGWC) relied on and built upon the evidence from OCCRI's sequence of Assessments in its own 2018 Biennial Report to the Legislature (Fig. 13.1):

> *"The Oregonian for Wednesday, August 15, 2018, led with the story of smoke that 'choked' the Portland airshed from forest fires 'filtering into Northwest Oregon from blazes in almost all directions . . . .' Less than a year earlier, Portlanders had awakened to a similar brownish haze obscuring the sky and the same public health advisory. DEQ [Department of Environmental Quality] said 2017 was 'different' from earlier bad fire years in that 'the entire state is . . . blanketed by smoke' coming not only from the Eagle Creek Fire in the Columbia Gorge but also from a dozen fires ranging from the Rogue River to Mt. Hood, as well as from fires in Canada and California. DEQ called the condition 'rare'.*
>
> *But it's not, anymore."*
>
> **"Climate change Comes to Oregon 2018, Biennial Report to the Oregon Legislature, Section 1" (Oregon Global Warming Commission, 2018)**

Figure 13.1 Mt. Washington from Black Butte Ranch near Sisters, Oregon. Smoke from Mili Fire, 2017. *Cover Photo, 2018 Biennial Report to the Legislature—2019 Legislative Session. Photo credit: Angus Duncan.*

This Report, coming just in advance of the 2019 session that would finally take up a statewide carbon cap, was intended to send a single overall message: those climate effects telegraphed by climate scientists for three decades, and foreshadowed in OCCRI's 2010 Climate Assessment, had arrived in Oregon.

Nothing was speculative anymore about drought and scarily scant snowpack. There was normal precipitation the winter of 2014–15, but warm ambient winter temperatures resulted in a record low snowpack and drought declarations in 25 of Oregon's 36 counties. Wildfire smoke blanketed the state in 2018, shutting down the summer tourist dollars that many communities lived on the rest of the year. Heat-related and respiratory illnesses were on the rise in the state. A "Blob" ranged upon the Oregon coast—"... an anomalous body having sea surface temperature ... 2.5°C above normal ... thought to affect ocean food web nutrient levels [Oregon Global Warming Commission, 2018 Biennial Report]." Other effects, predicted in 2010 and earlier, were being realized by 2018.

The Commission's report linked these emerging trends to parallel, often more pronounced climate effects around the globe (e.g., Hurricanes Katrina, Sandy, Harvey and Maria in the United States; Mangkut and Haiyan in the Philippines; extreme flooding in Japan, India, Thailand and elsewhere; all with serious loss of life and property).

A blog I posted on the Commission's website in April of 2020 observed the similarities between that year's onset of coronavirus and climate change: "the pandemic is like climate change on speed; and climate change seems a pandemic in slo-mo. In both cases we . . . have been given the grace of an early warning and time to prepare (Duncan, 2020)."

"there will be no vaccine for climate change"

In neither case did we made meaningful use of the grace periods, in Oregon or nationally, instead delaying and denying until the effects were unmistakable and locked in for some indeterminate but deeply damaging future. Of course, there will be no vaccine for climate change.

How and why this state of affairs prevailed, all the warnings of science and accumulating evidence of predicted effects notwithstanding, is the subject of the balance of this chapter.

Fitful progress: Oregon's coming to terms with climate change

The first global climate report from the IPCC was published in 1990. Two years later Oregon set itself a GHG goal: to not exceed its 1990 emissions levels. In 1997 the State adopted a requirement that new fossil fuel generating plants would have to partially offset their emissions or pay a modest penalty, but otherwise did not adopt any meaningful measures to arrest carbon emissions growth and achieve actual reductions. The state's emissions in 1990—its baseline—measured some 56 mm tons of carbon dioxide equivalent annually (Oregon calculates five other GHGs in addition to carbon dioxide, then converts the five to reflect their climate forcing effect as a carbon dioxide equivalency). By 1999, emissions had grown by almost 30%, to 72 mm tons, paralleling national trends.

While other countries, especially European nations, were setting reduction goals and imposing carbon caps in a first effort to implement the 1992 Kyoto Protocol, the Clinton Administration, while signing onto the agreement, was unable to get Senate approval and otherwise did little to affect national emissions. The George W. Bush Administration was generally uninterested, and President Barack Obama, while publicly supportive, was first distracted by the Great Recession of 2008 and then hamstrung by a Republican Congress unconvinced of the seriousness of climate change and bent on denying Obama any wins it could.

States and cities stepped into this policy vacuum. Seven Northeast states formed the Regional Greenhouse Gas Initiative (RGGI) to cap GHG's from their power plants. Three west coast Governors committed to economy-wide emissions reductions but only one—California—was able to approve carbon cap legislation before the multistate effort collapsed in the face of the 2008 recession.

Oregon's Governor Ted Kulongoski took the three-state commitment seriously by empaneling, in 2003, a Governor's Advisory Group on Global Warming of 25 exceptionally qualified citizens. The Advisory Group reported back to him in 2004 with recommendations for critical state climate actions (including a carbon cap) and three emissions reduction targets:

- by 2010, the state should have arrested the upward climb of GHG's and bent the trajectory downwards;
- by 2020, the state should be 10% below 1990 emissions levels; and
- by 2050, the state should be at least 75% below 1990 levels.

The 2004 Report recommended the Governor assemble a new stakeholders group focused only on designing a carbon cap that could link to comparable instruments in CA and other states, as the RGGI states earlier had done. That second group reported back in 2006 with a carbon cap framework. The next year the Legislature adopted the three emissions goals, and also established the OCCRI located at Oregon State University, charging it with developing the Oregon-specific science that would underpin mitigation and adaptation work.

The 2007 Legislature also established the OGWC, giving that body a broad hunting license to identify the state's climate policy needs and vulnerabilities, to track Oregon's emissions, and to recommend to the Governor and Legislature means for reducing emissions and mitigating their effects. Lamentably, the Legislature failed to give the Commission any authority other than offering recommendations for changing practices to reduce emissions. It failed to provide any budget that might allow for analysis to determine the efficacy or cost-effectiveness of the Commission's recommendations. It did enjoin other state agencies to cooperate with and support the work of the OGWC but gave that injunction no priority, no ability to hold agencies accountable, and no commitment of resources or deployment of tools such as subsidies and regulation. The Oregon Department of Energy was directed to provide staff support, which consisted of one-half of one professional analyst's time when the Department did not have higher priorities for that position.

The absence of Commission authority to hold agencies (or the Legislature) to account for failure to act decisively was a caution. So was the absence of the tools—staff and budget—that could fund analysis to support policy recommendations or communications with Oregon citizens and institutions. Commission members could construe the mandate as either a down payment on a more serious State effort to address climate issues, or as a pro forma gesture that would substitute the form for the substance of commitment.

The Governor asked me to chair this new Commission. Notwithstanding the limitations imposed by the originating statute, I agreed for two reasons. First, Oregon needed a voice focusing on climate policy and loud enough to be heard, at least occasionally, above the distractions of the short term demands on—and from within—agencies. Second, the Commission membership gave it throw weight: two utility CEO's, the head of the Port of Portland, the Executive Director of the state's largest environmental advocate, the heads of the major State agencies (on a nonvoting basis) and representatives of Oregon's University System including the OCCRI Director. While this interest-dominated membership presaged resistance to some needed actions (such as a carbon cap that would constrain the utilities), affirmative decisions would have some throw weight behind them. The membership and regular briefings from OCCRI would ensure the Commission would never stray too far from hard data and peer-reviewed science.

We were also encouraged by the local governments in many of Oregon's urban areas that were stepping up to assume emissions reduction responsibilities.

Oregon Global Warming Commission
Report to the Legislature
2011
Including Key Actions and Results from the Commission's Interim Roadmap to 2020
February 2011
KeepOregonCool
Oregon Global Warming Commission

Figure 13.2
2011 report to the legislature (cover).

Urban areas can leverage density and urban design to capture carbon efficiencies in transportation (public transit + bike/pedestrian options) and the built environment (building heating/cooling, lighting, and water efficiencies). The City of Portland, at the center of the state's largest urban area, adopted its first global warming containment strategy in 1993; and its current version, jointly with surrounding Multnomah County, in 2009. Their GHG emissions in 2009 were 2% below 1990 levels, whereas statewide emissions were up 15%. Per capita emissions in the rapidly growing metro area were down 20%. We understood that climate change was driven only by GHG parts per million in the atmosphere and was indifferent to per capita or economic growth numbers, comforting as they were to us. Thus, the State and the local governments have always set their goals to absolute emissions reduced (e.g., the 2009 combined City-County Climate Action Plan set goals of a 40% absolute reduction below 1990 levels by 2030 and 80% below by 2050) (Fig. 13.2).

The Governor's Advisory Group, in proposing goals and implementing measures to the Governor and Legislature in 2004, was working from data that showed a consistent 1%–2% emissions growth every year since the Oregon's 1992 commitment to hold the line at 1990 levels. We were obliged to work with emissions data that had been somewhat randomly collected and indifferently processed; our most recent numbers were from 5 years earlier and showed 1999 emissions reaching almost 28% above 1990 levels. Despite a 2000–01 recession that might have dampened emissions growth, we had to assume that the trend line holding since 1990 was continuing to do so, and urgent action was needed to arrest emissions growth and bend the curve downwards, as stipulated in our 2010 goal. In fact, Oregon's emissions peaked in 1999, reflecting growth in industrial emissions during the prosperous 90's, and the closure of a large nuclear plant that was replaced largely by new natural gas generation (Fig. 13.3).

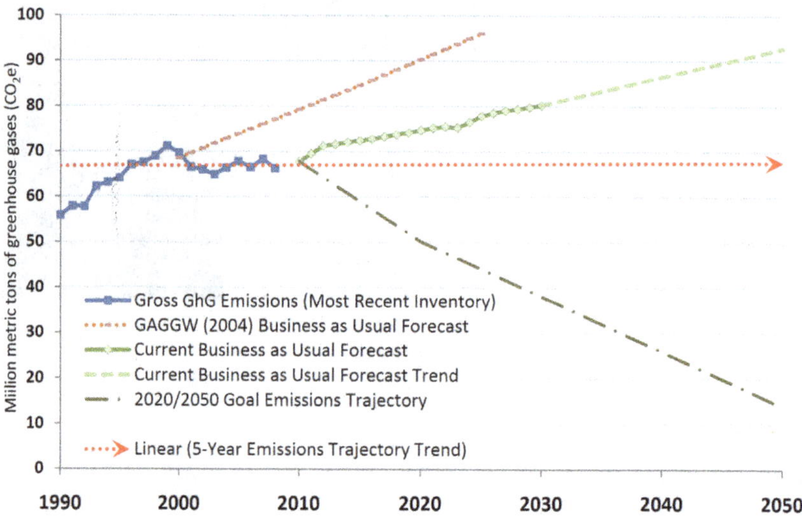

Figure 13.3 Oregon's GHG emissions history to 2010 with projections under Business-As-Usual compared to an emissions trajectory that would reach the state's 2050 emissions goal.

From 2000 to 2013, Oregon's emissions largely declined, led by reductions in electricity emissions. Two factors drove this trend: (1) continued investments in energy efficiency offset increases in population and associated demand growth; and (2) by 2010 the effects of a shift first to natural gas generation, and later to wind and solar energy displacing power dispatched from coal plants.

Oregon met its 2010 GHG goal, to arrest the upward trending emissions and begin to capture real emissions reductions. Emissions had declined from 72 mm tons in 1999 to 64 mm tons by 2010. Emissions continued to decline through 2014, to 60 mm tons, before beginning to rise again to an estimated 64 mm tons in 2017. We do not have the data for 2017–19 but emissions likely continued to grow until the rise was interrupted by the COVID pandemic and accompanying recession.

The emissions growth since 2013–14 was triggered by transportation recovering from the 2008 to 2009 recession. A key indicator, Vehicle Miles Traveled, had stayed level in the first decade of the new century, declined with the 2008–09 recession, then began again to climb. In the same early period the Bush Administration policies allowed vehicle efficiencies to level off (they would improve again under the Obama Administration, then be weakened by Trump). National GHG emissions did not peak until 2009, 10 years after Oregon's topped out. Emission levels have been going sideways since then, with a credible decline in power sector emissions offset by rising transportation emissions.

Electricity—the Boardman chapter

Oregon, Washington, Idaho, and Montana—the four Pacific Northwest states—had, until the mid-1960's, the cleanest and cheapest electricity in the country. The mid-20th century had seen the development of thirty-some large Columbia-Snake Basin mainstem hydropower dams and many more tributary projects, starting with the Depression-era construction of the Bonneville and Grand Coulee dams. Electrical load growth in the postwar era used up the hydro surplus, obliging the region's private investor-owned utilities (IOU) to begin building first coal and nuclear, then gas generation plants. Of the roughly 15,500 megawatts (MW) of regional coal generation, two-thirds belongs to Pacific Power and serves customers in six states including Oregon. The balance of coal capacity is shared among six other IOU's. Only one coal facility is physically located in Oregon (Boardman, in the north-central part of the state) and within the State's jurisdiction. Most coal generation is sited near the coal mines of the Intermountain West and sent over high-voltage transmission lines to regional loads. "Coal-By-Wire" comes to Oregon from Montana, Wyoming, Utah, and elsewhere.

Coal generation is dirty, its GHG emissions joined by smokestack clouds of NOx, SOx, mercury, particulate matter, and other unpleasant residues. While the fuel itself is not expensive, managing the air and water pollutants and coal ash waste has increased costs as pollution regulation, driven by human health and air visibility concerns, has become more stringent.

At the same time, alternatives to coal burning have become more available, less costly, and less polluting. Natural gas that used to be vented or flared from oil fields has become a marketable product that burns cleaner and produces half the GHG emissions per unit of output. "Fracking" extraction techniques, with their environmental controversies, have made gas generation less costly still. Concerns are building as data become available on fugitive gas (methane, or CH_4) losses from gas wellheads and pipelines.

The levelized costs (e.g., total capital and operating costs equalized by year across the facility operating life) of wind and solar generation have dropped dramatically in the last decade, down 68% and 88%, respectively (Bartlett, 2019). At the same time improved battery and other energy storage technologies have emerged to add capacity value to these resources and meet integration and reliability requirements. These renewables, especially when sited where the wind and solar resources are strongest and access to the transmission grid is accessible, are out-competing coal and gas for cost-effectiveness and environmental compatibility (Fig. 13.4).

Oregon took early steps to encourage renewable energy, adopting a Renewable Portfolio Standard (RPS) in 2007 that mandated utilities to serve 25% of their load from new renewables by 2025. That standard was increased in 2016 legislation to 50% new renewables by 2040. In its 2011 "Roadmap to 2020" Report, the OGWC recommended support for new transmission capacity to service new

Figure 13.4 Wind farm near Roosevelt, Washington. Photo credit: Angus Duncan, 2007.

renewables, Smart Grid and other new renewable energy integration tools, and an admonition to "Ramp Down Emissions Associated with Coal Generation."

These strategies—ramping down coal and ramping up renewables while maintaining priority development of energy efficiency—have been the central features of the Commission's, and Oregon's, GHG reduction efforts over the last decade. The third pillar of clean energy policy, energy efficiency, was consistently pursued during this most recent decade when efficiency became the region's second largest electricity resource, but Oregon was not the innovator and mover it had been in earlier years.

In mid-2009 PGE proposed to invest a half-billion dollars in its 585 MW Boardman coal plant in north-central Oregon, from which it derived 20% of its power supply. The plant, which was the largest single source of GHG's in the state, had been authorized in 1975, two years before passage of the federal Clean Air Act and so for years was excused from meeting that law's most stringent standards. Now it required substantial air quality retrofits to continue operating. The makeup investment, however, likely would have resulted in the plant operating for another 20 years or longer to amortize the new capital costs. The issue posed a perplexing challenge to Oregon's statutory clean air policies, which addressed airshed pollutants (SOx/NOx/Mercury, etc.) but not GHG emissions in its cost/impact analysis. It posed a challenge to the environmental community, which was finally making room for climate change in its agenda. It was challenging to PGE as it tried to reconcile its customers' desires for low power rates with their emerging preference for clean, climate-friendly power supplies.

The Boardman question also challenged the new Global Warming Commission as it sought for itself a meaningful role in driving emissions reductions, beyond only making policy recommendations and issuing reports. The Commission's predecessor Governor's Advisory Group had recommended capping carbon emissions from sources like Boardman and replacing the dirty power with clean renewables.

As Chair of the OGWC (and with 35 years of credentials in regional electric utility planning), I was able to enter into working relationships with several environmental advocacy groups, with the State air quality regulators, and with PGE policy managers. Separately and then collectively, we assembled an alternative to PGE's default reinvest-and-run strategy. In brief, PGE would agree to spending a much smaller amount of money—$50 million instead of almost $500 million—to make retrofits that would put the plant in compliance only through 2020. At that point PGE would stop burning coal at the plant. Although the proposal would keep Boardman operating and emitting GHGs for another 10 years, important environmental and consumer advocate groups came together to support it, which mandated a hard closure while also meeting the air regulators' minimum standards and enabling their approval. Opposition continued from some labor organizations representing Boardman work staff and resisting shutdown, and from the Sierra Club and some environmental groups advocating earlier closure dates.

But the agreement found broad community support and stuck. It made progress toward Oregon's environmental and GHG reduction goals, if not as much as a real sense of climate urgency might have recommended. It was a compromise of the sort that historically had exemplified problem-solving in the state, suggesting that that model was still viable. And it paved the way for important future climate negotiations between environmentalists and the large electric utilities.

Electricity—the end-of-coal chapter

The environmental community made clear to Oregon's two large electric utilities, PGE and PacifiCorp (PAC), that the Boardman agreement was a stepping-stone toward decarbonization of Oregon utilities. The utilities' coal-by-wire power deliveries to Oregon customers from out-of-state coal power plants would remain as targets. Internal polling by environmental groups suggested that a ballot measure to end coal-by-wire deliveries to Oregon customers and replace these with renewable wind and solar from a strengthened RPS, had potential to succeed at the polls. The presumption was that the two electric utilities which served 70% of Oregon's electricity were seeing similar results in their internal polling. An initiative battle would be costly and uncertain for all parties. And the Boardman agreement provided a basis for an environmental/customer advocate/utility conversation (in fact, the parties had been meeting informally several times a year to maintain the hard-won communications channel). Discussions begun in 2015 led to the introduction of Senate Bill 1547 in Salem's 2016 session. The bill was supported by the same cohort of environmental advocates and the two large utilities. It would require the utilities to largely cut off coal-by-wire not later than 2030, and conform to a revised RPS that would lead, in steps, to a standard that required 50% of each utility's load to be served by new renewable resources not later than 2040. The "new" was critical since an RPS was supposed to spur new renewables, not just count existing ones (mostly hydro, but some wind and solar already in the door). Together with existing renewables (including mostly federal hydro serving public power), the RPS had the potential to push the Oregon's 2040 electricity mix to an 80% or higher renewable energy content.

The "50% new renewables" requirement was also critical to discourage the utilities from simply replacing old coal with new gas combustion turbines that could have a life of 30 years or more. That would have locked Oregon into another extended and costly fight to terminate gas plants before they could be fully amortized.

The parties took their agreement to the Oregon Legislature's 2016 "short session" with the expectation that bringing a compromise with both utility and environmental buy-in would sufficiently grease the legislative skids (the Governor was also on board). There was some business community equivocation, and some state-level Republican legislator resistance to an anti-coal bill that reflected emerging rhetoric in the 2016 national Republican Presidential nominating process. Notwithstanding these headwinds, the bill passed out of committee with bipartisan support and was approved by the full House in a bipartisan vote that included five Republican Representatives.

The Senate had a more conservative Republican caucus that had been using obstructive tactics—requiring each bill be read in full; not attending committee hearings—all session to resist several majority measures including the coal bill (SB 1547). We still believed a compromise supported by utilities and environmentalists could draw bipartisan support. But the RPS requirement would also apply to a small number of the larger public power utilities, which mostly served rural Oregon. And R's claimed, contrary to testimony from the utilities, that power rates would be driven upwards. The R Senators balked (one R Member who was considering supporting the bill returned from his caucus meeting abashed and voted no). The bill, which even 5 years earlier might have expected the bipartisan support it won in its House vote, barely scraped by in the Senate in the closing hours of the session, and then only because the D's agreed to abandon other contentious bills.

"SB 1547 was a win but also a warning: politically-turbulent seas lay ahead."

SB 1547 was a major step forward for clean energy, a validation of the ability of utility and environmental parties to still find sufficient basis for a compromise well-grounded in science and economics. It also confirmed that processing these fraught and complex decisions through the legislative process was better than fighting them out as ballot measures. Yet SB 1547 also served as a very large, brightly colored caution flag that the politics of compromise and incremental progress was an increasingly high-risk proposition in Salem; and that the divisions in Salem reflected the increasingly polarized politics of Oregon, and the country. Divisions along party and geographic lines were hardly new, but uncompromising election day battles were usually left at the polling places, and problem-solving compromises were a more customary outcome by the time legislators reached Salem. Now the risk especially to R's of being "primaried"—that is, facing a more conservative challenger for their party's nomination—was increasing nationally and in state races, including in Oregon. Solution-seeking, compromise, and science were being outflanked by ideological lines drawn on taxes, "job-killing regulation" from big government overreach, and most significantly, environmental outcomes that parts of rural Oregon did not support. But Oregon was increasingly a Blue state, with most population and voting growth taking place in west-side urban areas. SB 1547 was a win but also a warning: politically turbulent seas lay ahead.

Electricity—the capping carbon chapter

With SB 1547, the two electric utilities indicated they had given all that should be expected of them. In fact, they and we could map out a plausible pathway forward in which electricity emissions declined in a manner that should meet or exceed the State's goal of 80% below 1990 levels (Fig. 13.5).

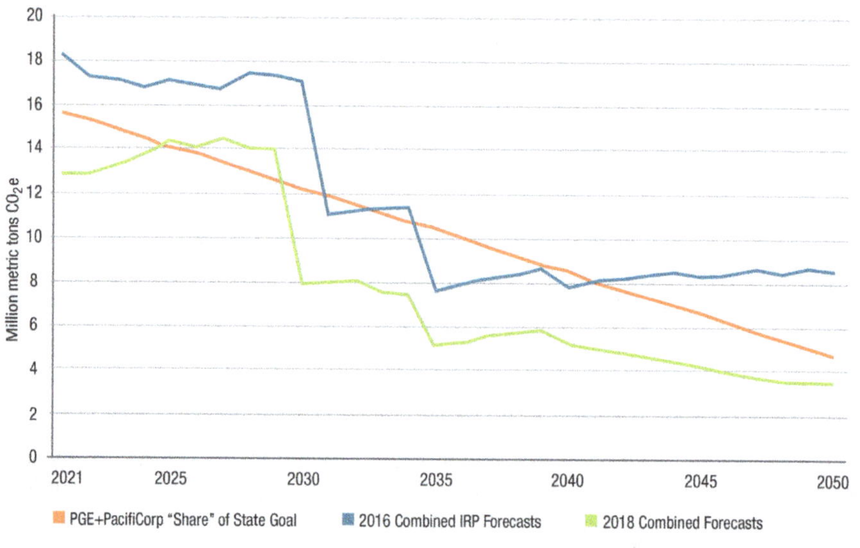

Figure 13.5 Estimated emissions reductions for PGE + Pacific Power.

That still left us with little control over emissions from other sectors—industry, agriculture, forestry, the retail gas utility, and especially transportation—that would swamp utility gains. And these sectors were harder to engage than were our utilities. The next tranche of emitters did not have easy access to the low-carbon technologies that could substitute with low or negative cost effect. And negotiating with two million individual, unregulated car drivers was a different proposition from getting to yes with two regulated utilities. To get at this remaining two-thirds of Oregon's emissions would require a new mix of incentives and regulatory tools.

Most of all, it would require the tool called for as far back as the Governor's Advisory group (2004), an economy-wide carbon cap. Because we had done the early work of sketching out how such a cap could be designed effectively to hold down costs, and because we had the examples of the Northeast states and of California to draw upon for lessons learned, we were confident we could shape a carbon cap that would be neither oppressive nor ineffective.

We also knew that accomplishing this by ballot measure was not a good bet. A carbon cap, especially one that took into account the differences among emitting sources, would be complex. And complexity does not usually do well at the ballot box. Skeptics had only had to look north to the State of Washington where two separate tries at pricing carbon—a carbon tax in 2016, and a carbon cap in 2018, both went down to crushing defeat. Both were easy to cherry pick for negatives: a flat cap without exceptions will contain myriad apparent injustices (e.g., regressive effects, where low income households pay the same as the one-percenters; or impacts on businesses that sold in national and global commodities markets without leverage to pass costs along). A cap with allowance trading, adjustments for hardship effects and so on would be criticized for complexity, and for who was included and exempted by where the compliance lines were drawn. Postelection court challenges were all but certain.

A carbon cap is akin to a tax code, building code, or utility regulation. Each presents a complex set of choices that demand complex solutions be found, and complex implementing tools devised to achieve a sellable balance of effectiveness and equity.

Another legislative solution was needed. This time it was not going to be a small group of environmentalists and emitters working solutions out among themselves and presenting a common package to legislators. For this, legislators would have to take the lead. Happily, there were exceptionally good chairs of both House and Senate Committees with jurisdiction, and substantial D majorities in both chambers. The committees conducted extensive outreach in advance of the 2019 session and were prepared with drafts from the outset. As it turned out, these early drafts still did not anticipate and address all the compromises this kind of bill entails. Not all the demands were on the table early. For some interests, their needs were unclear in advance, or arose only as other amendments were agreed to. For others, holding out afforded leverage when legislators truly need to count votes and an interest group names its price.

The electric utilities were not on board early this time and held out until they got a deal that acknowledged the trajectory they were already on from SB 1547; they should only need to buy emissions allowances, they argued, if their emissions rose above those curves. The gas utility, envious of that deal, wanted a similar one for itself. Rural areas claimed they had longer distances to drive (not necessarily) and should not be disadvantaged compared to city dwellers with access to transit. Certain industries claimed they were "energy-intensive and trade-exposed" or EITEs. A formula to determine who was in or out of that class engendered fractious discussion. And so on; over 100 amendments were accepted into the bill, and many more debated. The amendments themselves, offered or agreed to by Democrats in a spirit of sausage-making compromise, were then used by opponents to up their complaints of complexity. The Democrats going-in legislative strategy of swiftly moving a completed bill to the floor to avoid end-of-session horse-trading and games-playing was an early casualty; instead, a late-session bill that passed the House on a party line vote stalled in the Senate.

The debate in the House was itself instructive. Republican after Republican arose to condemn the bill and highlight the debating points of their preagreed strategy to shape public opinion. Yet nearly every one began with the phrase, "I'm not a climate denialist, but ...." The steady drip of evidence from the global scientific community, reinforced locally by the OCCRI and OGWC Reports and combined now with tangible evidence (wildfire smoke; sparse snowpack; public health effects) were finally moving public opinion past climate denial, "hoax" theory, and environmental conspiracy. Remnants of the denial cohort still existed, but Republican opposition, at this point more cultural than evidence-based, had to find new grounding. Now, climate might be a concern but solutions—and certainly any solution that involved government intervention—was ipso facto unacceptable. There was no Republican acknowledgment of the disruptive effects that climate change would bring to Oregon families and businesses, disruptions that were already causing serious damage. No looking down the road at gathering consequences.

Only if the Democratic majority would offer a carbon plan that involved no additional cost, no market intervention, no constraints on business or personal activities ... in short, no effective measures and no change in BAU, might Republicans consider supporting a bill.

When the D majority was clearly prepared to move their measure to a Senate floor vote, the Senate R's dropped their bombshell. With slightly more than a month remaining in the half-year session, the Senate R's walked out. And not only walked out, but in some cases journeyed across the State line and into Idaho where they would be safe from any Oregon state trooper sent by the governor to retrieve them.

> *"... when the Democratic majority decided to trade Oregon's economic free-market system for one of central government control—while ignoring our constitution and making a shambles of Oregon's rural and low-income economies—we walked."*
>
> **Bentz (2019) Why We Walked, Oregonian July 11.**

The Senate Ds still had their majority of course. But they did not have the quorum required to conduct business. In 47 US states a quorum is 50% plus 1. In Oregon it is two-thirds plus one, a peculiar rule Oregon borrowed back at statehood from the Indiana State Constitution, on which ours—for reasons lost to history—is based. When 12 Senate R's decamped, the chamber was two votes shy of a quorum. This time, unlike with SB 1547, there was no compromising and no bipartisan option.

The Senate R's, as one, opposed HB 2020, the carbon cap bill. Representative Cliff Bentz, their leader in opposition to the cap bill, explained in an OpEd: "In a democracy, the majority rules. But when the Democratic majority decided to trade Oregon's economic free-market system for one of central government control—while ignoring our constitution and making a shambles of Oregon's rural and low-income economies."

The R's advanced multiple explanations for their intransigence—cost, unfairness to rural Oregon, should let Oregonians vote on the cap, etc. The underlying reason, however, was Republican denial, across the country and in Washington DC, of climate change being real and needing a big government fix. For R's the risk of being "primaried" remained, but there was little need for that threat anymore. In an era of Trumpian denial, withdrawal from the Paris Agreement, abandoning auto fuel economy standards, and so on, Oregon Rs were not going to be outdone by their brethren elsewhere. The bill died.

A second bite at the apple was scheduled for Oregon's February 2020 short session. Again, leadership labored to graft additional fixes (concessions) into the bill. For example, most rural counties were exempted from the transportation carbon cap and associated costs. This and other concessions were a waste of air and effort; the Senate R's walked again, this time joined by House R's. No quarter was asked or offered by the walkers. And the short session ended on that deeply fractious note.

Transportation—Oregon loses its way

The mounting difficulties in fashioning an economy-wide carbon solution both paralleled and reflected the challenges in coming to grips with Oregon's largest emissions sector, transportation.

First, some framing data (from Oregon Global Warming Commission, 2018 Report to the Legislature): transportation energy use is about 30% of Oregon's total (measured as BTU's), almost all—over 90%—from petroleum-based fuels. Transportation's share of state GHG emissions is almost 40%, up from 35% in 2014. Emissions per vehicle had declined about 12% since 2005, reflecting greater vehicle efficiency. However, total emissions had been increasing in the years since the Great Recession, up 7% between 2012 and 2017 (and almost certainly continuing through 2019), reflecting more and larger vehicles, and 10% higher overall Vehicle-Miles-Traveled in the same period (Fig. 13.6).

About half of Oregon's GHG emissions come from light-duty cars and trucks—personal vehicles, delivery trucks and so on. A quarter of GHG emissions come

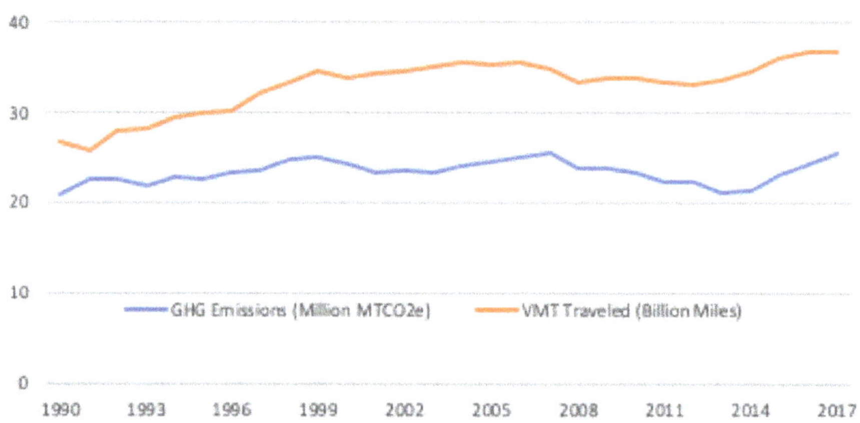

Figure 13.6 Oregon transportation vehicle miles traveled and transportation emissions 1990–2016 (Oregon Department of Transportation and Oregon Department of Environmental Quality).

from heavy-duty trucks, the long-haul semis; and the balance, from air/rail/shipping. Emissions from the heavy-duty truck sector have shown the most disturbing increases, reflecting stronger economic activity from the end of the 2008 Great Recession to the onset of the COVID pandemic. There have also been fewer regulatory and other incentives to spur greater vehicle efficiencies as those efforts stalled out in the Trump Administration.

A state Oregon's size has little market leverage, by itself, to demand more carbon-efficient cars and trucks. California has 12% of the US new light duty vehicles market; Oregon, a little over 1%. But because California has special status under the Clean Air Act to adopt emissions standards that differ from federally set levels, and other states can choose which standard to adopt, Oregon and another dozen or so states have joined with California—collectively equal to about 40% of the US new car market—to demand more efficient vehicles. This alliance was welcomed by the Obama Administration and enjoyed the general acquiescence of most auto manufacturers but was never accepted by national petroleum interests. The oil companies joined with the Trump Administration in declaring war on state-level efforts to mandate vehicle efficiency improvements, while the states allied with California have counter-sued. Progress toward meaningful national and state vehicle efficiency goals has stalled while the courts, and more likely the polls, sort out the mess.

Oregon also adopted a Clean Fuels Standard that requires vehicle fuels sold in the state to have reduced their carbon content (and therefore, emissions) from 2015 levels by 10% by 2025 (later raised to 20% by 2025 and 25% by 2030). Oregon offers a buyer's rebate of $2500 to $5000 per electric vehicle (EV).

There are programs to spread EV charging stations around the state. The three west coast states are among the top four in per capita EV sales, but the number of EV's in Oregon hovers at around 1% of the total fleet.

Meanwhile, gas and diesel small trucks and SUV's, with the poorest fuel efficiencies, take an increasing national and state market share. Transportation GHG emissions inch upward when for climate purposes they should be declining dramatically.

Oregon had hoped to push back against these trends with the 2019 and 2020 carbon cap efforts. The cap would have put the obligation on parties responsible for emissions—in the case of transportation emissions, on the fuel suppliers—reducing their allowable emissions each year and requiring them to purchase allowances for any excess. Revenue from the allowances would fund further transportation sector emissions efficiencies. The legislative failure left Oregon with fewer options going forward, and the state is now sorting through these, most importantly the substitution of an DEQ-administered cap-and-reduce regulation of fuel suppliers.

Forest carbon—an opportunity opens up

There are other emissions sectors within Oregon's GHG reduction strategy, including materials, waste management, nonfuels agricultural activities, and these also must be brought under control. But energy use produces 80% of Oregon's GHG emissions, and energy use by utilities, industry, and transportation has been the priority target. It remains so in 2020, but forest carbon has been identified as a sector of comparable significance. When the Commission first looked at forest carbon for its 2011 "Roadmap" we concluded there were not the data and analysis to warrant other than rules of thumb for this sector. We proposed that the wetter westside public forests (mostly federally owned), which seemed to be accruing net carbon, be left alone to do so. Oregon's drier forests extending east from the crest of the Cascades were less carbon dense and appeared to have more fire risk and fuels accumulation issues, so we deferred to federal managers to first tackle these tasks. Wood products, we proposed, would have to come from privately owned forests, mostly on the west side of the Cascades. We also proposed that State and Federal forest professionals commit to developing the data on carbon content and fluctuations needed for more tailored solutions to be devised.

They agreed, aand by 2016 we were seeing incomplete but more useful numbers coming through the US Forest Service Forest Inventory and Analysis (FIA) program. These data described the distribution of carbon in Oregon forests by geography, ownership, and "pool" within the forest (e.g., standing live and standing/fallen dead trees pools, substory vegetation, soils). More importantly, it described the flows of carbon among these pools, between forest and atmosphere, and between the forest and human extractive uses.

Most importantly, the data described a special status for Oregon westside forests, showing that they were among the most carbon-dense (carbon per hectare) forests in the world, more so than the Amazon, Indonesian, or Central African forests (Krankina et al., 2014). Taken together with their extensions south (into northern California) and north (up to and around the Gulf of Alaska), these forests constituted a global carbon sink comparable to that in Indonesia (which is approximately twice as large and half as carbon-dense). And Oregon forests were increasing their carbon content by tens of millions of tons annually.

We found ourselves stewards of a carbon resource of global significance. If the forests in the other states (and British Columbia) were acquiring and sequestering carbon at a comparable rate, the importance of holding onto both existing stores and these exceptional gains in a world where carbon losses to the atmosphere were the rule was hard to overstate. This was especially so when the data were showing the globe's other great forest regions beginning to shift from net carbon sinks to net sources.

The strident opposition of the forest products community mobilized against these possibilities (Fig. 13.7), and against any climate action that would affect prevailing levels of timber harvest from private industrial forestlands or the much more limited but still significant harvest from public—federal and state—forestlands. "Removals" (harvest) from the 20% of Oregon forestland controlled by industrial forest owners was the source of almost 70% of Oregon's live-tree harvest GHG emissions (USFS FIA, 2017). Owners and forest workers organizing themselves as "Timber Unity" joined to stage highly visible demonstrations at the State Capitol in opposition to the carbon cap bill. While that bill did not address forest carbon from timber harvest, it did address vehicle emissions from carbon-based fuels—gasoline and diesel. Timber Unity was not impressed when the bill exempted from the cap fuel use in the work of tree harvest. It called on its members and their log trucks to circle the capitol building for several days running, throttling their engines noisily, and sounding their hydraulic horns. The media were irresistibly attracted to the spectacle and filled their nightly newscasts with audibles and visuals from the demonstrations. The action unnerved the few coastal Democrats remaining in the Legislature (who asked for and got the fuel exemptions for specified logging activities) but did not undo the committed majority votes for the cap. [Note: The November, 2020 election saw all but one of the remaining coastal legislative seats shift from Democrat to Republican; the remaining (Senate) seat is held by a Democrat who votes consistently with Republicans on environmental questions.)

The Global Warming Commission submitted its forest carbon findings in a special 2018 report to the Legislature. The Report identified ongoing data gaps and analytic needs, but also vulnerabilities (overharvest) and opportunities

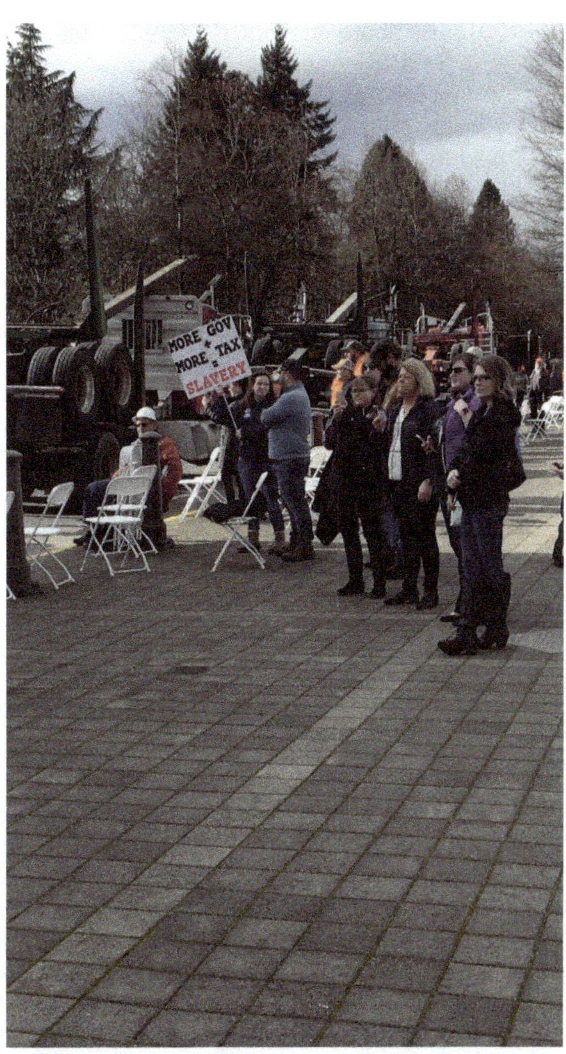

Figure 13.7 Timber unity demonstrators circle the Oregon Capitol, 2020. Photo credit: Angus Duncan.

(increased carbon sequestration through measures such as longer harvest rotations and reforestation). It cited research from Oregon State University suggesting the potential to increase forest carbon capture and sequestration by + 50% from reduced timber harvest in public forests, longer rotations for all forest owners, and other forest practice changes that would be challenging but achievable. Since the forest products industry was arguing for a policy of timber harvest and carbon sequestration in forest products (e.g., construction lumber), the Commission's report asked for additional analysis of the net carbon flux associated with harvest. Data from peer-reviewed papers and other analyses (Ingerson, 2009; Law et al,

2018; Hudiburg et al., 2019) suggested that as little as 15%–20% of the carbon in a living tree ended up in a durable product, and the average duration of the longest-lived such products (e.g., a 2X4 in a wood-frame building) was around 50 years. In the alternative, a mature 80-year old tree could be expected to live, and hold its carbon, for another several hundred years; whereas a stand of trees not subject to timber harvest might hold its carbon in perpetuity.

An option for the Commission in 2018 might have been to propose a forest carbon goal for the state that would set an interim target—as had been done in 2004 for the energy sectors—and refine it as the requested new data and analysis emerged. A larger and more challenging goal but one clearly within reach would be to (1) target overall state net carbon neutrality by the mid-2030's (net emissions offset by net sequestration), and (2) position the state serve as a growing carbon sink beyond that point as policies continue to drive down emissions and forest carbon uptake continues to capture and hold increasing amounts of atmospheric carbon. Because any such proposal would likely be contentious, and the State's carbon policymaking was likely to be quite contentious enough with a fossil fuels carbon cap coming before the Legislature in 2019, we chose only to put the question on the State's table without proposing targets and timelines. Any such strategic choice can be second-guessed, and in hindsight it is unlikely the forest carbon proposition would have altered the carbon cap outcome. At the same time, the question now had its place on the state's carbon agenda and would stay there, insistently, as the next chapter in Oregon's pilgrimage opened in 2020.

The Governor's executive order

At the end of February 2020, as Oregon's Senate Rs staged their second walkout (joined by their House colleagues) over the second iteration of a carbon cap-and-trade bill in a replay of 2019, climate advocates considered their options. The take-no-prisoners partisanship that gripped the US Congress had also locked Oregon politics into another round of climate denial. Notwithstanding new concessions and accommodations written into HB 1530 it was not going to see a single Republican vote. Due again to Oregon's quorum rule, the bill would prompt neither floor debate nor a vote in either chamber, only a bicameral walkout.

> *"If we adults don't take action right away, it is the next generation that will pay the price."*
>
> **Governor Kate Brown, 2020, announcing her Executive Order 20-04.**

The legislative leadership and Governor accepted, after two tries, that there was not a legislative path forward. On March 5 the leadership adjourned the 2020 session. On March 10 Governor Brown issued Executive Order (EO) 20-04. The Governor's statement read, in part:

> *"... climate action is crucial and urgent. If we adults don't take action right away, it is the next generation that will pay the price... we will pursue every option available under existing law ...."*

The EO proposed actions in a broad range of sectors, including:

- Declining carbon emissions caps in transportation fuels, natural gas and industrial emissions (but because the State had no authority to charge for allowances, they would be issued free and parties would be able to buy, sell, or trade for them);
- A stronger Clean Fuels standard for transportation fuels, requiring 25% reduction in emissions by 2025;
- Beefed up efficiency codes for appliances and new buildings;
- Requiring State agencies to integrate and prioritize carbon/climate outcomes in evaluating policy and programmatic actions, and to apply a carbon "lens" in such evaluations;
- Charging the Global Warming Commission with bringing recommendations back by mid-2021 on integrating the carbon potential of "natural and working lands" including forests into the State's climate policies (Fig. 13.8).

DEQ would be largely responsible for this carbon cap-and-reduce strategy, which would proceed on the same schedule the legislation would have prescribed, commencing application by January 1, 2022. While the agency may structure compliance obligations to accommodate some of the same sector circumstances—trade-exposed industries, for example—that the legislation would have, it is under no statutory obligation to do so. While there could be an accommodation for vehicle fuels in rural counties, there was no obligation to carve one out. Concessions to the business community and the constituents of

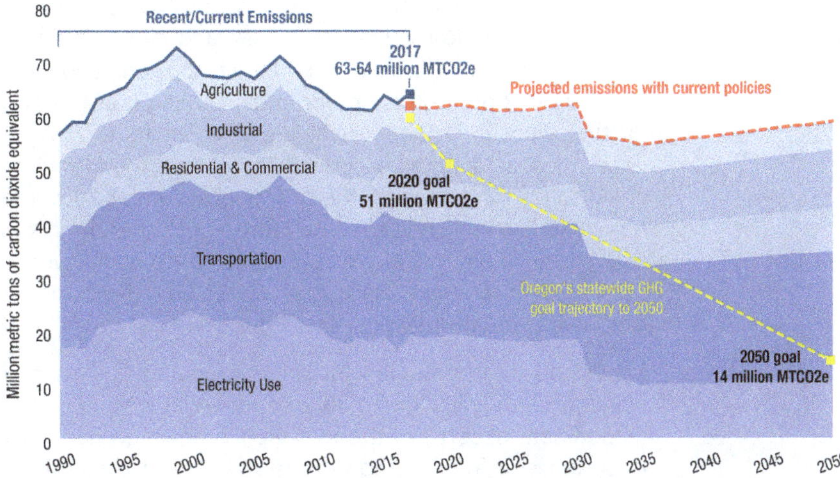

Figure 13.8 Oregon's emissions, past and projected 1990–2050 (State of Oregon DEQ, ODOE 2018).

rural Republican legislators that had been written into the failed legislation would be off the table unless DEQ, in its discretion, decided on the merits to resurrect them. In effect, the intransigence of Republican legislators has disadvantaged many of their constituents.

Important tools for advocates of the cap are missing as well in the administrative model. Because DEQ has no authority to sell allowances, revenues that might have been invested in carbon efficiencies are no longer generated. DEQ believes it has no authority to regulate emissions from out-of-state sources such as power plants delivering coal-by-wire to Oregon utilities (but SB 1547, ramping out such coal-generated deliveries make this less important). There is concern that free allowances can result in windfall profits to some recipients, as occurred in the European Community.

The jury will be out on the effectiveness of the DEQ-designed and administered cap-and-reduce approach until actual operating experience informs us. It is a second-best solution, but in the polarized and solution-averse politics of 2020, it appears the best we can do even in a deep blue state like Oregon, for now.

While the Governor's EO requires agencies to submit their plans for thoroughly integrating GHG's and climate into their agency agendas, only in the case of the carbon cap does it spell out a timeline and sequence of agency actions. For the rest, follow-up actions and reporting are unspecified. While no doubt the Governor expects every soldier to do his/her duty, there is ample precedent for BAU to quickly crowd out a new initiative like climate. The 2011 Legislature directed the Oregon Department of Transportation (ODOT) to develop an analysis-based Statewide Transportation (emissions reduction) Strategy—the STS. ODOT marshaled a convincing group of technical experts and policy stakeholders, hired excellent consultants and facilitators, and by 2013 had produced a complete and feasible greenhouse gas reduction strategy for the transportation sector. By then, however, ODOT and new leadership on the state's Transportation Commission had lost interest in climate and rediscovered its true vocation, fixing and building roads. Climate was still an apparent priority of the then-Governor, but there was no mechanism for implementing it at ODOT and no accountability for the failure to do so. Several legislative efforts to remedy this and other such lapses by structuring an accountability process found their way to that graveyard of legislative achievements, the deadpool known as "government reorganization." Absent such an accountability structure—goals, benchmarks, reporting—there was little chance a new priority would nudge aside so many older ones. The STS was consigned to a dusty back shelf at ODOT, there to remain until the 2020 Executive Order retrieved it.

Importantly, responsibility for recommending carbon policy for "natural and working lands"—including forestlands—was given to the OGWC. This was a recognition, of sorts, of the role the Commission had played in identifying the importance of forest carbon in the state's carbon toolkit. After the Commission published its Forest Carbon Accounting Report in 2018, the Governor directed that the lead on this topic would be transferred from the Commission to the Oregon Forestry Department (ODF)—a curious decision given that department's general disinterest in the subject until the OGWC began its examination. There was also concern in some quarters that ODF, as close as it has been historically to the timber industry, might not have either the independence or expertise to undertake this responsibility. The decision to shift responsibility back to the Commission improves prospects for a fair balance of carbon and economic outcomes that will be consistent with the state's carbon goals and priorities. It also gives ODF room to transition its own management and stewardship priorities by examining them through a carbon outcomes lens, as the EO directs.

Will the actions called for in the EO be carried out, or allowed to fade away from neglect? Will the many additional actions needed to attain state GHG goals but not captured in the EO find a way to attract the attention of Governor and legislators? Will the potential of Oregon's carbon-dense forests, and the State's obligation to leverage that carbon-capturing capability, be recognized and leveraged by aggressive state and federal forest policies? What is the mechanism for learning while implementing, adjusting strategies and tools, and embedding climate as a policy priority if it depends on the next Governor to adopt the same EO-level of commitment?

The why and wherefore: failures of governance and of imagination

So how did progressive Oregon, and the country, get to this stasis on climate change? Other countries have not locked upon the issue. Even the Brexit-inflected and flustered United Kingdom has a coherent and determined if stumbling-toward-solutions policy approach. Even China, as dependent as it is on coal, acknowledges its obligation to exit that fuel and is doing so, closing the oldest, least efficient plants first, still bringing new coal generation on-line but showing a net decline in coal GHG emissions.

Well, the United States is exceptional. Right? Perhaps not in all the admirable ways we once might have claimed, and leaving aside for the moment race relations, systematic destruction of native peoples, air and water pollution, etc. We have been that exceptional country in many ways; the country that imagined, and defined, the future of politics, technology, industrial production, mobility, and communications. When Nixon debated Khrushchev in that American kitchen replica exhibited in Moscow (I am dating myself here), the American kitchen won hands down.

So why cannot we get our heads and hands around this wicked climate issue? What was it about Oregon's long and winding and struggling road toward reasonable and science-based climate policies—and those same struggles for the nation—that should inform our efforts going forward? Here are some still-forming thoughts on the subject.

1. ***The center is not holding anymore***: American politics and intellectual life have never been static, but in the past there has usually been a center to it that most people subscribed to most of the time. Historically our political parties have been largely center-left and center-right. Movement of that center has been prodded and provoked by people and ideas from further left and right, radical and reactionary, often to useful effect, but their demands have been moderated by the center. We have resided confidence in our institutions: the military, the media, scientists, who have had access to more information in depth and have intermediated between those sources and the rest of us. This confidence was hugely reinforced by winning our meritorious war (WW II), by the country's postwar economic and technological successes, and by the broad sharing of the benefits of those successes. It has been subsequently narrowed and pummeled by the civil rights movement and Nixon's racist Southern Strategy, the Vietnam war, the recession and stagflation of the 1970's, the Reagan/Gingrich assault on the legitimacy of government functions, the interventions and interminable wars in Iraq and Afghanistan, and the Great Recession. I may have left a few body blows out of that list. All centrist institutions have lost credibility with someone in the process. Making policy progress, on issues from the social safety net to Black Lives Matter to climate change, has suffered as partisan divisions resist compromise and solution-building.
2. ***The erosion of confidence in science***: When we lost confidence in our institutions, we fell back on what we could see and we wanted to. These perceptions often aligned with the learnings and teachings of science, especially when that science threw up neat gadgets like computers and iPhones. Sometimes they did not align, as when the industries that brought us useful goods like gasoline and chemicals and jobs producing them also brought us air and water pollution.

> *"We have consistently preferred, over that 30 years, to deny the evidence of climate change the way we denied, for so long, the evidence that cigarettes kill."*

For the last 30 years scientists have been bringing us evidence of climate change and its causes, which are mostly ... us. They have warned that we needed to reduce the GHG emissions that accompany most of our daily lives, from our vehicles to our power plants and coal mining jobs. We have consistently preferred, over that 30 years, to deny the evidence of climate change the way we denied, for so long, the evidence that cigarettes kill. "Inconvenient Truth" was the title of Al Gore's take on climate change, and they were inconvenient enough that the video might have driven as many people away from the science as were persuaded. In fact, it might have driven away or drawn forward almost nobody, instead reinforcing each viewer's prior disposition (see confirmation bias discussions in Chapter 1, The Nuts and Bolts of Science-Based Advocacy; Chapter 2, When Scientists are Attacked: Strategies for Dissident Scientists and Whistleblowers; and Chapter 8, Why Advocate—and How?) while undermining the credibility of science as intermediary. While this is more an issue of political malpractice than scientific missteps, its solution will make demands on the science community to understand the differences in how the two disciplines use many of the same words and express outcomes (e.g., "certainties" versus "probabilities"). *Publishing* a paper is not enough; its evidence, logic, and findings must be *communicated* and *understood* in the translation from paper to policy (see Chapter 10, Out of the Ivory Tower: Campaign-Based Science Messaging for the Public).

3. ***Erosion of global cooperation***: One of the crucial legacies of WW II was the "western alliance" of democracies that won the war. Well, the USSR also had a modest hand in defeating Germany, but the myth of the successes of the West necessarily excluded that Eastern power, since it was the target of the Alliance. The west has had its successes and its missteps over the half a century from WW II to the fall of the Berlin Wall, but on balance it had to be counted a powerful success in protecting and improving lives. And it did so by building its own set of institutions: NATO, the UN, the OECD and its successor European Community, the World Health Organization, and many multistate treaties such as the one that phased out the hydrofluorocarbon aerosols and rescued the earths ozone layer. Now, 75 years after formation of the UN, these institutions are under assault not just by the nations then largely excluded from the western alliance and its offspring (e.g., Russia; China; various Middle East states and stateless forces) but by core nations, most emphatically but not only by the Trump-led United States. And this is occurring just as the need for global cooperation has become most crucial. Climate change is the ultimate Tragedy-of-the-Commons problem, made more vexing still by divergent views on responsibility and equity and technology sharing that invite a collapse of discourse, a kind of rearranging-of-deck-chairs-on-the-Titanic. We argue about who else should be doing what, and who should not be telling us what we can do, instead of finding a basis for global cooperation. No surprise that the same paralysis afflicts climate policy-making within the United States. This issue mostly resides in the world of foreign policy choices and tradeoffs. But while some of us might be scientists, or lawyers, or business owners or environmental advocates, we are all citizens who choose our political leaders. We can ask of them a commitment that our country be a responsible member of the global community, realizing in return

the added value of problem-solving when a truly global problem looms. The Paris Agreement beckons the Biden Administration.

4. ***Partisanship and tribalism in the US politics***: Winning while staying within sideboards of decency and reason used to be fundamental to American politics, even when it came with a dash of political cynicism. And winning did not exclude compromise, finding common ground, and problem-solving. Then, the Democratic Party abandoned its southern flank to become the advocate of civil rights for Blacks, and the Republican Party seized the opportunity offered by welding its business-first policies to a barely disguised racial pitch to the White south. The Southern Strategy typified, and accredited, an emerging win-at-any-cost mentality, that in turn demanded loyalty and penalized compromise. Problem-solving gave way to putting the opposition on the defensive as much as possible. This abandoning the middle was abetted by social media inviting partisans to self-select their news and interpretations, implicitly discrediting differing views ... and more, discrediting facts, however well-grounded. By the 20-teens, surveys were finding parents more appalled by their children marrying out of their party than out of their religion. Partisanship in public office was enforced by the practice of "primarying": running a candidate to the right of a moderate Republican, or to the left of a moderate Democrat. Finding common ground on substantive issues, whether health care or climate change, became subordinate to winning and maintaining partisan discipline. That has to be reversed. We have to restructure our electoral process to advantage the center and de-weight the extremes. We have to demand of our leaders less posturing and more problem-solving. This may mean—I think it will—that the Republican Party needs to rediscover its moderate center-right culture and separate itself from those who preach and practice racism and mindless partisanship. This may mean an extended absence for the Party from elective office, a high price to pay but it is a high "crime" being paid for. And perhaps not; after all, the Republican party in the 1850's created itself out of a similar stasis of southern slave defenders and a decaying Whig party. Within 10 years Abraham Lincoln occupied the White House.
5. ***Markets uber alles***: The market is a wonderful tool that has contributed innovation and economic efficiency when applied intelligently and bounded appropriately. But "markets" as a conceptual abstraction can distort thinking when sold as an ideology and an absolute value, an end and not a means, the epicenter of gravity of individual freedom and so exempt from regulation and boundaries.

"Greenhouse gas emissions and pandemic viruses may be the ultimate tests of whether individual rights and community values can be reconciled, with science acting in the critical intermediating role."

A free and prosperous society is one in which individual liberties and community values coexist, each qualifying and complementing the other. We do not assert our individual right to proceed through a red light when cross-traffic has a green one. Equally the private market elements of energy and transportation systems act to deliver optimum commercial and individual value, but are regulated to ensure that ***public*** goods—public safety, public health, clean air and water, and so on—are at least not compromised, and at best enlarged; that the dangers of a Tragedy-of-the-Commons are identified and averted. GHG emissions and pandemic viruses may be the ultimate tests of whether individual rights and community values can be reconciled, with science acting in the critical intermediating role. At least with a pandemic the penalties and the rewards are contemporary and broadly apparent. With climate-altering emissions, costs and limits on BAU are front-loaded, whereas rewards and penalties are deferred; but they are also unavoidable and all but irreversible once the atmosphere is sufficiently loaded with carbon. If the community—the country; the world—cannot assert the priority of community values over individual rights to emit pollutants and ignore science, there will be no vaccine to rescue us from climate catastrophe. There are tools, such as the Social Cost of Carbon, that can be used to force into our decision-making those costs carbon inflicts on the community when markets would otherwise defer those costs until too late. Utility "least cost planning" in the Pacific Northwest, in its original state, required that *all* costs be internalized in the power planning process and weighed up front, including costs that might be difficult to monetize or even quantify (e.g., species extinction). One result: energy conservation, required to be treated as a resource subject to such cost analysis, scored highly in this process and is now the second largest "source" of electricity in the region (after hydroelectricity).

6. ***Campaign funding reinforces partisanship and BAU***: As positioning on substantive issues was becoming more partisan and less fact-based, money has flowed unconstrained to the partisan groups that arise to do battle for and against. Campaign finance suffers from its over-dependence on free market ideas; the money both controls and validates policy positioning. Ostensibly the United States is wedded to a one-person/one-vote principle, and indifferent to the obvious capacity of money to amplify voices and votes. Campaign contributions so soil politics today that functioning democracy is imperiled. A Supreme Court allied to corporate interests is unwilling to constrain the financial power of those interests, whether in hostile corporate takeovers or intervention in political contests. The advent of Political Action Committees (PACs) had allowed consolidation of corporate, but also political, religious and institutional resources into still more powerful influences in political contests. Since big money almost by definition is the product of the prevailing distribution of power and interests, it prefers the profitable status quo to any uncertain future. The status quo includes existing energy sources and the companies that mine, pump, refine, distribute, sell, and burn them. These are not friends of

moving to address climate change. In effect, economic power reinforces political power; and political power, economic power. And because most voters are not looking much beyond the next election cycle, proposals to tinker with the machinery of elections cannot promise a desired next election outcome. Only when enough people outside the circles of wealth understand the money/power/money dynamic will the country be able to push back on it. There are more votes outside those circles than within them, if enough of us prioritize breaking that dynamic apart.

7. ***Discount rates and the perilous future***: The difficulties are not just with the companies making money from BAU that should not; BAU is part of our human wiring. Psychologically we tend to overvalue the present and discount the future. If today looks a lot like yesterday, we expect tomorrow will look much the same. Usually it does, too.

"Climate change is COVID-19 in slo-mo."

So, when we plan ahead 100 years, or even a decade, we almost always deeply discount the probability of significant and possibly abrupt, even cataclysmic, change. The present Administration discounted the probability of a major viral pandemic; discounted it in the abstract, before this year (2020), but continued to discount the probability that the coronavirus might look different in March or May than it looked in January. Climate change is COVID-19 in slo-mo, but with just as inexorable a progression absent strategies to hold it back. It is plain hard for us to visualize and act today on what appears to be distant and therefore seemingly diminutive risk. We highly value the here-and-now; we steeply discount future risk, or reward (although we do assume our lives will improve in an indefinite future, because we also are wired to be optimists). We especially discount outcomes that will not affect us personally, which by definition means outcomes beyond our lifetimes. As the great English economist John Maynard Keynes noted, "In the long run we are all dead." But that should not allow us to discount the future for our descendants. Haudenosaunee (Iroquois) philosophy offers a better guide: think and act as though your actions will have consequences for seven generations ahead.

In 2006, at the direction of the UK Government, Lord Nicholas Stern developed and published his "Review on the Economics of Climate Change." One of Lord Stern's most controversial observations was that the economic analysis of climate change might warrant a *negative* discount rate. That is, far from discounting future effects, we should weigh ***more*** heavily effects two or three generations hence of choices we make today, since we will see the near-term benefits (e.g., of low-cost energy), whereas future generations will bear the costs (of accumulating climate change). Our choices today, with their not-so-distant-anymore consequences, should bear the weight of this kind of rigorous economic analysis that values intergenerational fairness. The difference between COVID-19 and climate change, apart from velocity, is that there will be no vaccine for climate change. By the time we awaken to its imminent and most damaging effects, it will be too late to retrieve all those GHG emissions we blithely released over prior decades.

8. ***Failure of imagination***: This final observation is not much different from the preceding one, but the point is worth restating in another way. Imagine you are a British colonial subject in Singapore in 1942. You are aware that the United Kingdom is at war with the Axis powers including Japan. You are generally aware that the armed forces of Japan have subdued much of China and Indochina and are not so distant from Singapore. But life, and lawn tennis, continue in the colony—the Gibraltar of the East as it was known—much as they always have. It is not conceivable that life in our impregnable colony, hardly altered from one decade to the next, might be at risk. Then, in not much more than a week in early February 1942, Japanese forces overwhelm the British and colonial defenders. Britons are rounded up and deposited in camps where they live out the war (and are better off than some 130,000 Malay Peninsula defenders, Brits, Malays, and Australians, marched off to POW camps where many died). These kinds of stories, of disbelief in radical change, were ubiquitous also among German Jews and others swept up in the European theater of war. How could the status quo be so utterly upended, and so many otherwise capable and lively human intelligences be so flummoxed? Why did no one imagine how things could change?

Speaking truth to power: closing thoughts

Climate change is going to be like this. The most wicked of our world's large problems to solve. Somehow, we have to summon the better angels of our intelligences to look forward objectively and fearlessly. We have to contemplate the risks, allow our imaginations to explore them, posit alternative futures to BAU, and adjust our choices and behaviors to blunt the most destructive of these futures. We have to lift our eyes and imaginations up from today's puzzles and perplexities to that array of saving or savaging futures and the velocity with which they will be upon us. We must, without more delay, conceive of alternative defenses, weigh their relative costs and merits, recognize our shared responsibilities as scientists, advocates, leaders, voters, and members of the global community.

And act, we must.

References

Available from: https://www.portland.gov/sites/default/files/2019-08/cap_may_2010_web_0.pdf (2009).

Bartlett, J., What are the Costs and Values of Wind and Solar Power? Resources for the Future/Common Resources 10.08.19. Additional costs (e.g., for integrating variable generation into the grid) and benefits (no GHG's or other emissions) would be included in a utility comparative cost analysis. 2019. Available from: https://www.resourcesmag.org/common-resources/what-are-costs-and-values-wind-and-solar-power-how-are-they-changing/.

Bentz, C., Why We Walked, Oregonian July 11 2019.

Brown, G.K., quoted in The Oregonian, March 11 2020, 12.

Duncan, A., COVID-19 and Climate Change April 28 at Oregon Global Warming Commission web site KeepOregonCool.org. 2020.

Hudiburg, et al., 2019. Meeting GHG reduction targets requires accounting for all forest sector emissions. Environ. Res. Lett. 14, 095005.

Ingerson, A., Wood Products and Carbon Storage: Can Increased Production Help Solve the Climate Crisis? Washington, D.C.: The Wilderness Society. 2009.

Krankina, O., DellaSala, D.A., et al., 2014. High biomass forests of the Pacific Northwest: who manages them and how much is protected? Environ. Manag. 54, 112–121.

Kulongoski, G., GEN-1,GEN-2, GEN-2a, Final Report, Governor's Advisory Group on Global Warming, December 2004.

Law, H. et al., Land use strategies to mitigate climate change in carbon dense temperate forests, 2018. Available from: www.pnas.org/cgi/doi/10.1073/pnas.1720064115.

Oregon Climate Change Research Institute, First Climate Assessment, Executive Summary, 2010. Oregon State University. Available from: http://www.occri.net/media/1107/ocar2010_v12_executive-summary.pdf.

Oregon Climate Change Research Institute, Fourth Climate Assessment, Executive Summary, 2019. Oregon State University. Available from: http://www.occri.net/publications-and-reports/fourth-oregon-climate-assessment-report-2019/.

Oregon Department of Energy web site, Current state electricity resource mix, 2020. Available from: https://www.oregon.gov/energy/energy-oregon/Pages/Electricity-Mix-in-Oregon.aspx.

Oregon Global Warming Commission, Climate change Comes to Oregon 2018, Biennial Report to the Oregon Legislature, Section 1. 2018. Available from: https://keeporegoncool.org.

Stern, N., Stern Review on The Economics of Climate Change (pre-publication edition). Executive Summary. HM Treasury, London. Archived from the original on 31 January 2010. Retrieved 31 January 2010. 2006.

USFS FIA, Qin OGWC Forest Carbon Accounting Project Report to Oregon Legislature, 2018, page 12, and USFS FIA Table 2. Carbon stocks (Tg CO2e) and annual change per component (Tg CO2e/year) by owner group, all Oregon, 2001-2005 to 2011-2015. 2017. See interactive Oregon forest ownership map at: https://oregonforests.org/content/forest-ownership-interactive-ma.

The politics of conservation—taking the biodiversity crisis to the streets

David Johns

School of Government, Portland State University, Portland, OR, United States

A former warden in Chitwan (National Park) recalls how angry villagers often confronted him during periodic meetings with local communities. They demanded greater access to the park to satisfy their need for firewood and grazing areas for their livestock. Eventually, they would always ask, "Who is more important, people or rhinoceros?" And he would reply, "There are only four hundred rhinoceros in Chitwan, and there are over eighty thousand human residents. I am sorry, but it is my duty to protect the rights of the minority." (Dinerstein, 2003: 179)

But the saving of wild beings from obliteration cannot be expected to pay for itself in more than a sprinkling of cases. For most of the wild things on earth, the future must depend on the conscience of (hu)mankind (Carr, 1964: 172–173).

The disconnect

Conservation science meetings and publications are increasingly disconnected from the fate of biodiversity, which is spiraling downward. There is no evidence that attendance at meetings (before the Covid-19 lockdown) and journal impact factors stop extinctions, population declines, or ecological unraveling, even as they may advance careers. That few scientists still think that speaking truth to power is effective is progress. Many conservationists believed at one time that science would provide values in support of biodiversity that would be persuasive with others (Johns, 2000). But science does not generate values though scientists have values, as do those who fund it. When science did not live up to those expectations, many conservationists decided it was not worth the investment. For scientists this was doubly frustrating—not only would decision-makers often ignore the science but so also would some activists.

Conservation Science and Advocacy for a Planet in Peril.
DOI: https://doi.org/10.1016/B978-0-12-812988-3.00013-2

Yet both decision-makers and advocates need knowledge about how the world works that science generates if they are to set sound goals and make good decisions about how to reach them.

Both advocates and decision-makers will listen if they care. Caring is a strong motivator, whether it is from love for wildlife and the Earth or a desire to influence or be a decision-maker. Bruce Babbitt, a former US Interior Secretary, told conservationists when he came into office that they could not simply expect him to do the right thing, they had to make him. How to do that in the face of strong pressure from opposing interests? Doing so via activists is not enough; scientists need to act directly themselves. Although the science is about establishing the truth of the world, the task of scientists is much broader—as people who care as well as know about wildlife, ecosystems, and the threats to both they are uniquely suited to lead conservation efforts. Although often trained to appear dispassionate, they can engage their passion and learn how to use politics on behalf of life.

Even truth coded in compelling stories is not terribly persuasive with decision-makers. They respond to what they care about, and while some of them share conservation values and emotional connections with the natural world, many do not. What most do care about is staying decision-makers. To stay in power, they must be responsive to powerful organized interests, who are themselves important decision-makers: large banks, big high-tech firms, trade groups, mass media, energy companies, big law firms, big foundations, military establishments including weapons makers, religious elites, and those similarly situated—all groups that are embedded strategically in the very structure of societies. State officials, elected or otherwise, and the other institutions mentioned, are all able to manipulate fearful and poorly informed populations. They set the terms of the debate, define the issues and the choices people have, and then influence those choices.

The structure of these institutions as a whole, including their differences and conflicts, rests on a colonial relationship with other species and the Earth: one of domination and exploitation imposed by violence—from shark finning and factory farming, murdering rangers and tigers to enabling melting the arctic icecap and starving polar bears, turning the boreal forest into chopsticks, and bleaching corals. Decolonization is always a struggle—the powerful do not give up without a fight (Douglass 1985 [1857]; Johns, 2019).

This is why some conservationists have long recognized that safeguarding biodiversity and wilderness is revolutionary, requiring fundamental changes to existing societies' commitment to endless material growth and more people. Other conservationists shy away from this recognition because of the tremendous inertia, the controversy, and funder resistance. United Nations entities are not known for getting to the roots of things—the definition of radical—because they seek consensus or the lowest common denominator.

Nonetheless, the Science-Policy Platform on Biodiversity and Ecosystem Services issued a major report finding that "(g)oals for conserving and sustainably using nature and achieving sustainability cannot be met by current trajectories, and goals for 2030 and beyond may only be achieved through transformative changes across economic, social, political and technological factors." To meet goals requires "urgent and concerted efforts fostering transformative change" (IPBES, 2019: §§ C and D).

Magical thinking is sometimes described as continuing to do the same thing and expecting a different outcome, that is, not reality based. That has been the case with international conservation goals for at least the last two decades and will continue to be the case with the next round of targets if conservationists continue as they have. What can they do differently?

In this chapter, I argue that conservationists, including scientists as leaders, must bring irresistible pressure on decision-makers throughout the world to make biodiversity and wilderness a top priority. A recent editorial in Nature (2020: 337–338) observed that "Biodiversity is rarely allowed to stop or delay a new airport runway or power plant." The editorial then focuses on anemic recommendations—measurable targets, "action" plans, and more frequent government reporting on progress. These may help a bit but will not result in the far-reaching changes needed. The end of apartheid or slavery, decolonization, and the downfall of authoritarian regimes were the result of well-organized grassroots movements that fought the powerful to a standstill (e.g., Wood, 2000, 2003; Piven, 2004; Luders, 2010). These efforts went far beyond "speaking truth to power," as the term is usually meant. It involved speaking truth to power with power.

Forcing change

Former slave and abolitionist Frederick Douglass (1857/1985: 204) observed that "Power concedes nothing without a demand. It never did and it never will. Find out just what any people will quietly submit to and you have found out the exact measure of injustice and wrong which will be imposed upon them, and these will continue till they are resisted with either words or blows or with both." *How* do conservationists bring about decolonization of the natural world? Unlike ending slavery, halting the destruction of the wild requires that humans must act for nonhumans and the Earth. Although we have seen what disturbance a pandemic can bring to complex societies, it is unlikely one will act decisively and with the precision needed to topple the structures of domination. The economic dislocation caused by the virus will likely limit economic growth for next few years—which could result in less ecological damage but also make less available to support species recovery and protected areas (Naidoo and Fisher, 2020).

Ordinary and transgressive politics

The history of reform efforts in the modern world demonstrates that only sustained transgressive action results in deep-going change (McAdam et al., 2001)—action that breaks the rules and causes disruption to the point where it becomes more expensive for elites to hang on to the status quo than to make major concessions, including conceding power in some cases.

Radical goals are not completely dependent on radical methods or transgressive action, but on a strategy that incorporates these with insider tools: taking advantage of pressure from the streets and elite factions, lobbying, and electoral pressure. Indeed, most change comes as a function of punctuated equilibrium (Repetto, 2006: 24–46). Most of the time policy adjustments are just that—incremental adjustments around the edges. Significant changes such as cleaning up some forms of pollution or protecting public forests and species come after years of work often combined with the happenstance of crises. The US Endangered Species Act, despite many efforts to weaken it, is still one of the strongest such laws in the world and was passed (1) in the midst of a war being run by a besieged president (Richard Nixon) needing the recently enfranchised 18- to 20-year-old voters, and (2) on the heels of the first Earth Day which mobilized 10% of the US population, and without any strong opposition because its effects were unknown (Repetto, 2006: 1–23; Bevington, 2009). It would not pass today.

Ordinary or mainstream politics is sometimes described as the art of the possible. Conservationists find it hard to compete in this arena because opponents have such overwhelming resources (money, staffing, connections, media, expertise, jobs to offer) and they have written the rules so they will win almost all the time. They are also able to take advantage of many aspects of extant human personality—political advisor and theorist Nicolo Machiavelli (1996 [1531]: 18, I.5) observed that most people simply want to be left alone, but some seek great wealth and power and to obtain it they must take advantage of the majority. They will not give up their pursuit of either without a fight—it is part of their personality. Periodically, the large majority resists and even replaces an elite with what becomes a new one. Most uprisings fail, however. Humans are the cause of all conservation problems and the obstacles to conservation are mostly political. This includes laws and institutions that favor profits and human domination of the planet and its life, cultures that devalue other species and make exploitation or inaction in the face of exploitation acceptable, and institutionalized emotional immaturity. In short, opponents of conservation have societal predispositions on their side.

Scientists, like other conservationists, are inclined to want "access" and be "players," and too often settle for very little in exchange for their support, such as a seat that the table (which ≠ reform). The bigger organizations, especially, have become what Wilderness Society founder Bob Marshall (1934) warned against decades ago "We want no (fence) straddlers, for in the past they have surrendered too much...which should never have been lost." So, if an insider game of mainstream politics cannot halt or even appreciably slow the loss of biodiversity, what combination of outsider and insider actions can?

Transgressive action and asymmetric politics

Outsider action includes noncooperation such as not working (in/formal strikes), not buying (boycotts), not paying debts, building alternative institutions, and using them for social transactions, especially economic ones that include biodiversity compatible livelihoods. Scientists have often withheld cooperation from destructive projects to deny them legitimacy. Noncooperation is not inherently about sacrifice, but freedom from the treadmill of work-consume stuff-work, ad infinitum. As the corona pandemic demonstrated by slowing the frenetic busyness, much of the Earth can become quiet and nature come out of hiding. Why go back?

There are political enforcers of the extant economy and they will try to impose cooperation as the French did in Algiers or Selma authorities did in Alabama. To succeed noncooperation must be disciplined and collective, not individual (Luders, 2010). Links to political organizations will help ensure strategic coordination of a variety of actions: vigils, petitions, marches, demonstrations, parades, celebrations or protests of holidays and birthdays, sit-ins and occupations (which can include civil disobedience), blockades; hit and run actions such as posting satirical cartoons of leaders. Pirate radio and television can effectively penetrate societies, even with jamming.

Strategic integration rests on broad shared goals, coming at targets from different directions and in different ways but with apparent independence to make repression more difficult. Conservation's opponents generally have economic and political institutions on their side from media to police and prisons, so conservationists must think like guerillas. The disadvantaged can win asymmetric conflicts if they focus on opponents' weaknesses, guard their own, and are innovative strategically and tactically (Arreguin-Toft, 2005). It is not just a matter of relative power, but resolve, unity, sound strategy, leadership, patience and the long game, sometimes technology, and luck. It is as important to understand the political landscape as to understand the biological.

Scientists may lead, partly staff, or influence activist organizations; in any case, they will depend on extensive support networks that are not formally public. Such groups offer moral and material support, provide recruits, resting places, safe houses and often mass statements by means of peaceful demonstrations, attendance at trials and hearings, bearing testimony by their presence, quiet noncooperation, and so on. Many groups decide before an action who will and will not take risks such as arrest, beatings, or worse. Legal, food, medical, and moral support is arranged ahead of time, as is media coverage. Some groups—students in some countries, the middle class in others, peasants in yet others—are presumed loyal or enjoy a status that gives them some impunity from punishment for protest. With local and federal police in the United States using violence against those protesting the murders of unarmed Blacks and others by police and vigilantes, mothers in Oregon have organized and taken to the streets to stand between protestors and police in battle gear to try to stop police violence. Women, who are often victims of violence in private, nonetheless enjoy some protection in public. Scientists, whether or not academics, may enjoy similar protection in some societies, barring waves of anti-intellectualism. The rules are not written or enforceable and students protesting at Tiananmen Square in 1989, though not threatening the state and enjoying popular support, were violently repressed by an aging, frightened party leadership.

There is historically little overlap between the social justice movements of the 60s and the present and the conservation movement (distinguished from the environmental movement), so the latter lacks much street experience. They would do well to get some experience by joining current protests.

Self-defense

Sometimes movement factions are divided by those who are nonviolent in principle, those who are nonviolent strategically, that is, because using violence will diminish support or will be used to justify more official violence against the movement, and those who regard violence as necessary and effective. Many in the US civil rights movement were willing to shoot back in self-defense but would not initiate violence against the Klan or police (often the same). Nimtz (2016) and Cobb (2014) argue that threats or fear of violence often resulted in concessions by elites to movements. Most anti-colonial struggles are violent because dealing with colonial violence made it necessary (e.g., Fanon, 1963; Wood, 2000, 2003), although there are always nonviolent or moderate factions that authorities seek to legitimate and work with. There are also those who confine themselves to soft actions—creating encrypted guerilla communications, disrupting cyber systems, generating publicity, and seeking diplomatic support from other powers. Even if those seeking change cannot prevail by force, they can impose costs that force regimes to the table.

The Organization of African Unity recognizes the right of people to act militarily to gain self-determination or in self-defense against severe abuses (Gross, 2016). But the group so acting must show it legitimately represents those oppressed. There is little question that the treatment of other species is grossly abusive and denies to them their lives and livelihoods, but currently the law does not grant people the right to act in their defense. Rights, however, are usually not initially granted; they are taken and then receive legal recognition over time.

Many anti-colonial struggles result in regime opponents seizing territory and administering it while conflict continues. An analogy might be made with parks and wilderness areas not under state administration for one reason or another. Protected area authorities typically employ rangers, usually authorized by the state, but sometimes enforcing the law under regional authority when weak states cannot or will not. Many nongovernmental organizations (NGOs) aid states or protected areas by supporting and supplying rangers. And when protected areas come under control of corrupt states, states that bow to exploitative interests, warlords, or others who seek to use protected area land, water, and wildlife for their own ends, enforcement by NGOs is appropriate. Sea Shepherd frequently enforces international law on the high seas or even in exclusive zones at the request of states against pirates or poachers; they also act against states violating treaties they have signed to protect the oceans.

Targeting and campaigns

To achieve progress in biodiversity protection and recovery, it is important for the conservation movement and scientists to target decision-makers carefully. Which ones are sympathetic? Which ones might be enticed to defect from the status quo? Which ones are susceptible to pressure? Which ones must be isolated for supporting bad policies? Who are the most effective opponents of conservation and how can they be undermined? When elites are united, they almost invariably prevail because of the resources they possess. When they can be divided over substantive interests, by mistrust, or on other bases, then conservation can prevail. Security-force defections are a good indicator that power is slipping from the status quo and opportunities for change are growing.

The US dolphin-safe tuna campaign is a good example of a multifaceted campaign. It targeted the United States and other countries' decision-makers with political pressure, the US tuna canners with boycotts, and tuna fisherman with exposure and loss of markets. With the United States consuming half the canned tuna and only three major brands the retail market was vulnerable to boycott. Canners feared school children would appear on television protesting Flipper (a bottle-nose dolphin from a 1960 TV series) being killed. There was an undercover film showing dolphins drowning in nets, being crushed in fishing blocks, and beaten by fishermen. Lawsuits against the US executive branch forced them to require importing countries to abide by "dolphin friendly" regulations and to do battle with and ultimately prevail over World Trade Organization objections to blocking imports (Layzer, 2012: 348–366).

In another effective campaign the Zuni Tribal Council, seeking to halt coal mining of land they considered sacred, reached out not just to regulators and the New Mexico congressional delegation but to white churches who responded by joining the struggle as allies. The Zuni were successful in part because they were innovative. When billboard companies, under pressure, would not rent to them, they painted their ads on the side of semitrailers and parked them at high volume locations or drove them around. The Zuni also sent out runners to communicate with other tribes, mainstream church leaders, and others—a traditional form of contact that attracted media coverage and raised the issue's profile (Cox, 2006: 243, 267–273).

As the Earth has filled up with people and the demand for material stuff has grown exponentially, conservationists and scientists find themselves more and more in conflict with governments and growth interests. Although conservationists have been instrumental in achieving protected area status for many places and species, securing and safeguarding life requires about half the planet in strict protection (Noss, 1992; Noss et al., 2012; Roberts, 2007; Wilson, 2016). Conflict is likely and sometimes could be intense. Even when protection is achieved conflict with poachers, agricultural and industrial encroachment, infrastructure, and over recovery of predators is highly likely—even in societies where the law is typically obeyed and enforced.

Not always a fight but always hard work

Conservation is not always an open fight, however. Some governments at times are supportive of conservation efforts for a variety of reasons, including national pride and popular demands. Governments may also respond to visionary people who combine a heartfelt commitment to the wild with resources brought strategically to bear. Tompkins Conservation purchased over 810,000 hectares in Patagonia (southern South America) and deeded it along with infrastructure to the governments of Chile and Argentina leveraging over 5.2 million hectares of parks in the two countries. This effort also aided restoration of several native species that had been extirpated. Land purchases and negotiations with the government took place over many years as did building relationships with regional populations and countering anti-conservation propaganda (Jimenez, 2014; Johns, 2019).

Gorongosa National Park lies toward the southern end of the Great Rift Valley, in Mozambique. Founded in 1960 by colonial authorities, it was neglected and worse during the war to gain independence and then devastated by a longer civil war fueled in part by the apartheid regime of South Africa. Over a million people died and wildlife was reduced to food. With major support from the Carr Foundation and the vision of Joaquim Chissano, President of Mozambique, and Nelson Mandela of South Africa, reestablishing the park became a priority: they believed that strong ties to the natural world contributed to peace and would bring prosperity to the surrounding communities (Johns, 2019). The park hires locals, offers training, funds schools and medical care and conservation compatible economic activities. It also sponsors girls' clubs to keep them in school and ensures that they have economic opportunities. The park itself is about 405,000 hectares, but with buffers and connectivity zones north and south, it is over 3 million hectares.

Others are undertaking similar efforts around the globe, but not many wealthy people have the vision or passion for the natural world of Kris and Doug Tompkins or Greg Carr. That there are not enough of them to secure half of the Earth, that the ocean cannot be owned and privately protected though it can be privately plundered, means that only political movements can do the job.

What segments of the population can bring meaningful pressure on decision-makers to secure and heal half the Earth? Who will constitute a movement that can bring the necessary economic, cultural, and political clout? Identifying those segments of the population is critical to making the best use of limited conservation resources. Mobilization involves recruiting people and their networks into conservation organizations and creating alliances with other groups, especially scientists. Who will lead and mobilize others? There is no obvious conservation equivalent of the working class in some theories of social change. Many conservation leaders have training in biology and related sciences; their emotional connection to the wild has drawn them to both conservation action and science to try to better understand what they care about. But they lack political experience and that must change.

Mobilization

Morality and material interests

Conservation is both a moral fight—seeking justice, protecting emotional attachments, and meeting species' needs—and a material fight—ending exploitation and domination and securing healthy ecosystems. Recruitment and alliances will be both moral and material but the admixture will vary based on the audience. Appeals to justice and the self-transcendent will be primary—appeals for the good of the biological community. Mere material appeals rest on ever-shifting ground and tend to trigger calculation and narrow self-interest, which undermines conservation action and consideration of other species. Rockefeller funding supported protecting the coastal redwoods (*Sequoia sempervirens*) in a National Park, but only if its size was limited and business was not offended (Gonzalez, 2001: 79–94). Yosemite Park was ceded by California to the United States only with the approval of the Southern Pacific Railroad, which wanted ~4047 hectares for a rail line. It is fair to say that Gifford Pinchot, first head of the US Forest Service, prevailed over John Muir, wilderness advocate, and Sierra Club founder, because he had the backing of industry (Gonzalez, 2001: 30).

People join causes or actions as part of cohorts or even broader networks, rather than as individuals. Grievances are generally felt more intensely than are aspirations and that may explain why defensive actions more easily arouse people and are more likely to generate coalitions. Overlapping networks and shared interests and identities are also major factors underpinning coalitions. Where networks do not overlap, brokers often play a central role; movements can also act to generate threats and a sense of urgency (Staggenborg, 2010: 100–120). It is also easier to create coalitions in the same issue area, such as conservation, than to reach across issue areas, where ideological differences are greater.

Allies

Yet, conservationists alone are unlikely to be strong enough anytime soon to win; they need allies across issue areas. Among them:

- animal rights, animal welfare NGOs
- anti-factory farming groups
- NIMBY (not in my backyard) groups
- public health and medical professionals
- birth control groups
- anti-consumption, anti-plastic groups, and climate change groups
- some civic leaders
- some youth groups
- criminal defense and conservation bar
- religious leaders who value creation
- labor leaders with vision
- cultural leaders (filmmakers, songwriters, music performers, humorists, novelists, dramatists, journalists)
- prodemocracy advocates and agitators
- progressive businesses
- veterans' groups
- and last but not least—scientists.

Some anti-racism advocates see the link between how human groups without power and nature are treated, although solving most intrahuman problems (Kopina and Washington, 2020) does not mean resolving the domination of nature. Absent from this list are hunters who have sometimes joined conservation alliances only to turn on them over predator protection and roaded access. Alliances require understanding each group's values, emotional connection to the wild, their disposition to action, and willingness to take risks; then a strategy can be crafted to mobilize them.

Movement, identity, and story

Successful movements link individual identity with a collective vision. When Martin Luther King gave his 1963 March on Washington speech he roused people with "I have a dream…" not "I have a strategic plan!" (In part, the difference between a leader and a manager.) Mobilization and sustaining momentum, especially in the face of obstacles, are about justice-seeking and establishing the moral high ground. Justice resonates with aspects of the existing sense of self and identity, and through action gives expression to the dream and goals. Stories and actions describe what is broken, what must be fixed, what obstacles must be overcome, and demonstrate the fallacies of opposing stories and moralities. Effective stories say why fixing what is broken is urgent and personal, and why it is right (See Chapter 10, Out of the Ivory Tower: Campaign-Based Science Messaging for the Public herein; Sachs, 2012). They appeal to both the universal and the local. They provide the path to healing and wholeness, reweaving what global institutions have wrecked in trying to standardize the world and bring it under control, ignoring local ecologies.

New stories and myths usually find partial roots in existing stories; the morality is new. They contribute to the emergence of a new identity and cohesion among those mobilized that may help scientists in bringing behavior change to institutions and individuals. Public displays of emotion and beliefs, keeping together in time—all of which can be part of ritual—reinforce this new identity, and bond people to each other and to the cause (McNeill, 1995; Kertzer, 1988). Often holidays and other celebratory events can be rebranded, with different meanings attached to them—Columbus Day in North America is no longer about the "discovery" but about conquest and oppression which must be thrown off. The Feast of St. Francis celebrates the wild rather than human domination. Some celebrations seek to address the ambiguity in the extant: should the US holiday honoring Martin Luther King inspire activism and change or merely public service: end hunger or merely feed the hungry. Creating new holidays or celebrating anniversaries of victories can advance conservation, for example, Earth Day (Etzioni, 2004: 2–40).

In many respects, conservationists have won the overarching rhetorical battle—there is minimal criticism of the scientific underpinnings of biodiversity. Few among global decision-makers are openly critical of efforts to safeguard and recover species or maintain ecological integrity; though some groups seek to sideline conservation and focus on human problems. Indeed, the very existence of paper parks, exaggeration, and other forms of misrepresentation indicate that governments want to appear biodiversity-friendly. Most governments have signed the Convention on Biodiversity, Millennium Goals, and Aichi sustainability targets. But the targets have not been met. A recent editorial in Nature (2020: 578: 337–338) noted that "Biodiversity is rarely allowed to stop or delay a new airport runway or power plant." There is, after all, money to be made and power to be enhanced. Airports and power plants and other projects that comprise the activities of endless growth have well-organized interests backing them, many of whom are decision-makers themselves. Specific conservation objectives still find many enemies and many countervailing stories that must be overcome. *Science* just announced in its news section (Servick, 2020: 688) that India was proceeding with a coal mine in an elephant (*Elephas maximus indicus*) reserve, an oil project in a wildlife sanctuary, and considering a uranium mine in a tiger (*Panthera tigris tigris*) reserve.

Campaigns

In Chile opponents of protected areas expressed fear of foreign control, losing access to water, and sheep depredation from reintroduced predators. Lobbying and agitating against new parks included anti-conservation science stories (Jimenez, 2014). Conservationists took care to reassure locals that parks were about preserving their biological and cultural heritage. The parks "produced" nature that people craved. Each potential protected area had a local leader that stayed on top of the issues and politics, which were always fluid. Oates (1999) found in his African and Indian projects that without someone on the ground in each place, success was not possible.

In the arid Western United States, where cattle and sheep ranching routinely degrades the landscape and results in persecution and local extirpation of many predators and the slaughter of wildlife competing with introduced livestock (e.g., *Bison bison* that wander outside of Yellowstone National Park), ranchers have successfully protected their narrow interests because they are well entrenched in state-level political parties, county government, and federal agencies. Critics of arid lands ranching often refer to the US Bureau of Land Management as the Bureau of Livestock and Mining for its emphasis on exploitation and note that every cow has two Senators, giving them outsized leverage. Ranchers also have learned to successfully use the romanticism of the cowboy and the Old West. Conservationists have never developed a compelling competing story (Layzer, 2012: 140–173; Repetto, 2006: 232–252). Some private ranchers such as Ted Turner and Diana Hadley have made their lands predator friendly, but they are a minority.

Although people care about the natural world, it is not the issue they vote on and there are few political candidates leading on the issue. Indeed, broad public support for strong conservation policy can rarely compete with a well-organized minority such as ranching, oil and gas development, or housing subdivisions. For one thing, "the public" does not really exist; it is not organized and thereby able to take focused, coordinated action and sustain that action, or follow the policy actions of officials and reward or punish them on that basis. The fishing industry routinely prevails over sound science and broad public support for marine protection. And nation-state claims to fishing areas have less to do with protein to feed the world's poor than with feeding terrestrially grown meat, securing jobs and international allies, and maintaining naval access to parts of the oceans (Finley, 2017: 3, 142–154).

The most effective means of spurring many audiences to action is via personal and professional networks of those already committed to conservation and seeking to strengthen relationships and shared understandings. Opinion leaders are also effective, and these may include cultural icons of various sorts, from musicians and actors to editors, pastors and charismatic officials, humorists, and various forms of street theater. Mainstream television is a poor source of information, but many say that it is where they get their information. Most media outlets are businesses or parts of government that can put them at the mercy of advertisers and circulation or censors. Conservation does not control significant media outlets other than those associated with advocacy NGOs (websites, newsletters) and professional societies (scholarly journals); they all have limited circulation and many are less oriented to outreach than to internal communications. As noted, stories that resonate most effectively use film, theater, dance, oratory, and nonwritten forms (See Chapter 10, Out of the Ivory Tower: Campaign-Based Science Messaging for the Public herein, Sachs, 2012). Narrative film is particularly good for conveying myth—stories about the sacred—the most important and fundamental purposes of a group (Gottschall, 2012: 60–64; Rappaport, 1999: 272, 280–310). But for these media to work beyond simple awareness, they need to feed people, including scientists, into organizational structures. Without coordinated collective action on behalf of clear wildlife and ecological goals—which science must play a major role in informing—there can be no conservation.

The television serials of Miguel Sabido produced for Televisa in 1970s Mexico and the radio and television serials of Population Media Center produced globally more recently have been successful at changing individual behavior, especially with regard to human reproduction. Their success hinges on sound research into the lived experience of the target audiences, reflected in the characters, recognizing the difficulty of change and reality of some backsliding, the rewards of change, and the availability of societal support institutions such as birth control device dispensaries (Ryerson, 2014). These tools have not been used to raise general conservation awareness nor in support of direct conservation campaigns—but the impact on the causes of biodiversity decline due to human population is undeniable.

Organization

Conservation success hinges on sustained mobilization, creation of an organizational structure into which resources such as time, energy, scientific knowledge, strategic skills can be focused to achieve movement purposes and goals. It is the difference between many acting separately and many acting as a group. Too many modern conservation NGOs, led by scientists or not, want check writers to support professional staff who lobby, circulate petitions, and the like when what is needed to overcome political obstacles is lots and lots of activists and scientists to fill the street and hearing rooms and make decision-makers feel the heat. Similarly, alliances and coalitions need an organizational framework that generates a multiplier effect via coordinated goals and action. Often a charter document is important. Overlapping memberships, experience with coalitions, and shared identity make coordination more effective. The Student Nonviolent Coordinating Committee (SNCC) and the Black Panther Party both had national organizations that the United Farm Workers lacked. Together they implemented a national grape boycott; SNCC and some Panther leaders saw the farmworkers as economically exploited like southern black sharecroppers and so with a common cause.

Tension always exists between growing membership and building coalitions on the one paw and staying focused on the vision on the other. It is not a simple matter. Winning specific battles can take allies who do not share all of conservation's goals of Half-Earth, biodiversity compatible human societies, or ecocentrism. The path to Half-Earth may be protected area by protected area, geographic region by geographic region. Existing protected areas were not created all at once. People with different motivations may fight for similar ends. Successful movements must be able to incorporate these groups at different times and in different ways—for example, join forces to overcome endless growth or meat factories. Conservationists, whatever their training and experience, must be prepared to deal with differences among groups they ally with and keep their eyes on the prize (See Chapter 10, Out of the Ivory Tower: Campaign-Based Science Messaging for the Public herein, Sachs, 2012). A big tent usually means a compromised tent. Big tents may be appropriate at times—to overcome specific obstacles—but if it sidetracks from fundamental conservation goals, those goals are never reached. That is why those deeply committed to conservation must keep focused on the vision and create a strategy to get there—rather than one focused on strengths, weaknesses, threats, and opportunities that mire groups in the here and now and defined by others.

Cesar Chavez (2002: 66) said there is never enough money to organize; if one focuses on money, the organizing never happens. The Ford Foundation used its funding to try and turn the US civil rights movement and the anti-apartheid struggle away from basic economic change to black capitalism (INCITE, 2007). The southern Oregon Klamath Basin collaboration, in the words of independent biologist Higgins (2012), did not result in "a deal that looks at the ecological needs of the Klamath (River). This is a social deal where people have come together and found out where they can agree. Unfortunately, where they have agreed doesn't overlap with the needs of the ecosystem." Successful coalitions require sound leadership, with clear strategic vision and goals, and keen awareness of the investment costs and likely outcomes. Egos also need to be put aside—not always an easy thing.

Social movements often fail because they do not embody what they are trying to create—in this case biodiversity compatible human societies. That vision is yet to be fully defined. Nature-focused scientists will play a role in defining the vision along with others. It is also important to understand what structural and other factors are standing in the way of bringing that vision about—what groups are in the way of change and what groups are supportive and can be mobilized to attain the vision. Those leading the effort must understand the step-wise process of mobilization. Groups start with different initial levels of commitment and willingness to act and embrace different levels of risk taking. There are costs to opposing the status quo for individuals and organizations—the temptations of timidity offered by elites who proffer (meaningless) seats at the table or threaten job loss or worse. The danger of being treated like a grizzly or a tiger is real, that is, violent repression. Conservationists, including scientists, must be prepared for repression if opponents feel threatened, and know how to document and expose it, how to spot infiltrators and provocateurs and isolate them, and how to defend themselves. Media, lawyers, infiltrating the police, and self-defense are important tools.

Crises

Generally, conservation success hinges on higher levels of mobilization than opponents—bringing more resources to bear more skillfully, being smarter, persevering, and staying focused. Undermining opponents can achieve the same thing. Dividing elites and opponents, helping to create and exacerbate crises among them can contribute to progress. Insight into incipient crises is extraordinarily useful in making use of these events. Human behavior will continue to generate crises ranging from mere scandals to economic collapses or food shortages via climate change. During the 2020 coronavirus outbreak—which is causing great anxiety among the global middle class because it affects them (no vaccine at the time) and not just the global poor—some have written about nature striking back (Gallagher, 2020; Carrington, 2020; Rees, 2020). No epidemic, however, has reduced human numbers enough to make much difference (Snowden, 2019), though slowed economic activity has given aspects of nature a respite. Nature did come out of hiding a bit as a result of corona, but the human rubber band is snapping back. Rees (2019) and others have suggested that climate change is blowback due to stupid human behavior and could make industrial agriculture impossible, leading to the potential deaths of 6 billion. Catastrophe, however, is not a sound basis for policy. Being flexible and taking advantage of systemic dysfunction is reasonable and often offers some predictability, but to halt the extinction underway requires human action on behalf of the natural world that reverses the human trajectory. Humans (including scientists) must "strike back" against the purveyors of exploitation and domination because only they can change the existing order and create a new one with some precision.

Such work is never easy, especially for scientists. How can organizations forge ecologically adapted societies and undermine the status quo while simultaneously surviving in the status quo? In a world of highways, how does one get around without motors until they are not needed? Tactical innovation is constantly needed as is keeping the vision always front and center.

Larger organizations and some smaller ones must deal with the problem of hierarchy and the challenge of partial success, both of which can lead to timidity; historically those at the top of reform and revolutionary movements do not lead the same lives or face the same risks as the rank and file and can lose their fervor and urgency (Michels, 1962[1915]). In the case of conservation, most humans do not face the grim challenges wildlife face from habitat loss and fragmentation and persecution. Partial successes can also give NGO leaders a greater stake in the status quo, with similar results. Managers replace leaders as organizations become "professionalized" to better compete within the system that needs to be fundamentally changed. Organizations that stay with the vision often have a long-serving, small, and homogenous leadership circle. High turnover weakens leadership in relation to funders and others that would drive strategic direction from outside.

Staying focused and leadership

In the case of human-centered social movements, it is possible for individuals to push their leaders or withdraw support based on their actions. But other species cannot speak in ways that are easily understood by many humans, even though we know that evolution is continuous (Darwin, 1989[1890]). At the same time, our egocentricity makes us prone to ignore the cries of others and distance ourselves from the victims of domination if we *do not feel the domination directly, or in an otherwise personal and urgent way.* Our understanding of the world is limited, and we often lack adequate empathy and wisdom. Our lack of understanding and even the capacity to understand limits us (Ehrenfeld, 1978; Wright, 2004). Biologists, ecologists, naturalists, and some activists have immersed themselves in the natural world and have a good knowledge of other species, ecosystems, and the whole, but more important than knowing is caring, the capacity to bond with a place or another species or individuals of that species.

Nonhuman life does not vote or organize and cannot give (or withdraw) permission to represent them in decision-making that affects them (Gray and Curry, 2020, 2016; O'Neill, 2006). If actual representation is impossible, can virtual representation (relying on human judgment) be made to work? What practices must be incorporated into the conservation movement to integrate the needs of other species and *create institutions and practices that do so?* Experience has shown, as noted, that if a movement does not incorporate practices by which it seeks to order the larger society, then such practices are unlikely to be realized in a new order.

Adequate representation of other species will depend on conservation leaders, or a specialized subset of leaders that exercise a veto over policy affecting other species and protect conservationists from self-serving or expedient denial. Nature-focused scientists have a major role in this, but so do those who are ecocentric and who have strong emotional ties to the Earth and other species. Human societies have always had those appointed to police ethics, but in large-scale societies, they almost invariably come to rationalize domination in its many forms. Much work needs to be done to institutionalize nonhuman representation.

Leaders have other critical roles to play. Selecting good deputies is imperative—those who will check their self-aggrandizing tendencies and encourage their self-effacing tendencies, reminding leaders that the movement is bigger than them. Often leaders choose those who will stroke their fragile egos. When groups or movements have real leaders—those who are bold, inspire, are inventive, grasp opportunities, stimulate action by taking action, make members feel worthy, good and important, and breathe life into the cause—deputies must provide sound management and demand the space and authority to carry out clear goals (Aminzade and Goldstone, 2001: 39, 129–130, 153). One task of leaders is to look as far down the path as possible, keeping their eyes on the vision. Often, however, they are easily distracted by the next shiny, bright thing. Managers in the top position, by contrast, are often too narrowly focused on the near term and resist innovation and risk. Good leaders model emotion and its appropriate expression as well as emotional development. Those in leadership positions are also critical to initiating coalitions and building on their personal and professional connections and networks.

Speaking truth to power: final thoughts

Societal change is a messy, circuitous business. It is often difficult to tell if progress is being made. Speaking truth to power is more than about giving voice to the interests of nonhumans, as the failure to realize most biodiversity targets in the past. The sort of speech that stops runways, power plants, and the hubris of endless growth is the speech of action–when power in the citadel is met with power from the streets. Conservation leaders, whether trained in science or with experience in advocacy, must keep the vision central and follow where it leads. Some have counseled abandoning a strong conservation vision because achieving it has proven difficult. What is required is a different strategy, not a compromised vision. If opponents think conservationists will tire, then we will be ignored. A vision can become an uncritical faith, but this is not inevitable. A much greater threat comes from professionalizing the conservation movement, relying on organizational experts, and replacing the vision with the organization. Movements can also become unhinged and cause great misery as they fail—but that comes from trying to increase human control or human perfection, not from dismantling control. Reinforcing a sense of efficacy is as important as real progress. Many conservation organizations depend on trying to foster hope. "But hope is only man's mistrust of the clear foresight of his mind. Hope suggests that any conclusion unfavorable to us *must* be an error of the mind" (Valery, 1919: letter of 11 April). We will do better to focus on doing what is right, regardless of circumstances and a dubious sense about the future. Conservation best serves life on earth when the method of science is combined with emotional bonds with Nature, an advocates' ability to seize opportunities, and an uncompromising vision.

References

Aminzade, R., Goldstone, P., 2001. Leadership dynamics and dynamics of contention. In: Aminzade, R.R., Goldstone, J.A., McAdam, D., Perry, E.J., Sewell, W.H., Tarrow, S., Tilly, C. (Eds.), Silence and Voice in the Study of Contentious Politics. Cambridge University Press, Cambridge UK, pp. 126–154.

Arreguin-Toft, I., 2005. How the Weak Win Wars. Cambridge University Press, Cambridge.

Bevington, D., 2009. Rebirth of Environmentalism. Island Press, Washington DC.

Carr III, A., 1964. The Land and Wildlife of Africa. Time-Life, New York.

Carrington, D., 2020. Coronavirus: 'Nature is sending us a message', says UN environment chief. The Guardian. Wed 25 Mar 2020 03.00 EDT; Last modified on Wed 25 Mar 2020 16.15 EDT(US) <https://www.theguardian.com/world/2020/mar/25/coronavirus-nature-is-sending-us-a-message-says-un-environment-chief.> Downloaded 25 March 2020.

Chavez, C., 2002. The Words of Cesar Chavez. Texas A&M University Press, College Station TX.

Cobb, C.E., 2016. This Nonviolent Stuff'll Get You Killed. Duke University Press, Durham NC.

Cox, R., 2006. Environmental Communication for the Public Sphere. Island Press, Washington DC.

Darwin, C., 1989. The Expression of the Emotions in Man and Animals. Vol 23, The Works of Charles Darwin. New York University Press (Pickering and Chatto), New York (London).

Dinerstein, E., 2003. The Return of the Unicorns. Columbia University Press, New York.

Douglass, F., 1857/1985. The significance of emancipation in the West Indies. In: Blassingame, J.W. (Ed.), The Frederick Douglass Papers. Series One: Speeches, Debates, and Interviews, Volume 3. Yale University Press, New Haven, pp. 1855–1863. Speech, Canandaigua, New York, August 3, 1857; collected in pamphlet by author. P 183-208.

Ehrenfeld, D., 1978. The Arrogance of Humanism. Oxford University Press, New York.

Etzioni, A., 2004. Holidays and rituals. In: Etzioni, A., Bloom, J. (Eds.), We Are What We Celebrate. New York University Press, New York, pp. 3–40.

Fanon, F., 1963. The Wretched of the Earth. Grove Press, New York.

Finley, C., 2017. All the Boats on the Ocean. University of Chicago Press, Chicago.

Gallagher, D., 2020. Pope Says Coronavirus Pandemic Could be Nature's Response to Climate Crisis. CNN, Rome, Updated 9:52 AM ET(US), Wed 8 April. <https://www.cnn.com/2020/04/08/europe/pope-francis-coronavirus-nature-response-intl/index.html>.

Gonzalez, G.A., 2001. Corporate Power and the Environment. Rowman & Littlefield, Lanham MD.

Gottschall, J., 2012. The Storytelling Animal. Houghton Mifflin Harcourt, Boston.

Gray, J., Curry, P., 2016. Ecodemocracy: helping wildlife's right to survive. 37 ECOS 1, 18–27.

Gray, J., Curry, P., 2020. Ecodemocracy and political representation for non-human nature. In: Kopnina, H., Washington, H. (Eds.), Conservation: Integrating Social and Ecological Justice. Springer Verlag, Cham, Switzerland, pp. 155–166.

Gross, M.L., 2016. The Ethics of Insurgency. Cambridge University Press, Cambridge UK.

Higgins, P., 2012. The Klamath Basin: A Restoration for the Ages. KVIE, Sacramento, Minute 48 et seq.

IPBES (Intergovernmental Science-Policy Platform on Biodiversity and Ecosystem Services), 2019. Summary for policymakers of the global assessment report on biodiversity and ecosystem services. Plenary of the Intergovernmental Science-Policy Platform on Biodiversity and Ecosystem Services. Seventh session, Paris, 29 April–4 May 2019.

INCITE: Women of Color Against Violence, 2007. The Revolution Will Not Be Funded. South End Press, Boston MA.

Jimenez, I., 2014. Personnel Communication.

Johns, D., 2000. Biological science in Conservation. Wilderness science in a time of change conference—Volume 2: Wilderness within the context of larger systems. USDA Forest Service, Rocky Mountain Research Station, Ogden UT.

Johns, D., 2019. Conservation Politics: The Last Anti-Colonial Struggle. Cambridge University Press, Cambridge UK.

Kertzer, D.I., 1988. Ritual, Politics and Power. Yale University Press, New Haven CT.

Kopina, H., Washington, H., 2020. Conservation: Integrating Social and Ecological Justice. Springer Nature, Cham Switzerland.

Layzer, J.A., 2012. The Environmental Case, 3rd Congressional Quarterly, Washington DC.

Luders, J.E., 2010. The Civil Rights Movement and the Logic of Social Change. Cambridge University Press, Cambridge UK.

Machiavelli, N., 1996. Discourses on Livy. University of Chicago Press, Chicago.

Marshall, R., 1934. Letter from Bob Marshall to Harold Anderson, 24 October 1934.

McAdam, D., Tarrow, S., Tilly, C., 2001. Dynamics of Contention. Cambridge University Press, Cambridge UK.

McNeill, W.H., 1995. Keeping Together in Time. Harvard University Press, Cambridge MA.

Michels, R., 1962. Political Parties. Free Press, Glencoe IL.

Naidoo, R.,Fisher, B., 2020. Sustainable Development Goals: pandemic reset. 583 Nature 198–201 (9 July).

Nature Editors, 2020. New Biodiversity Targets Cannot Afford to Fail. 578 Nature: 337–338 (20 Feb).

Nimtz, A.H., 2016. Violence and/or nonviolence in the success of the civil rights movement: The Malcolm X-Martin Luther King Jr Nexus. New Poli. Sci. 38 (1), 1–22.

Noss, R., 1992. The wildlands project land conservation strategy. Wild Earth Spec. 1, 10–25.

Noss, R.F., Dobson, A.P., Baldwin, R., Beier, P., Davis, C.R., Dellasala, D.A., et al., 2012. Bolder thinking for conservation. Conserv. Biol. 26, 1–4.

Oates, J., 1999. Myth and Reality in the Rain Forest. University of California Press, Berkeley.

O'Neill, J., 2006. Who speaks for nature? In: Haila, Y., Chuck, D. (Eds.), How Nature Speaks. Duke University Press, Durham NC, pp. 261–277.

Piven, F.F., 2004. Challenging Authority. Rowman & Littlefield, Lanham MD.

Rappaport, R.A., 1999. Ritual and Religion in the Making of Humanity. Cambridge University Press, Cambridge UK.

Rees, W.E., Yes, the climate crisis may wipe out six billion people. The Tyee. 18 Sep 2019. <https://thetyee.ca/Analysis/2019/09/18/Climate-Crisis-Wipe-Out/>. Downloaded 4 Oct 2019.

Rees, W., 2020. The Earth is telling us we must rethink our growth society: why COVID-19 previews a larger crash. We must do ourselves. <https://thetyee.ca/Analysis/2020/04/06/The-Earth-Is-Telling-Us-We-Must-Rethink-Our-Growth-Society/>. downloaded 9 April 2020.

Repetto, R. (Ed.), 2006. Punctuated Equilibrium and the Dynamics of U.S. Environmental Policy. Yale University Press, New Haven.

Roberts, C., 2007. Unnatural History of the Sea. Island Press, Washington DC.

Ryerson, W.N., 2014. Effectiveness of Entertainment Mass Media in Changing Behavior. Population Media Center, Shelburne VT.

Sachs, J., 2012. Winning the Story Wars. Harvard Business Review Press, Boston MA.

Servick, K., 2020. India greenlights development. Science 688 (15 May).

Snowden, F.M., 2019. Epidemics and Society. Yale University Press, New Haven CT.

Staggenborg, S., 2010. Social Movements. Oxford University Press, New York.

Valery, P., 1919. The crisis of the mind. letter to Athenaeum (London), April 11 and May 2. From the French "La Crise de l'esprit" in *La Nouvelle Revue Française*, August 1919.

Wilson, E.O., 2016. Half Earth. Liveright, New York.
Wood, E.J., 2000. Forging Democracy from Below. Cambridge University Press, Cambridge UK.
Wood, E.J., 2003. Insurgent Collective Action and Civil War in El Salvador. Cambridge University Press, Cambridge, UK.
Wright, R., 2004. A Short History of Progress. Anansi, Toronto.

When science is silenced: scientists fighting back against the politicization of their work

*Augusta C.F. Wilson**
Climate Science Legal Defense Fund, New York, NY, United States

I am a lawyer who practices in a specialized area: I work with climate scientists who find themselves the targets of politically motivated attacks by those who do not like the work they are doing. It is fairly common, when I tell people what I do, for them to express a combination of dismay and disbelief that my organization, the Climate Science Legal Defense Fund (CSLDF), needs to exist.

Politicization of science is, of course, hardly new; it goes back at least as far as Galileo in the 17th century (see Chapter 16, Speaking Truth to Power for the Earth). In the modern era, the playbook was, in many ways, written during the so-called tobacco wars. A Surgeon General's report released in 1964 effectively put an end to any legitimate scientific debate about whether smoking cigarettes cause cancer. But the tobacco industry, and outside interests associated with and funded by it, would continue to seek to create the appearance of scientific controversy (Brandt, 2012) and to sow doubt about the validity of the science in the public's mind for decades (Heath, 2016; Johns and Levy, 2019).

Climate science is the latest major front in this ongoing war. Many of the same tactics—and even some of the same individuals—that were deployed to call the science around cigarette smoke into question have been brought to bear in an effort to create uncertainty about the fundamental science or a lack of consensus among climate science when, in fact, there is no disagreement.

Climate researchers are not alone in this, of course—researchers in various fields find themselves the targets of harassment or attempts at censorship or intimidation or

**CSLDF has worked with or represented some of the scientists whose stories are described in this chapter.*

Conservation Science and Advocacy for a Planet in Peril.
DOI: https://doi.org/10.1016/B978-0-12-812988-3.00017-X

caught up in manufactured controversies around their work when it touches on topics that are politically sensitive or important to industry interests. In many cases, these are scientists who have chosen not to pursue potentially lucrative careers in the private sector but have opted instead for a career in government service or at a public university. In some instances, they have taken incredibly brave stands—sometimes at great personal cost—to ensure that their research reaches the public, for whose benefit it was intended. This chapter will tell some of their stories.

Weaponized use of open records laws against scientists

Reaction to the "hockey stick" graph

Michael Mann is a Distinguished Professor of Atmospheric Science at Penn State University and the director of the University's Earth System Science Center. He became a well-known figure in the climate science community in the late 1990s, when he and coauthors Malcolm Hughes and Raymond Bradley published two studies in which they reconstructed the average temperature of the northern hemisphere going 600 years in one study and 1000 years in the other. These reconstructions showed global mean temperature beginning to rise so rapidly in the 20th century that, when plotted on a curve, the data resembled the shape of a hockey stick (Cho, 2017b).

The "hockey stick" graph, as it became commonly known, rapidly became iconic in scientific and even public discourse around climate change. It demonstrates in a remarkably visceral way the impact that human use of fossil fuels since industrialization has had on the climate. This made it a potent symbol and, while it brought Dr. Mann and his coauthors considerable recognition from many of their scientific peers, it also quickly made them targets of those opposed to taking action on climate change.

In 2009, near the time of a United Nations Climate Conference in Copenhagen, an unknown external attacker hacked a server at the Climatic Research Unit at the University of East Anglia in Britain, an event frequently referred to as "climategate." Numerous emails were taken from the server and released, including some that included Dr. Mann or referenced his work. Climate contrarians took some portions of these emails out of context and used them to create a false impression of data manipulation or falsification. Despite the fact that numerous independent investigations found no evidence of fraud or scientific misconduct, a tremendous amount of damage was done. Dr. Mann became the target of deeply troubling personal attacks. His safety and that of his family was threatened—in 2010 he received a letter laced with white powder and an email that said, "You and your colleagues who have promoted this scandal ought to be shot, quartered, and fed to the pigs along with your whole damn families." The white powder, fortunately, turned out to be cornstarch.

In addition to these deeply upsetting personal attacks and threats, Dr. Mann's work on the "hockey stick" graph had also made him the subject of legal harassment, some of it orchestrated by public officials. Although, as referenced earlier, he is now at Penn State University, until 2005, Dr. Mann was on the faculty at the University of Virginia (UVA). In May 2010 the Virginia Attorney General, Ken Cuccinelli, who is a vocal climate skeptic, filed a civil investigative demand—effectively a subpoena intended to force the UVA to turn over a large quantity of Dr. Mann's email correspondence and other documents and materials related to grant applications that Dr. Mann submitted while at the UVA. Cuccinelli invoked the by-then debunked "climategate" emails in attempting to justify the need for an inquiry into Dr. Mann as he sought government funding for his research (Helderman, 2010).

There was a strong response from the academic and scientific communities. The American Civil Liberties Union (ACLU) of Virginia and the American Association of University Professors (AAUP) wrote a letter imploring the university to resist providing the documents Cuccinelli was seeking. Nature magazine published an editorial condemning Cuccinelli's actions and referring to them as a "witch-hunt" (Natures Editorial Board, 2010). The American Meteorological Society and the University Corporation for Atmospheric Research also wrote a letter to the president of UVA asking him to resist the subpoena (UCAR, 2010). The UVA petitioned a Virginia circuit court to set aside the civil investigative demand, and a legal battle officially began.

A Virginia judge dismissed the investigation. However, not long afterward, a group called the American Tradition Institute—now known as Energy and Environment Legal Institute (E&E Legal)—submitted a Freedom of Information Act request to UVA seeking essentially the same personal correspondence that Cuccinelli was seeking (Cho, 2017b, *supra*; Union of Concerned Scientists, 2010). E&E Legal is a coal-funded interest group affiliated with the Competitive Enterprise Institute, Exxon Mobil, and the Koch brothers (American businessman Charles Koch and his brother, David Koch, now deceased, have actively supported organizations that lobby against climate change mitigation or that promote climate denial for years). It has made a practice of targeting climate scientists with harassing open records requests, seeking to sift through huge numbers of scientists' emails searching for anything potentially embarrassing.

Impact and response from the scientific community

Perhaps the single most concerning aspect of these high-profile attempts by both a Virginia government official and E&E Legal to obtain significant quantities of Dr. Mann's email correspondence is the deeply chilling effect this could have on scientists' willingness to engage in free and open discussion and debate with their scientific colleagues.

The emails released from the East Anglia attack illustrated how common it is for scientists from different institutions and, often, different countries to collaborate on research and author papers together. Unsurprisingly, these scientists rely heavily on email as the most efficient means to work together as they develop research projects while being sometimes physically at great distance from one another. In these exchanges, scientists often use jargon and linguistic shortcuts common in their fields—readily understood by colleagues, but easy to misrepresent if presented to a lay-audience without appropriate context. Scientists—even, and perhaps especially those who are collaborating on a project and therefore presumably have high levels of professional regard for one another—frequently disagree with, challenge, and question each other. This is an important part of the scientific process, intended by all involved to ensure that ultimately the work is done with the highest level of rigor and that published results are as accurate and scientifically sound as possible.

If scientists who work at public research universities fear that these discussions may be made public, taken out of context, and used by a special interest group to embarrass or discredit them, they will inevitably begin to self-censor. They will hesitate to openly discuss potential weaknesses in their research or ask one another hard questions—typically done with an aim toward improving the work and striving for the most accurate and sound results—because such discussion might be twisted to falsely suggest that they are incompetent, or even intentional bad actors, and that their work is unreliable.

This chilling effect will be detrimental to the scientific endeavor as a whole. And, since even scientists who do not work for a publicly funded institution may have their emails made public if they correspond with a colleague who does, and who happens to become the subject of one of these harassing records requests, the chilling effect will spread far beyond publicly funded science. The ACLU and the AAUP expressed a similar sentiment in a letter to the UVA president. They wrote, "If scientists refrain from novel methodological approaches because they may be characterized as 'fraudulent,' then scientific research, and, by extension, society as a whole, will be the loser" (ACLU Press Release, 2010).

Further "hockey stick" open records abuse

Dr. Mann was not the only scientist who worked on the "hockey stick" graph who became a target. His coauthor, Malcolm Hughes, is a dendroclimatologist—he studies tree rings and what they can tell us about environmental conditions in the past as reflected in tree growth. Like Dr. Mann, Dr. Hughes was the target of multiple open records requests filed by E&E Legal. E&E Legal submitted these requests to the University of Arizona, where Dr. Hughes was the director of the Laboratory of Tree Ring Research before his retirement. This series of requests also targeted Jonathan Overpeck, who was a colleague of Dr. Hughes' at the University of

Arizona and was also one of the lead authors of the 2007 Report of the International Panel on Climate Change (IPCC). The "hockey stick graph" formed an important part of the basis for the IPCC's conclusion in 2001 that the 1990s were likely the warmest decade of the millennium that preceded them (Cho, 2017a).

E&E Legal's open records requests sought thirteen years' worth of Drs. Hughes and Overpeck's documents and emails from their time at the University of Arizona. The requested documents included, among other things, prepublication data and drafts. What E&E Legal was seeking to do was not only an incredible invasion of these scientists' privacy, but also triggered an array of other significant concerns ranging from academic freedom to intellectual property. Not least of all, both Dr. Hughes and Dr. Overpeck were forced to spend huge amounts of time searching for the requested documents which, as already mentioned, were in some cases more than a decade old. There were even some significant technical difficulties in getting some of the requested records out of the by-then-outdated format in which they were stored and making them compatible with the current technologies.

The University pushed back to some degree on E&E Legal's requests, producing some of the requested documents but withholding several thousand others, and a legal battle ensued. CSLDF joined with the two scientists, as well as the American Meteorological Society and Pfizer, to file two "friend of the court" or amicus briefs supporting the University in its attempt to resist the overly broad and intrusive requests. We focused on trying to explain to the court how this case was part of a growing trend of misuse of public records laws by activists across the political spectrum to target scientists. We described how they stifle collaboration between scientists, divert scientists' time and energy away from their work, discourage scientists from working in controversial fields, and hinder efforts by public universities and government agencies to recruit new scientific talent.

Unfortunately, after a confusing series of rulings and a reversal by the judge of his own previous decision (Kurtz, 2016), in June 2016, the Arizona Superior Court ultimately held that the University needed to turn over the contested emails. While this was a disappointing result, it yielded at least two important lessons. First, although E&E Legal trumpeted that the emails it had so avidly sought through multiple appeals would surely reveal the scientists' activist climate agenda (E&E Legal Press Release, 2016), they were remarkably quiet once the emails were actually produced. E&E Legal had been posting a fairly regular string of press releases, news articles, and commentaries about the progress of the case on its website, but there is a conspicuous lack of follow-up discussion about the actual content of the emails once the court ordered the University to turn them over. This is strong evidence that the emails did not, in fact, reveal anything of much note other than dedicated scientists trying to do their jobs.

Second, the open records law in place in Arizona simply was not working as the legislators who drafted it intended it to, and was confusing both for institutions trying to implement it and for judges trying to interpret it. The legislative history of the Arizona public records law at issue indicated a clear intent to ensure that scientists in labs at research universities, as well as at the various pharmaceutical companies that had set up shop in the state, could collaborate without fear that prepublication and draft work, or intellectual property such as trade secrets, would be made public through an open records request. Yet, in this case, the statute worked a very different result.

The Superior Court judge handling the case was initially persuaded that the release of the emails could do substantial public harm but had to revisit the ruling after a disagreement with the appeals court about statutory interpretation. In ordering that the emails in question be produced, the judge went out of his way to acknowledge concerns about the potential chilling effect that the release of the emails might have (Energy & Environmental Legal Institute Plaintiff vs. Arizona Board of Regents, et al., 2016).

This episode pointed to a more overarching problem with the patchwork of state laws that dictate the handling of records at many publicly funded universities and labs throughout the United States: these laws vary considerably in the comprehensiveness of their approach to protecting scientists against politically motivated fishing expeditions into their email accounts, as well as in their fundamental clarity and quality. Navigating this maze of different laws can be so complex that the CSLDF has developed a guide that explains the open records laws in each of the 50 states, as well as the District of Columbia, highlights ambiguities and weaknesses in those laws, as well as categories of research records that may be particularly vulnerable to invasive open records requests (Climate Science Legal Defense Fund, 2018).

In many instances, a concerted effort to get state legislatures to revisit open records laws to clarify and strengthen them, as well as to make them more consistent across state lines, is clearly called for. In the University of Arizona case, Judge Marner seemed to say as much when he entreated the scientists and their supporters to turn to the Arizona legislature to address what he clearly agreed was a significant problem with the Arizona open records statute (Marner Under Advisement Ruling, *supra*, at 4).

Open records requests as an intimidation tactic

Regrettably, the kind of weaponized use of open records laws experienced by Drs. Mann, Hughes, and Overpeck is not new, and climate researchers are by no means the only scientists who have been targeted.

Take, for example, the story of environmental chemist Deborah Swackhamer. Dr. Swackhamer is now retired from the University of Minnesota, but in 1996 she was a full-time faculty member there investigating why alarmingly high levels of a chemical called toxaphene—typically used as a pesticide, and banned in the 1980s—were appearing in Lake Michigan (Hemphill, 2016). One possibility she and other researchers were investigating was whether it might be a byproduct of the operations of pulp and paper mills situated on rivers that emptied into the Great Lakes.

Then, the University of Minnesota received a Freedom of Information Act request that, as of 2016, still held the record for the largest such request the University has ever received according to reporting in Agate Magazine. "They wanted to see records of my grants, my teaching materials, phone calls, all my data, for a 13-year period," Swackhamer told Agate. "We shipped off container after container of papers; they kept coming back for more information." The attack deeply affected Dr. Swackhamer's personal life as well as her professional life, because a similar request was sent to the Environmental Protection Agency (EPA), where her husband was administering research projects on toxaphene.

Although there was never an absolute confirmation of who was behind the requests, Dr. Swackhamer and others believe that it must have been the paper industry. She told Agate, "I think their motive was not only to stop me doing research, but they really wanted my husband to stop funding it. I think basically they were trying to intimidate us." Industry officials denied involvement.

Despite the significant impediments that resulted from being targeted in this way, Dr. Swackhamer pressed on with her research. A few years later, that research showed that at least one paper mill in Minnesota was *not* in fact producing high levels of toxaphene.

Censorship of scientists for political reasons

Censorship of scientists' congressional testimony

This invasive open records request in the 1990s would, unfortunately, not be the last time Dr. Swackhamer was subjected to attempts to bully and silence her. In 2017 she was the chair of the EPA's Board of Scientific Counselors (BOSC), a key advisory committee that provides advice and recommendations to EPA's Office of Research and Development regarding EPA's research programs.

In May 2017 Democrats on the House Science Committee invited Dr. Swackhamer to testify at one of the Committee's upcoming hearings. She agreed to testify, but said that she would speak as a private citizen and not on behalf of the agency. She discussed this approach with senior officials in EPA's science office, who raised no objection (Mervis, 2017), and submitted her testimony to the committee. That testimony discussed the importance of a robust scientific foundation to inform environmental policy in the United States, and the importance of independent peer review (Swackhamer, 2017). In this context she raised concerns about EPA Administrator Scott Pruitt's decision to not renew half of the 18 BOSC Executive Committee members for a second term, noting for the Committee that Pruitt had stated through a spokesman that "more representation from the regulated community is needed on the committee" (Swackhamer, 2017, Testimony at 4). Dr. Swackhamer was far from alone in being concerned that large numbers of nonrenewals on important federal advisory committees were intended to make room for industry representatives (Reilly, 2017).

The day before the hearing, after Dr. Swackhamer believed her testimony was already embargoed, she received emails from EPA Chief of Staff Ryan Jackson pressuring her to change her testimony (Davenport, 2017; Mervis, 2017). As reported by the New York Times, Jackson wanted her to "stick to the agency's 'talking points'" about the dismissals of BOSC members.

This kind of pressure from an administration to get a witness to alter already-submitted testimony was shocking to Dr. Swackhamer as well as to many others. She refused to change her testimony and made her emails with Mr. Jackson available to the science committee's Democrats.

The concerns Dr. Swackhamer expressed in her testimony continued to be borne out, as in the weeks following that May 23, 2017 committee hearing all of the members of the BOSC subcommittees who were up for a second term were also told their appointments would not be renewed (Testimony of Swackhamer, 2019).

Subcommittees were left without chairs or vice chairs, and their planned meetings were canceled (Reilly, 2017). Unfortunately, all evidence suggests that Dr. Swackhamer faced swift retribution for her unwillingness to bend to pressure from the administration: a few months after her testimony, on October 31, 2017, she was notified that she was being removed from her position as head of the BOSC (Testimony of Swackhamer, 2019). Dr. Swackhamer described this move to media outlets as "not normal," and indicated that, although she did not know the "official reason" why she lost her BOSC chairmanship, she suspected it related to the controversy over her testimony (Bogardus, 2017).

Censorship of climate science during the Trump Administration

What happened to Deborah Swackhamer turned out to be only one example of how, as the Trump Administration took power, attempts to intimidate and silence scientists that had previously commonly come from outside interest groups began increasingly to come from within the administration itself. Since the 2016 presidential election, censorship has been an increasingly common concern for federal scientists, as they have all too frequently been forced to stand up to their own employers to ensure that information gleaned from taxpayer-funded research was allowed to reach the public. Climate scientists have been a special target.

Deep concerns about censorship of information about climate change began even before President Trump's inauguration, triggered by the then-President-elect nominating a series of climate contrarians to high-level positions, including to his cabinet—President Trump's first EPA Administrator, Scott Pruitt, was one prominent example. Scientists were so concerned that crucial data could disappear that they began organizing to download and preserve federal data to ensure that it remained available to the public after Mr. Trump took office (Dennis, 2016).

The scientists' fears turned out to be warranted. Censorship of information about climate change began with startling rapidity as President Trump transitioned into the White House. At approximately the same moment that he took the oath of office in January 2017, all references to climate change vanished from the White House website (Gunaratna, 2017). This scrubbing of information about climate change from various federal government website continued apace as the administration progressed. In the summer of 2019 the Environmental Data and Governance Initiative (EDGI) published research in which it analyzed close to 6000 pages from the websites of 23 federal agencies. It found that the appearance of the terms "climate change," "clean energy," and "adaptation" had dropped by 25% since Mr. Trump's inauguration, going from 6552 in 2016 to 4912 in 2018 (Nost et al., 2019; Baynes, 2019). At the same time, "catch-all" terms that tend to obscure clear analysis such as "resilience" and "sustainability" increased by 26%, according to the EDGI study.

Unfortunately, the silencing of climate scientists continued unabated throughout the course of the Trump Administration (Holden, 2019). In some ways the most concerning aspect of this trend was that career managers at scientific agencies appear in a number of instances to have absorbed the lesson that their departments and the scientists who work in them should avoid working on or speaking about climate change too overtly to avoid provoking the ire of political officials in the administration. In June 2020 Lisa Friedman of the New York Times reported that efforts to undermine federal climate science, "once orchestrated largely by President Trump's political appointees, are now increasingly driven by midlevel managers trying to protect their jobs and budgets and wary of the scrutiny of senior officials" (Friedman, 2020).

Federal scientists have reported instances of their superiors denying them the right to publish valid scientific work because it addressed greenhouse gas emissions, or putting them under intense pressure to remove references to climate change from their work. There have also been reported instances of officials going even further and directly changing, or attempting to change, language in scientific work product to eliminate references to climate change or to falsely suggest uncertainty around climate science that does not exist.

For example, a long-time Department of Interior employee, Indur Goklany, led an effort to insert misleading language about climate change into scientific reports after his promotion to the office of deputy secretary in 2017 (Tabuchi, 2020). This language erroneously claims that some studies have found that the earth is not warming, thus falsely suggesting that there is a lack of scientific consensus that the earth is warming—there is absolutely consensus in the scientific community that the earth is warming.

Nonetheless, this inaccurate language, which became known as "Gok's uncertainty language" after Goklany's nickname, was pushed on scientists at Interior, reportedly appearing in at least nine agency scientific reports. As described by the New York Times, Goklany also instructed department scientists to add that rising carbon dioxide—the main force driving global warming—is beneficial because it 'may increase plant water use efficiency' and 'lengthen the agricultural growing season.' These are highly misleading assertions. The scientific consensus is that climate change will have the net effect of reducing crop yields and causing significant disruption to agriculture world-wide.

This kind of direct effort to interfere with scientific work product to downplay climate change, or to suggest a level of uncertainty around the issue in the scientific community that simply does not exist, was regrettably not unique to Indur Goklany during the Trump Administration. Even worse, scientists who pushed back against these efforts and insisted on being outspoken about their work on climate change were frequently subjected to retaliation from within their institutions. For example, Joel Clement, a science and policy expert who was the director of the Office of Policy Analysis at the Department of the Interior (DOI) at the beginning of the Trump Administration, resigned from federal service when he was involuntarily reassigned in what he believed was retaliation for being vocal—both publicly and with White House and senior DOI officials—about the dangers climate change poses to native communities in Alaska (see Chapter 11, Essays From the Trenches of Science-Based Activism).

Direct interference with scientific work as a form of censorship

Other scientists experienced different forms of retaliation under the Trump Administration when they tried to convey scientifically valid information about climate change. Maria Caffrey is a climate scientist who was a partner at the National Park Service (NPS)—which falls under the umbrella of Interior—from 2012 until 2019. In 2013, she was named Principal Investigator on a research project assessing the impact of future sea level rise on coastal national parks. Dr. Caffrey dedicated several years to this work, and in the summer of 2016, she completed the first draft of a scientific report detailing the results of her research (Shogren, 2019).

That report referenced anthropogenic or human-caused climate change in several places. This was not commentary but rather was scientifically relevant since the models Dr. Caffrey used in her work were based on projections of future greenhouse gas emissions and resulting warming—in other words, understanding the assumptions about anthropogenic climate change Dr. Caffrey was using was important to understanding the scientific methods she used to conduct her research.

Dr. Caffrey later stated in testimony to the House Natural Resources Committee that she expected the NPS to publish her report in early 2017, around the time the Trump Administration came into power. As the year progressed, however, publication was repeatedly pushed back, with explanations from her supervisors at NPS about the reasons for the delay becoming increasingly unclear (Testimony of Caffrey, 2019).

Dr. Caffrey told the House Committee that the real reason for the delay became clear to her when, while she was on maternity leave in early 2018, she learned that supervisors at the NPS had edited her scientific work without her knowledge or permission and removed all references to anthropogenic climate change from her report.

Dr. Caffrey described further interactions with officials and supervisors at the NPS that confirmed her conclusion they were attempting to censor her scientific work to avoid drawing the ire of a new political administration that was opposed to action on climate change. A senior official at NPS told her that it was now "verbal policy in NPS that the term 'anthropogenic climate change' should not be used in scientific reports" and that "this is just the way it is right now." That same official also expressed concern that if the report was published with the references to anthropogenic climate change included, he would be reassigned, and the entire program that Dr. Caffrey worked in might be terminated.

The pressure on Dr. Caffrey to change her scientific work for political reasons came from other sources within NPS as well. She described to the House Natural Resources Committee how one supervisor suggested that if she, Dr. Caffrey, refused to remove the climate references, NPS might not release her report at all. Another NPS employee, who was eventually brought in to attempt to mediate the dispute that ensued, threatened to release the report without Dr. Caffrey's name on it unless she agreed to whatever changes he deemed appropriate.

Either of these outcomes would have been professionally devastating. Dr. Caffrey had spent years on this research. For those years to fail to yield any scientific publication—or for her work to be published, but without her being appropriately credited—would have left a significant gap in her publication record that would have damaged her prospects for future career advancement, either inside the NPS or with other employers.

Dr. Caffrey believed that what she was experiencing was a direct violation of the scientific integrity of her work, and she was not alone in thinking so. Another colleague who was originally a coauthor on the paper ultimately withdrew because "he did not wish to have his name associated with what he saw as a violation of scientific integrity." Only by being willing to enter into what was clearly a series of escalating conflicts with her NPS supervisors, and also by being willing to risk speak publicly about what was happening and drawing unwanted negative publicity to the NPS, was Dr. Caffrey able to ensure that her report was in fact released unedited (Caffrey et al., 2018).

Distressingly, Dr. Caffrey paid a steep personal price for standing up for her science. She testified to House Natural Resources Committee that her unwillingness to alter her scientific work for political reasons ultimately resulted in her being pushed out of the NPS. She was removed from her sea level rise project, even though she was still actively working on it and colleagues she was working with continued to seek and need her advice and input. Funding for her to continue to work on other projects mysteriously dried up, and she was prevented from pursuing new grant opportunities, even though her immediate supervisors valued her and wanted to keep her on. Dr. Caffrey ultimately filed a whistleblower complaint with the Office of Special Counsel in July of 2019 describing how the NPS had retaliated against her for resisting pressure to scrub mentions of climate change from her work.

Restricting publication as a form of censorship

Regrettably, Dr. Caffrey is far from the only scientist working on climate change issues during the Trump Administration who had to push back against censorship of scientific work for political reasons.

John Crusius is a research chemist with United States Geological Survey (USGS). He wrote an article on the topic of "natural" climate mitigation—in other words intentionally storing carbon in tropical forests, soils, and wetlands (Crusius, 2020). The article argued that "future changes in climate, land use, or natural resource policies could cause natural mitigation to cease and emissions to increase," negating much of the potential benefit of the natural carbon sink solutions. Dr. Crusius advocated for prioritizing "natural mitigation pathways that are least vulnerable to disruption by climate change or human activities," such as minimizing tropical deforestation, as well as identifying additional, safe, and effective ways to capture and store carbon.

Dr. Crusius received informal approval from USGS to publish his article in the well-regarded journal Earth's Future, published by the American Geophysical

Union. However, as the article progressed through peer review and approached publication, supervisors at the USGS abruptly pulled the plug on the project, refusing to allow Dr. Crusius to publish (Friedman, 2020, *supra*). To ensure that his research was published he was forced to engage with some uncomfortable conflict with his agency. He appealed the decision to deny publication and ultimately succeeded in getting the agency to agree to allow him to publish the article, albeit without his USGS affiliation.

Silencing climate science through elimination of funding

The general chill under the Trump Administration surrounding climate science extended beyond just those scientists actually employed by the federal government. The same New York Times piece that described what Dr. Crusius experienced when he tried to publish an article addressing greenhouse gas emissions also told the story of Noah Diffenbaugh, a climate scientist at Stanford University (Friedman, 2020, *supra*). In April 2017 Dr. Diffenbaugh and a group of coauthors published a study examining links between climate change and extreme weather events (Diffenbaugh et al., 2017). The paper acknowledged support from the Department of Energy (DOE), which, during the Obama Administration, had provided more than $1.3 million for the project. This kind of disclosure of funding is generally required by scientific journals as well as by universities. Although it was common practice, the acknowledgment triggered additional review by the DOE, something Dr. Diffenbaugh described as alarming.

According to the New York Times piece, a series of emails showed that managers in DOE's biological and environmental research program, known as BER, felt that their program was "under attack internally" and were concerned about the use of what had become "red flag" terms such as "extreme event attribution," "Clean Power Plan," and "Paris Agreement." The concern was purportedly to avoid giving the impression that the DOE was supporting research focused on policy evaluation. This is disconcerting, since that was precisely what BER had funded Dr. Diffenbaugh and his colleagues to do.

The DOE presented Dr. Diffenbaugh with a choice: either remove those terms or remove the acknowledgment of the agency funding. He and Stanford decided to do neither and published the paper with both the scientifically relevant "red flag" words and ethically appropriate disclosure of the DOE as one of the funding sources for the research. Officials from the DOE subsequently cut overall funding for the project that Dr. Diffenbaugh had been working as part of in half and eliminated funding for his specific project entirely.

More than anecdotal evidence of a change in culture

These anecdotal pieces of evidence that climate science and scientists were silenced during the Trump Administration are now also supported by data. A study published in the journal PLoS in April 2020 surveyed government scientists (Goldman et al., 2020). It found that at the EPA over a third of the scientists surveyed "agreed or strongly agreed that they had avoided working on climate change or using the phrase 'climate change,' though they were not explicitly told to avoid them...." It is important to note that a roughly equal number of EPA scientists surveyed responded that they disagreed or strongly disagreed with the same statement. Even so, this study reveals that a shocking number of scientists at the nation's primary environmental agency—an agency whose stated mission is to "protect human health and the environment"—have felt that they should avoid directly discussing or openly working on climate change, arguably the single biggest current threat to both the environment and human health.

Since January 2018, the CSLDF and the Sabin Center for Climate Change Law at Columbia Law School have jointly maintained an online resource called the Silencing Science Tracker. It is a public database that tracks news reports of attempts by the government to restrict, censor, undermine, or misrepresent science since the November 2016 presidential election. Although it initially focused on federal government actions, the Tracker has expanded to include antiscience actions by state governments as well. At this writing, the Tracker has documented close to 500 instances of government actors seeking to silence science.

The Tracker has done more than simply bear witness to the silencing of science. It has provided useful datapoints that help us understand important trends in where that silencing has been targeted, how it has worked, and how it has changed over time. For example, as CSLDF Executive Director Lauren Kurtz, attorney Susan Rosenthal, and Sabin Center attorney Romany Webb wrote in the Environmental Law Institute's Environmental Law Reporter in September 2020, out of 126 documented instances of federal government censorship of scientists, and 20 reported instances of scientists engaging in self-censorship, almost 80% involved information related to climate change (Webb et al., 2020). The Tracker thus supports what otherwise might be only anecdotal impressions with concrete data.

The data gathered in the Tracker is also helpful in identifying which agencies within the federal government have been hot spots for attempts to silence science. For example, as Fig. 15.1, reprinted from Webb, Kurtz, and Rosenthal's paper, shows the EPA was the single most represented federal agency in the Tracker from November 2016 to May 2020. The DOI and the White House also stand out as federal entities with high rates of antiscience actions during this time. By contrast, the Department of Homeland Security and the Department of Defense had relatively few reported instances of antiscience actions during that same time period.

This is useful information about which government scientists have been most affected by antiscience actions. Not only can it inform how we

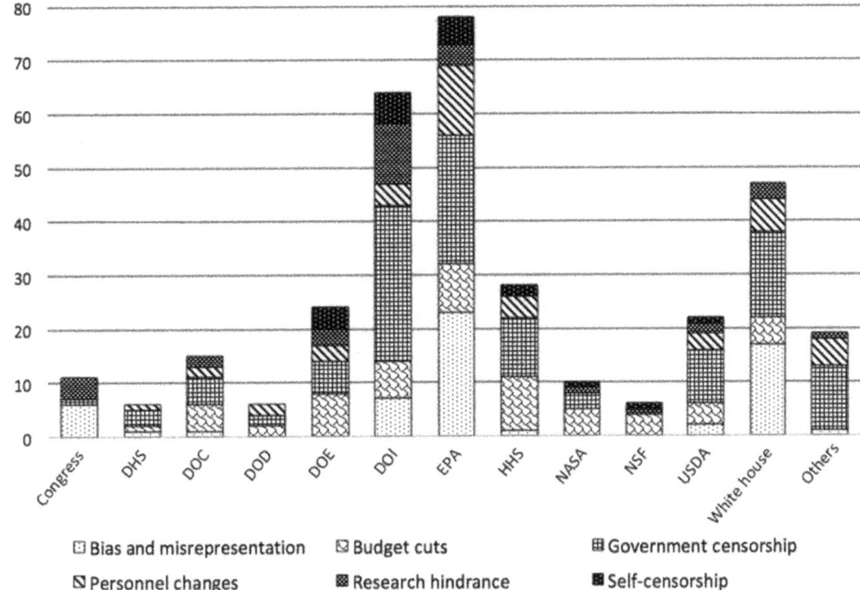

Figure 15.1 Federal antiscience actions by agency (Nov. 8, 2016 to May 7, 2020). *Copyright© 2021, Environmental Law Institute®, Washington, DC. Reprinted with permission from ELI®. Webb et al., 2020, Fig. 1.*

respond to immediate threats to science, but it can also help us understand where to focus resources and efforts in restoring scientific integrity at federal institutions if and when an administration less hostile to science takes power.

Congressional abuse

Another institution that has been a frequent source of politicized attacks on scientists is the United States Congress (also see Chapter 1, The Nuts and Bolts of Science-Based Advocacy, and Chapter 5, Blowing the Whistle on Political Interference: The Northern Spotted Owl). While congressional authority has often been used to protect and further scientific research, it has also unfortunately all too frequently been used to do exactly the opposite: to attack, intimidate, or undermine the credibility of legitimate scientists whose research is politically inconvenient (Climate Science Legal Defense Fund, 2020).

Abuse of the Unilateral Subpoena Power

During the 114th Congress in 2015, multiple House committees adopted new rules allowing their chairs to issue subpoenas without a vote. The Science, Space, and Technology Committee was one of the committees that adopted such a rule.

The Chair of the Science committee during the 114th Congress was Representative Lamar Smith of Texas, a well-known climate contrarian. Smith embraced this new unilateral subpoena power with impressive zeal—from January 2015 to January 2017 he issued 25 subpoenas, more than the committee had issued during its entire 54-year history prior to that. Many of them were targeted at undermining climate science and climate scientists.

Smith's decision to use his unilateral subpoena power to go on an anticlimate crusade was, unfortunately, an illustration of how misuse of congressional power can not only harm individual scientists but can also do considerable damage to entire scientific institutions and to the scientific endeavor as a whole.

One of Smith's chosen targets was the National Oceanic and Atmospheric Administration (NOAA). In 2015 NOAA scientist Tom Karl and coauthors published research in *Science* in which they presented an updated global surface temperature analysis. This analysis argued that, if known biases in the surface temperature records maintained by NOAA were corrected for, there was not, as had been suggested by an analysis by the International Governmental Panel on Climate Change, any discernable "slowdown" in the increase in global surface temperature from 1998 to 2012 (Karl et al., 2015).

The Karl et al. (2015) paper caused a significant stir in the scientific community, and debate and disagreement ensued. Some researchers presented evidence that a "slowdown" had indeed occurred (Tollefson, 2016), while others found that Karl and his coauthors had it right and validated their results (Mooney, 2017). This is how the scientific process actually works: ongoing research efforts coupled with healthy debate within the relevant community of researchers resulting in gradual moves toward consensus (Lewandowsky et al., 2016).

The notion of a "pause" or "hiatus" had previously been seized upon by climate contrarians, including Lamar Smith (Gerken, 2014). And unfortunately, Smith appeared to receive the Karl et al. (2015) paper as a considerable threat to the viability of his dubious talking point. He claimed that NOAA employees had raised concerns about the timing of the publication of the Karl et al. (2015) paper, suggesting that it had been rushed to publication to advance the Obama Administration's climate policies (Schwartz, 2015; Gramling, 2016). Smith never actually provided any documentation supporting this assertion, so it was never clear what the specific concerns were or who raised them (Cornwall, 2015). Nonetheless, Rep. Smith pursued a fairly relentless investigation of NOAA. He sent letters and subpoenas requesting "all documents and communications" related to the Karl et al. (2015) paper. He pressured NOAA to release, among other things, individual communications among the scientists involved in the work.

This inquiry and the accompanying aggressive use of the unilateral subpoena power against a federal scientific agency was so plainly politically motivated that scientific organizations felt compelled to express their deep concern. For example, in November 2015, the American Association for the Advancement of Science (AAAS) led a coalition of eight leading scientific organizations in sending a letter to Rep. Smith decrying this politicization of science and emphasizing that "scientists should not be subjected to fraud investigations or harassment simply for providing scientific results that some may see as politically controversial" (Pinholster, 2015). The coalition's letter went on to say, "We are concerned that establishing a practice of inquests directed at federal scientists whose findings may bear on policy in ways that some find unpalatable could well have a chilling effect on the willingness of government scientists to conduct research that intersects with policy-relevant scientific questions."

The controversy resurfaced in 2017, when the *Daily Mail* published an article reporting that a former NOAA scientist named John Bates had come forward with a suggestion that the Karl et al. (2015) paper was based on "misleading, 'unverified' data" (Rose, 2017). Bates himself later publicly contradicted significant portions of this story, stating that he had not meant to suggest that Karl and his coauthors had committed scientific misconduct (Cornwall and Voosen, 2017). The Independent Press Standards Organization ruled that the article included significant misleading statements, and the *Mail* eventually had to publish a description of that ruling at the top of the story on its website (Schipani, 2017).

Nonetheless, House Science Committee Chairman Smith continued to make considerable hay of the situation; he issued new subpoenas, and repeated the false accusations during a February 7, 2017 hearing, even though Bates had already given interviews to E&E News and the Associated Press in which he made clear that he was not alleging any data tampering (Schipani, 2017). While Smith's investigation around the Karl et al. (2015) paper ultimately went nowhere, it cost taxpayer money, wasted NOAA-employee time that could have been much more productively spent on the agency's actual scientific work, and—most concerningly—probably did considerable damage to the public's trust in one of our most important scientific agencies without any basis for doing so.

Abuse of congressional oversight authority: targeting of individual scientists

It is not only federal agencies or other institutions that have found themselves in the crosshairs of politically motivated attacks on science by members of Congress—in some of the more concerning instances, members have also targeted individual scientists. This kind of behavior is especially problematic: seeing an individual scientist dragged into the public spotlight, their credibility questioned, and their professional reputation tarnished because their scientific work was politically inconvenient for a member of Congress undoubtedly deters other scientists from pursuing valid and perhaps important research in politically charged areas, and may deter talented researchers from considering federal service, or even from entering the field at all.

Further fallout from the hockey stick graph

Michael Mann, whose experience being targeted with weaponized open records requests by the Virginia Attorney General and an outside interest group were discussed earlier in this chapter, was also subjected to a far-reaching congressional inquiry spearheaded by Representative Joe Barton (R-TX), who in the early 2000s was the chair of the House Committee on Energy and Commerce. Barton was closely associated with the fossil-fuel industry and a fierce opponent of legislation intended to address climate change.

In 2005 Representative Barton and his fellow committee member Ed Whitfield (R-KY) sent letters to Dr. Mann, as well as to his coauthors Drs. Bradley and Hughes and a few other scientists, citing a *Wall Street Journal* article as the basis for opening a review of their work around the "hockey stick" graph and requesting a considerable amount of information from the scientists (The Committee on Energy and Commerce, 2005). Both the tone and the content of the letters alarmed many in the scientific community, who saw this as an attempt on the part of members of Congress to fish for information they could use to discredit individual scientists (Schmidt and Rahmstorf, 2005). Despite the backlash, Barton commissioned a statistician named Edward Wegman to do a report on the validity of the work of Dr. Mann and his coauthors (Wegman et al., 2006). A later assessment by a noted computer scientist found that neither Wegman nor any of his team had any expertise in climatology, a Barton staffer provided considerable amounts of the "source material," and significant portions were plagiarized (Mashey, 2010). At the same time, an independent panel convened by the Republican-led Science Committee produced a peer-reviewed report that supported the findings of Dr. Mann and his colleagues (North et al., 2006). Neither this investigation nor any of the numerous other investigations Dr. Mann and his coauthors endured ever found any wrongdoing.

Dubious invocations of antilobbying restrictions

This kind of abuse of congressional oversight authority against scientists has not been directed only toward climate scientists. In December 2017 Linda Birnbaum, an eminent toxicologist and the Director of an executive branch entity called the National Institute of Environmental Health Sciences (NIEHS), published an editorial in the journal *PLoS Biology* entitled *Regulating Toxic Chemicals for Public and Environmental Health*. Dr. Birnbaum and her coauthor concluded the piece by saying, "[c]losing the gap between evidence and policy will require that engaged citizens, both scientists and non-scientists, work to ensure our government officials pass health-protective policies based on the best available scientific evidence" (Gross and Birnbaum, 2017).

Neither that sentence nor any other part of the editorial advocated for any specific piece of legislation or policy. Dr. Birnbaum also noted that she had received no specific funding for writing the article. The final sentence, calling for public engagement and science-based policies that are protective of public health, appeared uncontroversial to many—one law professor described it as "endorsing something akin to motherhood and apple pie" (Kaiser, 2018). Nonetheless, two members of the House science committee, both of whom had received money from industry donors with financial interests in impeding research into the harmful health and environmental effects of chemicals (Lerner, 2018) decided to use their powers to target Dr. Birnbaum.

Science committee chair Lamar Smith—whose abuse of his subpoena power to target climate scientists has already been described—and Representative Andy Biggs (R-AZ) sent a letter in early 2018 to the Inspector General of the Department of Health and Human Services, NIEHS's parent agency, in which they argued that the final sentence of the editorial represented a potential violation of the Anti-Lobbying Act (Committee on Science, Space, and Technology, 2018). They characterized Dr. Birnbaum as having "pressured" citizens to persuade them to communicate with their elected representatives.

Given that Dr. Birnbaum did not use any federal funds to write the piece, and that it did not champion any specific piece of legislation, there was very little basis for finding a violation of the Anti-Lobbying Act. In the final assessment the congressmen's investigation appeared to be a highly politically motivated attack on a scientist whose publicly expressed views were counter to the interests of their industry donors, and it rapidly lost steam.

Pro-climate members of congress wrestle with the line on oversight

The use of congressional powers to investigate individual scientists has not been the exclusive domain of Republican members of Congress. In 2015 the New York Times published a story about a researcher at the Harvard-Smithsonian Center for Astrophysics named Wei-Hock "Willie" Soon. The Times reported that Soon had received $1.25 million from fossil-fuel companies that he had failed to disclose (Gillis and Schwartz, 2015). Arizona Representative Raul Grijalva responded by asking Universities to turn over documents going back several years related to seven individual scientists who had testified at congressional climate hearings (Guillen, 2015).

The move garnered a quick response from the scientists themselves, and from various corners of the scientific community. Public reaction focused in particular on the portion of the requests that asked for records related to preparation of public testimony; there was concern that such a request could chill candid discussion and frank debate among scientists.

Markedly unlike his Republican counterparts, however, Representative Grijalva seemed to take concerns that some parts of his requests had gone too far to heart. He told the *Atlantic* that he was most interested in information about the scientists' funding sources—something the American Geophysical Union said in a blog post it believed was appropriate (Leinen, 2015)—and that he was willing to eliminate the request for scientists' communications that was causing significant concern in the scientific community (Geman, 2015).

Speaking truth to power: final thoughts

The stories in this chapter illustrate the steep personal price scientists have sometimes paid to stand up for their work and to defend their credibility when they have found themselves and their research in the political crosshairs as other authors of the chapters in this book have shown as well.

But the politicization of science has other deeply negative consequences: not only does it impede our ability to deal with pressing human health and national security issues like climate change (and, more recently, COVID-19), it also does harm to the scientific endeavor in the United States as a whole (see Chapter 7, Scientific Integrity and Advocacy: Keeping the Government Honest). In particular, scientific agencies have experienced a deeply concerning exodus of expertise during the course of the Trump Administration. The Washington Post reported that in the first 2 years of the Trump Administration, more than 1600 federal scientists left government, compared with an 8% increase during the same period of the Obama Administration (Gowen et al., 2020, also Chapter 7, Scientific Integrity and Advocacy: Keeping the Government Honest). This thinning of the scientific ranks has been driven by overt actions, such as the abrupt relocation of a Department of Agriculture Office from Washington, DC to Kansas City that caused two-thirds of employees to decide to leave their jobs rather than move (Natanson, 2019); it has also been propelled more subtly by an overall erosion of morale that has caused many scientists to quit in frustration (Vartan and Morber, 2020).

The good news is that there are some straightforward steps that could do a great deal to repair and restore scientific integrity in government agencies while protecting researchers. For example, improvements to State open records laws ensuring that scientists and universities do not have to produce prepublication data or drafts would offer meaningful protection to scientists from overly intrusive, weaponized open records requests and help ensure the free flow of scientific discourse. Strengthening scientific integrity policies at our scientific institutions to ensure that they clearly protect scientists against attempts at censorship, harassment, intimidation, or other political interference would likewise do much to protect the scientific endeavor (Chapter 7, Scientific Integrity and

Advocacy: Keeping the Government Honest). The Scientific Integrity Act, which was reintroduced in February of 2021, would go some distance toward requiring just that if it were to be enacted. The end of the Trump Administration certainly does not mean an immediate end to the kinds of challenges described in this chapter, but it does offer a tremendous moment of opportunity for us to take steps to protect science and scientists from political interference. Let us seize that moment.

References

ACLU Press Release, May 6, 2010. Groups say academic freedom will be chilled if school doesn't stand up for teachers. <https://www.aclu.org/press-releases/aaup-and-aclu-ask-uva-resist-ags-attempt-subpoena-records-climate-science-professor>.

Baynes, C. (2019). 'Trump administration removes quarter of all climate change references from government websites', *The Independent*, Aug. 17. Available at: https://www.independent.co.uk/news/world/americas/us-politics/trump-climate-change-government-websites-global-warming-a9020461.html.

Bogardus, K. (2017). 'Pruitt demotes critic as he remakes science boards', *E&E News*, Nov. 1. Available at: https://www.eenews.net/stories/1060065333.

Brandt, A.M., 2012. Inventing conflicts of interest: a history of tobacco industry tactics. Am. J. Public Health 102 (1), 63–71. Available at: https://www.ncbi.nlm.nih.gov/pmc/articles/PMC3490543/.

Caffrey, M., 2019. House Committee on Natural Resources. <https://naturalresources.house.gov/imo/media/doc/Dr.%20Caffrey%20-%20Written%20Testimony%20-%20FC%20Ov%20Hrg%2007.25.19%20(Scientific%20Integrity).pdf>.

Caffrey, M.A., et al., 2018. Sea level rise and storm surge projections for the National Park Service, Natural Resource Report Series NPS/NRSS/NRR—2018/1648. <https://www.nps.gov/subjects/climatechange/upload/2018-NPS-Sea-Level-Change-Storm-Surge-Report-508Compliant.pdf>.

Cho, R., 2017a. Perspectives of Scientists Who Become Targets: Malcolm Hughes. Climate Science Legal Defense Fund. Available at: https://www.csldf.org/2017/05/05/f-scientists-who-become-targets-malcolm-hughes/.

Cho, R., 2017b. Perspectives of Scientists Who Become Targets: Michael Mann, Climate Science Legal Defense Fund. Available at: https://www.csldf.org/2017/07/20/perspectives-scientists-become-targets-michael-mann/.

Climate Science Legal Defense Fund, 2018. A guide to open records laws and protections for research materials. <https://www.csldf.org/resource/a-guide-to-open-records-laws-and-protections-for-research-materials/>.

Climate Science Legal Defense Fund, 2020. How the abuse of congressional oversight powers harms science. <http://csldf.org/wp-content/uploads/2020/12/CSLDF-How-the-Abuse-of-Congressional-Oversight-Powers-Harms-Science.pdf>.

Committee on Science, Space, and Technology, Jan. 17, 2018. Letter from Rep. Lamar Smith, Chairman, Committee on Science, Space, and Technology, and Rep. Andy Biggs, Chairman, Subcommittee on Environment, to Hon. Daniel R. Levinson, Inspector General, US Dept. of Health & Human Services. <https://www.eenews.net/assets/2018/01/17/document_pm_02.pdf>.

Cornwall, W. (2015). 'Researchers dispute lawmaker's allegation that NOAA rushed climate study', *Science Magazine*, Nov. 25. Available at: https://www.sciencemag.org/news/2015/11/researchers-dispute-lawmaker-s-allegation-noaa-rushed-climate-study.

Cornwall, W., Voosen, P. (2017). 'How a culture clash at NOAA led to a flap over a high-profile warming pause study', *Science Magazine*, Feb. 8. Available at: https://www.sciencemag.org/news/2017/02/how-culture-clash-noaa-led-flap-over-high-profile-warming-pause-study.

Crusius, J., 2020. "Natural" climate solutions could speed up mitigation, with risks. Additional options are needed. Earth Future 8 (4). Available at: https://agupubs.onlinelibrary.wiley.com/doi/full/10.1029/2019EF001310.

Davenport, C. (2017). 'E.P.A. official pressured scientist on congressional testimony, emails show', *The New York Times*, Jun. 26. Available at: https://www.nytimes.com/2017/06/26/us/politics/epa-official-pressured-scientist-on-congressional-testimony-emails-show.html?_r = 1.

Dennis, B. (2016). 'Scientists are frantically copying U.S. climate data, fearing it might vanish under Trump', *Washington Post*, Dec. 13. Available at: https://www.washingtonpost.com/news/energy-environment/wp/2016/12/13/scientists-are-frantically-copying-u-s-climate-data-fearing-it-might-vanish-under-trump/.

Diffenbaugh, N.S., et al., 2017. Quantifying the influence of global warming on unprecedented extreme climate events. Proc. Natl. Acad. Sci. U.S.A. 114 (19), 4881–4886. Available at: https://www.pnas.org/content/early/2017/04/18/1618082114.abstract?sid = 30ed1648-a5b2-44fd-a3e5-172d425a678c.

E&E Legal Press Release, Jun. 16, 2016. AZ superior courts sides with E&E Legal and orders UofA to disclose 'climate' related public records including correspondence of longtime activists Hughes and Overpeck. Available at: https://eelegal.org/press-release-3/.

Energy & Environmental Legal Institute Plaintiff vs. Arizona Board of Regents, et al., Jun. 14, 2016. Under Advisement Ruling, Hon. James E. Marner, Arizona Superior Court, Pima county, case no. C20134963. <http://blogs.law.columbia.edu/climatechange/files/2016/06/2016-06-14-decision-EELI-v-U-of-A.pdf>.

Friedman, L. (2020). 'A war against climate science, waged by Washington's rank and file', *The New York Times*, Jun. 15. Available at: https://www.nytimes.com/2020/06/15/climate/climate-science-trump.html.

Geman, B. (2015). 'Grijalva: climate letters went too far in seeking correspondence', *The Atlantic*, Mar. 2. Available at: https://www.theatlantic.com/politics/archive/2015/03/grijalva-climate-letters-went-too-far-in-seeking-correspondence/446778/.

Gerken, J. (2014). 'House science chair says latest climate report is "clearly biased," but he only read the summary', *Huffington Post*, Nov. 7. Available at: https://www.huffpost.com/entry/lamar-smith-climate-change-denial_n_6123638?guccounter = 1.

Gillis, J., Schwartz, J. (2015). 'Deeper ties to corporate cash for doubtful climate researcher', *The New York Times*, Feb. 21. Available at: https://www.nytimes.com/2015/02/22/us/ties-to-corporate-cash-for-climate-change-researcher-Wei-Hock-Soon.html.

Goldman, G.T., et al., 2020. Perceived losses of scientific integrity under the Trump administration: a survey of federal scientists. PLoS One 15. Available at: https://journals.plos.org/plosone/article?id = 10.1371/journal.pone.0231929.

Gowen, A., Eilperin, J., Guarino, B., Tran, A.B. (2020). 'Science ranks grow think in Trump administration', *Washington Post*, Jan. 23. Available at: https://www.washingtonpost.com/climate-environment/science-ranks-grow-thin-in-trump-administration/2020/01/23/5d22b522-3172-11ea-a053-dc6d944ba776_story.html.

Gramling, C. (2016). 'House science committee demands that NOAA widen its internal search for climate change emails', *Science Magazine*, Feb. 26. Available at: https://www.sciencemag.org/news/2016/02/house-science-committee-demands-noaa-widen-its-internal-search-climate-change-emails.

Gross, L., Birnbaum, L.S., 2017. Regulating toxic chemicals for public and environmental health. PLoS Biol. 16. Available at: https://journals.plos.org/plosbiology/article?id = 10.1371/journal.pbio.2004814.

Guillen, A. (2015). 'Dem's climate probe brings "witch hunt" accusations', *Politico*, Feb. 27. Available at: https://www.politico.com/story/2015/02/climate-change-study-funding-raul-grijalva-115568.

Gunaratna, S. (2017). 'All climate change references deleted from WhiteHouse.gov as new chapter begins', *CBS News*, Jan. 20. Available at: https://www.cbsnews.com/news/all-climate-change-references-have-been-deleted-from-the-white-house-website/.

Heath, D. (2016). 'Contesting the science of smoking', *The Atlantic*, May 4. Available at: https://www.theatlantic.com/politics/archive/2016/05/low-tar-cigarettes/481116/.

Helderman, R.S. (2010). 'State attorney general demands ex-professor's files from University of Virginia', *Washington Post*, May 4. Available at: https://www.washingtonpost.com/wp-dyn/content/article/2010/05/03/AR2010050304139.html.

Hemphill, S. (2016). 'Researcher crusades for policies to protect water: profile of Dr. Deborah Swackhamer', *Agate Magazine*, Mar. 24. Available at: http://www.agatemag.com/2016/03/researcher-crusades-for-policies-to-protect-water-profile-of-dr-deborah-swackhamer/.

Holden, E. (2019). 'War on science: Trump administration muzzles climate experts, critics say', *The Guardian*, Jul. 26. Available at: https://www.theguardian.com/us-news/2019/jul/26/war-on-science-trump-administration-muzzles-climate-experts-critics-say.

Johns, D.M., Levy, K. (2019). 'How Trump's war on science is borrowing from the tobacco industry playbook', *Washington Post*, Dec. 13. Available at: https://www.washingtonpost.com/outlook/2019/12/13/how-trumps-war-science-is-borrowing-tobacco-industry-playbook/.

Kaiser, J. (2018). 'Republicans on House science panel suggest top environmental health scientist broke antilobbying law', *Science Magazine*, Jan. 23. Available at: https://www.sciencemag.org/news/2018/01/republicans-house-science-panel-suggest-top-environmental-health-scientist-broke.

Karl, T.R., et al., 2015. Possible artifacts of data biases in the recent global surface warming hiatus. Science 348 (6242), 1469–1472. Available at: https://science.sciencemag.org/content/sci/348/6242/1469.full.pdf?_ga = 2.239715907.1909031520.1603647909-203807853.1603203020.

Kurtz, L. (2016). 'Arizona court reverses protection for climate scientists', *Climate Law Blog*, Jun. 22. Available at: http://blogs.law.columbia.edu/climatechange/2016/06/22/az-court-reverses-protection-for-climate-scientists/.

Leinen, M., 2015. Protecting Academic Freedom and Holding Ourselves Accountable. American Geophysical Union, From the Prow. Available at: https://fromtheprow.agu.org/protecting-academic-freedom-holding-accountable/.

Lerner, S. (2018). 'House science committee wants to investigate a government scientist for doing science', *The Intercept*, Jan. 22. Available at: https://theintercept.com/2018/01/22/linda-birnbaum-niehs-house-science-committee/.

Lewandowsky, S., Risbey, J.S., Oreskes, N., 2016. The "pause" in global warming: turning a routine fluctuation into a problem for science. Bull. Am. Meteorol. Soc. 97 (5), 723–733. Available at: https://journals.ametsoc.org/view/journals/bams/97/5/bams-d-14-00106.1.xml?tab_body = fulltext-display.

Mashey, J.R., 2010. Strange Scholarship in the Wegman report. <https://deepclimate.files.wordpress.com/2010/09/strange-scholarship-v1-02.pdf>.

Mervis, J. (2017). 'Scientist tells her story in latest partisan battle on House science panel', *Science Magazine*, Jun. 28. Available at: https://www.sciencemag.org/news/2017/06/scientist-tells-her-story-latest-partisan-battle-house-science-panel.

Mooney, C. (2017). 'NOAA challenged the global warming "pause." Now new research says the agency was right', *Washington Post*, Jan. 4. Available at: https://www.washingtonpost.com/news/energy-environment/wp/2017/01/04/noaa-challenged-the-global-warming-pause-now-new-research-says-the-agency-was-right/.

Natanson, H. (2019). 'The USDA relocation to Kansas City is ripping apart the lives of its employees. Here are some of their stories', *Washington Post*, Sept. 7. Available at: https://www.washingtonpost.com/local/social-issues/the-usda-relocation-to-kansas-city-is-ripping-apart-the-lives-of-its-employees-here-are-some-of-their-stories/2019/09/07/9108a3b0-c935-11e9-a1fe-ca46e8d573c0_story.html.

Natures Editorial Board, 2010. Science subpoenaed. Nature 465, 135–136. Available at: https://www.nature.com/articles/465135b#/.

North G.R. et al., 2006. Surface Temperature Reconstructions for the Last 2,000 Years, National Research Council of the National Academies. Available at: https://www.nap.edu/read/11676/chapter/1.

Nost, E., et al., 2019. The New Digital Landscape: How The Trump Administration Has Undermined Federal Web Infrastructures For Climate Information. Environmental Data and Governance Initiative. Available at: https://envirodatagov.org/wp-content/uploads/2019/07/New_Digital_Landscape_EDGI_July_2019.pdf.

Pinholster, G. (2015). 'AAAS leads coalition to protest climate science inquiry', *AAAS News*, Nov. 24. Available at: https://www.aaas.org/news/aaas-leads-coalition-protest-climate-science-inquiry.

Reilly, S. (2015). 'EPA axes 38 more science advisers, cancels panel meetings', *E&E News*, June 20 (republished by Science Magazine). Available at: https://www.sciencemag.org/news/2017/06/epa-axes-38-more-science-advisers-cancels-panel-meetings).

Rose, D. (2017). 'Exposed: how world leaders were duped into investing billions over manipulated global warming data', *The Daily Mail*, Feb. 4. Available at: http://archive.is/hmXLn.

Schipani, V., 2017. No Data Manipulation at NOAA. FactCheck.org. Available at: https://www.factcheck.org/2017/02/no-data-manipulation-at-noaa/.

Schmidt, G., Rahmstorf, S., 2005. Scientists Respond to Barton. RealClimate. Available at: http://www.realclimate.org/index.php/archives/2005/07/barton-and-the-hockey-stick/.

Schwartz, J. (2015). 'Chief of house science panel picks battle over climate paper', *The New York Times*, Dec. 4. Available at: https://www.nytimes.com/2015/12/05/science/chief-of-house-science-panel-picks-battle-over-climate-paper.html.

Shogren, E. (2019). 'Scientist who resisted censorship of climate report lost her job', *Reveal*, Feb. 14. Available at: https://www.revealnews.org/blog/scientist-who-resisted-censorship-of-climate-report-lost-her-job/.

Swackhamer, D.L., 2017. House subcommittee on the Environment Committee on Science, Space, and Technology, U.S. House of Representatives "Expanding the role of states in EPA rulemaking," Room 2318 Rayburn House Office Building. <https://www.congress.gov/115/meeting/house/106025/witnesses/HHRG-115-SY18-Wstate-SwackhamerD-20170523.pdf>.

Swackhamer, D.L., 2019. Subcommittee on investigations & oversight Subcommittee on Environment U.S. House of Representatives: addendum to the hearing charter "EPA Advisory Committees: how science should inform decisions," at 3. Committee on Science, Space, and Technology. <https://science.house.gov/imo/media/doc/Swackhamer%20Testimony.pdf>.

Tabuchi, H. (2020). 'A Trump insider embeds climate denial language in scientific research', *The New York Times*, March 2. Available at: https://www.nytimes.com/2020/03/02/climate/goks-uncertainty-language-interior.html.

The Committee on Energy and Commerce, Jun. 23, 2005. Letter from Rep. Joe Barton and Rep. Ed Whitfield, Chairmen, House Committee on Energy and Commerce, Subcommittee on Oversight and Investigations, to Dr. Michael Mann, Department of Environmental Sciences, University of Virginia. <https://web.archive.org/web/20050624174711/https://energycommerce.house.gov/108/Letters/06232005_1570.htm>.

Tollefson, J., 2016. Global warming "hiatus" debate flares up again. Nature. Available at: https://www.nature.com/news/global-warming-hiatus-debate-flares-up-again-1.19414.

UCAR, May 14, 2010. Letter from the American Meteorological Society and the University Corporation of Atmospheric Research to John T. Casteen III, President, University of

Virginia. <https://www.ametsoc.org/index.cfm/ams/about-ams/ams-position-letters/letter-to-president-of-university-of-virginia-concerning-virginia-ag-investigation/>.

Union of Concerned Scientists, 2010. Timeline: legal harassment of climate scientist Michael Mann. <https://www.ucsusa.org/resources/legal-harassment-michael-mann>.

Vartan, S., Morber, J. (2020). 'As election day nears, taking stock of an expertise exodus', *Undark Magazine*, Oct. 26. Available at: https://undark.org/2020/10/26/trump-expertise-exodus/.

Webb, R., Kurtz, L., Rosenthal, S., 2020. When politics Trump science: the erosion of science-based regulation. In: Environmental Law Reporter, 50 E.L.R. 10709-10713. Available at: https://papers.ssrn.com/sol3/papers.cfm?abstract_id = 3698921.

Wegman, E.J., Scott, D.W., Said, Y.H., 2006. Ad Hoc Committee report on the hockey stick global climate reconstruction. <https://climateaudit.files.wordpress.com/2007/11/07142006_wegman_report.pdf>.

Further reading

Climate Science Legal Defense Fund & Columbia Law School Sabin Center for Climate Change Law, Silencing Science Tracker. <https://www.csldf.org/resource/silencing-science-tracker/>.

Speaking truth to power for the Earth

Dominick A. DellaSala
Wild Heritage, A Project of Earth Island Institute, Berkeley, CA, United States

Conservation Science and Advocacy for a Planet in Peril. DOI: https://doi.org/10.1016/B978-0-12-812988-3.00010-7

https://eoimages.gsfc.nasa.gov/images/imagerecords/4000/4882/AS11-44-6548_lrg.jpg (Royalty free public domain).

> *"If you could see the earth illuminated when you were in a place as dark as night, it would look to you more splendid than the moon." Galileo Galilei Dialogue Concerning the Two Chief World Systems, 1632* (https://en.wikipedia.org/wiki/Dialogue_Concerning_the_Two_Chief_World_Systems#:~:text=The%20Dialogue%20Concerning%20the%20Two,with%20the%20traditional%20Ptolemaic%20system).

Science as denial's antidote

In 1609, pioneering astronomer Galileo Galilei spied celestial bodies through his rudimentary telescope and had an epiphany about the relative position of the Earth in the Cosmos (Britannica.com/biography/Galileo-Galilei). His contrarian discovery that our home planet was in fact not the center of the solar system (*Geocentrism*) was met with swift condemnation from the Catholic Church that declared *Heliocentricsm* (Earth and planets revolving around the Sun) "heretical." Toward the end of his life, Galileo was placed under house arrest and forced to recant or face persecution. At his trial, he was alleged to have uttered—"and yet it moves" (Earth circling the Sun).

Not finished with muzzling Galileo, the Church went after fellow astronomer, Nicolaus Copernicus, in banning *"On the Revolutions of the Celestial Spheres,"* which likewise placed the Earth in orbit around the Sun. Religious condemnation took a toll on Giordano Bruno (burned at the stake), Johannes Kepler (excommunicated from the Lutheran Church), and Sir Isaac Newton (forced to recant to avoid persecution), along with many other leading philosophers and medical doctors cast aside as heretics, tortured, and isolated (for a summary of scientific persecution, see https://gcallah.github.io/history_of_science/ReligiousPersecution.html).

Condemnation of scientists and disrespect for their work was not unique to Galileo's time. In third century BC, the Greek philosopher and director of the Great Library of Alexandria (Egypt), Eratosthenes, mapped and accurately calculated the circumference of the Earth—an amazing feat at the time—positing that the Earth circles the Sun (https://www.britannica.com/topic/Library-of-Alexandria). As one of the first universities, the library was a treasure trove of humanity's breakthroughs in medicine, natural sciences, and mathematics including Euclidean and Pythagorean geometry. It cataloged intellectual and philosophical scrolls of the most influential scholars of the times: Aristotle, Homer, Plato, and Socrates. Sadly, the Great Library died a slow death from centuries of neglect (lack of funding and support), antiintellectual rebellions, periodic conquests, and an accidental fire set by Julius Caesar in his quest for power and domination (https://en.wikipedia.org/wiki/Library_of_Alexandria).

Artistic rendering of the Library of Alexandria (Wikipedia.com)

Economic collapse, social unrest, and violent conquests plunged Europe into the Dark Ages for centuries, even beyond the Renaissance and Age of Enlightenment. Nonetheless, in spite of persecutions and atrocities, science flourished (particularly outside of Europe), and in 1620 it would see the light of day as its parent, Sir Francis Bacon, introduced the world to the scientific (Baconian) method (https://www.biography.com/scholar/francis-bacon).

For centuries, society benefited from breakthroughs of discovery and innovation. Powerful telescopes would plot the path of comets on their timeless celestial journey, casting aside cosmic fears of comets as "harbingers of doom" and the "menace of the Universe." Microscopes and chemistry flasks ushered in new medicines that reduced infections and infant mortality, extended human life expectancy, and put snake oil salesman on notice. And the Haber-Bosch process (chemical process for nitrogen fertilizer) phenomenally increased agricultural production, instrumental in reducing world hunger but triggering a human population explosion and concomitant land conversion.

Despite these advancements, scientists are often victimized when speaking out. In some parts of the globe, they are jailed for "inconvenient" research. In the United States, scientists get death threats and sometimes need personal security escorts when their work is politically charged and in the public light (e.g., pandemic expert, Dr. Anthony Fauci, requiring body guards to attend White House pandemic briefings). We face threats to personal and familial safety, reprisals, censorship, canceled funding, and branding as "minority opinions" when facts are inconvenient (e.g., have an economic or lifestyle impact). We are attacked by industry-employed scientists and other scientists funded by corporations or government agencies with an economic or political stake in the outcome. Consider the concerted effort by chemical companies to destroy the ground-breaking pesticides work of Rachel Carson on chemical toxicity effects on the environment and efforts to defund and bar from publication the seminal work of Tyrone Hayes in documenting hormone interference in frogs and humans of widely used pesticide atrazine (https://www.youtube.com/watch?v = mP-6Gp5RbjQ).

Another case in point—the downplaying of the coronavirus pandemic and climate change by the former US President Donald Trump. By his account, the virus will simply go away if we stop looking for it and his medical advisors just shut up.

> *"If I listened totally to the scientists, we would right now have a country that would be in a massive depression,"*
>
> **Donald Trump speaking at a campaign rally in Carson City, Nevada.**

And Trump had this to say about his top medical advisor, Dr. Anthony Fauci, on the coronavirus:

> *"People are tired of hearing Fauci and these idiots, all these idiots who got it wrong."*

Meanwhile, ignorance, denial, and misinformation spread as fast as the virus in taking hundreds of thousands of lives in the United States as the science of wearing a mask, getting vaccinated, and social distancing became politically charged. This is a hard lesson to learn—consider—if we cannot convince the public to simply wear a mask to save lives, how in the world can we change behaviors enough to stop global climate chaos?

And, how about the draconian and insane threat issued by media mogul Steve Bannon, the former White House chief strategist who was also arrested and charged with fraud in the "We Build the Wall Campaign," referring to Dr. Anthony Fauci and FBI Director Christopher Wray:

> *"I'd put the heads on pikes. Right. I'd put them at the two corners of the White House as a warning to federal bureaucrats. You either get with the program or you are gone."*
>
> **Bannon's comment was immediately deleted and account suspended on Twitter but not Facebook!**

Trump and his minions also routinely denied climate science with outlandish retorts about how the planet is getting cooler, "*I don't think science knows.*" In his actions against humanity and science, President Trump is a contemporary Julius Caesar in excerbating the climate and biodiversity emergencies (Ripple et al., 2017, 2019). Even though he recently departed the White House, his absence by no means marks the end of a growing antiscience agenda. As scientists, do we let him and other deniers get away with this?

Speaking truth to power vs. remaining complicit

Latin is the rudimentary language of science. Most scientists have dealt with it in some form. These particular phrases call out the choices we face as conscientious beings and planetary citizens when the evidence calls us to act:

Qui Tacet Consentire Videtur = They (gender neutral instead of he) who remain silent are understood to consent.

Qui tacet consentire videtur, ubi loqui debuit ac potuit = They (gender neutral instead of he) who are silent, when they ought to have spoken and were able to, are taken to agree.

Latin proverb *Qui tacet consentit* = Silence gives consent.

When deciding whether to speak or remain silent, consider the planetary emergency.

For decades, we have known that the planet is overheating. Droughts and land conversion have intensified desertification, hurricanes in the Gulf of Mexico are more violent and frequent, tundra thawing is displacing Indigenous peoples from Arctic hunting grounds, rising seas are forcing Pacific island nations to abandon, and intense heat waves have become Europe's Grim Reaper (see Intergovernmental Panel on Climate Change IPCC, 2019 for summaries). The impoverished, elderly, people of color, indigenous, health-compromised, nations of the Global South, and nonhuman species are the unfortunate victims of abject climate denial. But it does not have to be this way, as the solution to a safe climate is within our reach if we quickly transition to a carbon-free energy economy and protect the vast forests from boreal to tropical as nature-based climate solutions (Griscom et al., 2017, Moomaw et al., 2019). To do otherwise is *Qui Tacet Consentire Videtur.*

Human and planet health are intertwined

Our health depends on getting this right. Humanity's ever-expanding ecological footprint has forced wild animals (due to poaching and habitat loss) into close contact with people in "wet markets" that can become super-spreaders of deadly viruses. One of the best inoculants we have for avoiding costly pandemics is to protect what remains of wild places and the indigenous cultures that have thrived within them for millennia (DellaSala and Baumann, 2020). Simply put, whatever we do to the planet, we do to each other.

We must end the hemorrhaging of biodiversity to survive. Over 1 million species, some of which may hold the cure for pandemics, are doomed with extinction (Intergovernmental Science-Policy Platform on Biodiversity and Ecosystem Services IPBES, 2019). An unprecedented discombobulation of the biosphere-atmosphere feedback system is setting us up for a deadly collision with the planet's life-support systems (Barnosky et al., 2011, 2012, Ripple et al., 2017, 2019) that will dwarf the Covid-19 pandemic and come with longer lasting set backs then even the loss of the Great Library. And like Caesar's burning-down-the house conquest, *we are burning down Nature's prodigious library from which we are but a single page in the vast encyclopedia of interconnected life.*

A cosmic perspective

www.shutterstock.com-1608962641

In spite of the pernicious treatment by science deniers, science builds on the long arc of history: from rudimentary telescope to spectroscopic images of deep space, from the double helix to the latest mRNA vaccines, from academic scrolls to Big Data, and from imagination of what worlds may exist beyond our own to exploratory probes that set sail into deep space with a desire to no longer be alone in the vast universe (https://voyager.jpl.nasa.gov/galleries/images-on-the-golden-record/). We all have a little Copernicus and Galileo in us as innovators, cultural creatives, and global problem solvers.

In 2015,18 astronauts sent a video to world leaders at the United Nations COP21 in Paris. Through the actions of "Planetary Collective," they had this to say about their experience and perspective gained from above the Earth:

1. humanity's effect on the planet is undeniable;
2. you get an eyewitness view on the environment;
3. you realize how blessed we are;
4. we are a very small part of a very big picture;
5. we are citizens of space and stewards of the planet; and
6. it is our responsibility to protect the Earth.

(https://www.upworthy.com/5-quotes-that-show-why-you-go-into-space-a-scientist-and-come-down-an-environmentalist).

Upon returning from the moon and speaking to *People* magazine in 1974, Astronaut Edgar Mitchell delivered this prophetic message to us Earthlings:

> *"You develop an instant global consciousness, a people orientation, an intense dissatisfaction with the state of the world, and a compulsion to do something about it. From out there on the moon, international politics look so petty.*
>
> *You want to grab a politician by the scruff of the neck and drag him a quarter of a million miles out and say, Look at that, you son of a bitch."* **https://www.indy100.com/article/one-incredible-quote-from-edgar-mitchell-the-6th-man-to-walk-on-the-moon--Zklu22ptj6e.**

It has been nearly 60 years since Soviet cosmonaut Yuri Gagarin and the US astronaut Alan Shepard first orbited the Earth, viewing our fragile blue planet from the tiny capsule of their floating tin cans. And in spite of prescient proclamations of these brave celestial travelers, the latest climate projections are a sobering reminder that we have precious little time before climate chaos is locked in and humanity is potentially plunged back into the Dark Ages.

Speaking Truth to Power: Epilogue

Recognizing our duty as planetary citizens, more scientists are marching in force with young people like Greta Thunberg, with Black Lives Matter, and with environmentalists, parents, and school teachers. Our ability to explain complex scientific constructs has vastly improved by scientists willing to speak truth to power, the likes of Carl Sagan, Neil DeGrasse Tyson, David Attenborough, James Hansen, David Suzuki, Rachel Carson, and Jane Goodall to name a few.

Science march on Olympia Washington Capital, March 2018, with permission B. Zeigler

Given the planetary emergency what is needed is a revolution in science-based advocacy on par with the breakthrough discoveries that pioneering astronomers and early scientists brought to society. Science-based conservation can help ensure that Earth is treated as the center of our personal universe, and that succeeding generations are blessed with a safe climate, an ecologically vibrant world, and a just society. We simply have no other choice.

In Carl Sagan's (1980) prophetic *Cosmos* and amazing TV series, he calls on each of us to *Speak for the Earth*:

> Our loyalties are to the species and to the planet. We speak for earth. Our obligation to survive and flourish is owed not just to ourselves but also to that cosmos ancient and vast from which we spring. http://www.cooperative-individualism.org/sagan-carl_who-speaks-for-earth-1980.htm

David Bowie's poetic song Space Oddity reminds me of how alone we are in the Cosmos when we fail to hold in reverence the magnificence of life on Earth:

> *"...This is Ground Control to Major Tom You've really made the grade And the papers want to know whose shirts you wear. Now it's time to leave the capsule if you dare.*
>
> *This is Major Tom to Ground Control I'm stepping through the door And I'm floating in a most peculiar way And the stars look very different today. For here Am I sitting in a tin can Far above the world. Planet Earth is blue And there's nothing I can do."*
>
> **David Bowie.**

Despite the sobering close of Space Oddity—as scientists, WE can and must act in unison to save planet Earth from ourselves.

In an interview with WBEZ (national public radio, https://www.wbez.org/stories/neil-degrasse-tyson-on-the-politics-of-saving-the-world/5f1a57fb-d23a-4f91-b28f-93a9bfe7f7c1), Neil DeGrasse Tyson had this to say about speaking truth to power:

> *"So there are people saying, 'Oh why are the scientists marching? What do they need?' It's not even about us. Yeah, I mean, it could be about us because it's great to have more research money than less research money and I do research. OK, so hold that aside for a moment.*
>
> *Whether or not you enjoy science, at the end of the day, it enhances your health, your wealth and your security. Oh, so that's the special interest that scientists are marching for: your health, your wealth and your security. It is unlike any other special interest group there is.... Science touches everyone, especially society. So if you were neither marching or you were not spiritually part of that march, then you don't deserve the future."*

I close this book with another issue raised by Neil deGrasse Tyson in his interview about "our bias problem" (https://www.facebook.com/watch/?v=410877186610168). He talks about three truths: personal, political, and objective. Objective is based on the scientific method pioneered by the likes of Sir Frances Bacon, Sir Isaac Newton, and Galileo Galilei. For instance, the gravitational acceleration of a free-falling object is 9.8 m/s^2 no matter what you may believe. The Earth is not flat and the Apollo moon landing was not staged even if you believe in conspiracies. The denial of objective truth that permeates the politics of climate change and biodiversity loss is the shocking reality we face if we care about saving life on Earth. As Albert Einstein once put it, "it is the duty of every citizen according to his best capacities to give validity to his convictions in political affairs." As citizens of planet Earth, and with the Hippocratic

Planetary Oath as our touchstone, scientists and advocates need to rise up and speak with objective truth in political affairs to avoid plunging us into planetary chaos in the decades ahead. I hope that I have made a compelling argument in the pages of this book for you to speak out—and if you already are doing so—thank you—please keep it up! The Earth needs you!

And finally, here are some must read publications (Blockstein, 2002; Hansen, 2009; Hayes and Grossman, 2006; Olson, 2009; Schneider, 2009; Hunter et al., 2010; Parsons, 2016; Noss et al., 2012) and supportive resources for speaking truth to power.

Advocacy and societal change

Americans for Carbon Dividends; http://www.afcd.org/

Carbon Tax Center; https://www.carbontax.org/

ClientEarth; https://www.clientearth.org/

EcoHealth Alliance; https://www.ecohealthalliance.org/about

Environment America; https://environmentamerica.org/

Environmental Investigation Agency; https://eia-global.org/

Global Footprint Network; http://www.footprintnetwork.org/

Global Forest Watch; https://www.globalforestwatch.org/

2000-Watt Society; https://www.2000-watt-society.org/act

Center for Advancement of the Steady State Economy; https://steadystate.org/

Wellbeing economy alliance; https://wellbeingeconomy.org/

Post growth institute; https://www.postgrowth.org/

Alliance for zero extinction; https://zeroextinction.org/

Center for Biological Diversity; https://biologicaldiversity.org/about/

Population connection; https://www.populationconnection.org/

Indigenous environmental network; https://www.ienearth.org/

Climate sources and training

350.org; https://350.org/

Climate Reality Project (Al Gore); https://www.climaterealityproject.org/

Carbon Disclosure Project; https://www.cdp.net/en

(*Continued*)

(Continued)

The Carbon Tracker Initiative; https://www.carbontracker.org/

Climate Action Tracker; http://climateactiontracker.org/

Climate Analytics; http://climateanalytics.org/

Climate Change Performance Index; https://germanwatch.org/en/CCPI

Climate Impact Lab; http://www.impactlab.org/

Climate Interactive; https://www.climateinteractive.org/

CO_2—Earth: Are we stabilizing yet? https://www.co2.earth/global-co2-emissions

Copernicus Climate Change Service; https://climate.copernicus.eu/about-us

Global Carbon Atlas; http://www.globalcarbonatlas.org

Global Carbon Project; https://www.globalcarbonproject.org/carbonbudget/

News organizations

Greenwire and E&E News (subscription based), The Narwhal; https://the-narwhal.ca/;
Mongabay; https://news.mongabay.com/

Bulletin of the Atomic Scientists; https://thebulletin.org/#

Canadian Broadcasting Corporation (CBC) What on Earth; https://links.lists.cbc.ca/v/443/6bfb647e3a526fec0a4f1bf9d8e6303034090e300c703e6323f1f694ea285081

Carbon Brief; https://www.carbonbrief.org/

CleanTechnica; https://cleantechnica.com/cleantechnica/

Climate Home News; http://www.climatechangenews.com/

DESMOG: Clearing the PR-Pollution that clouds Climate Science; https://www.desmogblog.com/

Grist; https://grist.org/

The Guardian Climate Change; https://www.theguardian.com/environment/climate-change

Heated; https://heated.world/

Inside Climate News; https://insideclimatenews.org/

(*Continued*)

(Continued)

The New York Times Climate; https://www.nytimes.com/newsletters/climate-change

Real Climate; http://www.realclimate.org/

Resources for the Future; https://www.rff.org/

The Washington Post Energy and Environment Newsletter; https://www.washingtonpost.com/climate-environment/?itid = nb_hp_climate-environment

National Public Radio Living on Earth; https://www.npr.org/podcasts/381444261/pri-living-on-earth

Thinktanks

Carnegie Climate Governance Initiative (C2G); https://www.c2g2.net/

Center for Climate and Energy Solutions; https://www.c2es.org/about/

The Center for Climate and Security; https://climateandsecurity.org/

Center for Global Development; https://www.cgdev.org/

Center for Research on Energy and Clean Air; https://energyandcleanair.org/

Center for International Environmental Law (CIEL); https://www.ciel.org/

Center for Strategic and International Studies (CSIS); https://www.csis.org/

Center for Sustainable Development; Earth Institute, Columbia University; http://www.earth.columbia.edu/articles/view/1791

Energy Watch Group; https://energywatchgroup.org/

Global CCS Institute; https://www.globalccsinstitute.com/

Global Center on Adaptation; https://gca.org/our-work

Global Challenges Foundation; https://globalchallenges.org/

The International Council on Clean Transportation (ICCT); https://theicct.org/

International Renewable Energy Agency (IRENA); http://www.irena.org/

Mauna Loa, Hawaii Observatory Earth System; https://www.esrl.noaa.gov/gmd/obop/mlo/

(*Continued*)

(Continued)

Mercator Research Institute on Global Commons and Climate Change (MCC); https://www.mcc-berlin.net/ueber-uns.html

Met Office (United Kingdom); https://www.metoffice.gov.uk/about-us/who

MIT Center for Energy and Environmental Policy Research; http://ceepr.mit.edu/

NASA, Global Climate Change; https://climate.nasa.gov/

National Academies of Sciences; http://sites.nationalacademies.org/sites/climate/index.htm

Rocky Mountain Institute; https://rmi.org

World Climate Research Program; https://www.wcrp-climate.org/

World Green Building Council; http://www.worldgbc.org/

World Resources Institute; http://www.wri.org/

Worldwatch Institute; http://www.worldwatch.org/mission

The World Weather Attribution Project; http://www.climateprediction.net/weatherathome/world-weather-attribution/

World Wildlife Fund (WWF); https://www.worldwildlife.org

Yale Program on Climate Change Communication; https://climatecommunication.yale.edu/visualizations-data/

Activist and lobby organizations

Center for Progressive Reform; http://progressivereform.org/

Climate Investigations Center; https://climateinvestigations.org/

Climate Liability News; https://www.climateliabilitynews.org/

Coalswarm; http://coalswarm.org/about-coalswarm/

EarthJustice: Because the Earth needs a good Lawyer; https://earthjustice.org/

EndCoal – Coal Plant Tracker; https://endcoal.org/global-coal-plant-tracker/

Fossil Free Divestment; https://gofossilfree.org/divestment/commitments/

Fossil Free Indexes; https://fossilfreeindexes.com/about-us/

Rainforest Action Network; https://www.ran.org/

(Continued)

(Continued)

Global covenant of Mayors for Climate and Energy; https://www.globalcovenantofmayors.org/

RE 100; http://there100.org/

Renewable Energy Policy Network for the 21st Century (REN21); http://www.ren21.net/gsr-2018/pages/ren21/ren21/

Under2Coalition; https://www.under2coalition.org/about

Scientists support sources

The Alliance for World Scientists; https://scientistswarning.forestry.oregonstate.edu/

Union of Concerned Scientists; https://www.ucsusa.org

Public Employees for Environmental Responsibility; help for government scientists; https://www.peer.org/

Climate Legal Defense Fund; https://www.csldf.org/; probono support to scientists under attack

Wild Heritage, a project of Earth Island Institute maintains a scientist sign on list; https://wild-heritage.org/our-work/scientist-sign-on/

American Association of the Advancement of Science provides advocacy information; https://www.aaas.org/resources/workshop-advocacy-science-advocacy-initiatives.

The Society for Conservation Biology (www.consbio.org) maintains a global membership of some 4000 professionals with dozens of local chapters in seven geographic regions (Africa, Asia, Europe, Latin America and Caribbean, Marine, North America, Oceania).

* The above list was prepared in consultation with all chapter authors of this book especially Franz Baumann.

References

Barnosky, A., et al., 2011. Has the Earth's sixth mass extinction already arrived? Nature 471, 51–57. Available from: https://doi.org/10.1038/nature09678.

Barnosky, A., et al., 2012. Approaching a stated shift in Earth's biosphere. Nature 486, 52–58.

Blockstein, D.E., 2002. How to lose your political virginity while keeping your scientific credibility. BioScience 52, 91–96. Available from: https://doi.org/10.1641/0006-3568(2002)052[0091:HTLYPV]2.0.CO;2.

Griscom, B.W., et al., 2017. Natural climate solutions. PNAS 114:11645-11650. www.pnas.org/cgi/doi/10.1073/pnas.1710465114.

DellaSala, D.A., Baumann, F., 2020. April. Public health depends on a healthy planet. The New Republic 20. Available from: https://newrepublic.com/article/157361/public-health-depends-healthy-planet.

Hansen, J., 2009. Storms of our grandchildren. The Truth About the Coming Climate Catastrophe and Our Last Chance to Save Humanity. Bloomsbury, New York, NY.

Hayes, R., Grossman, D., 2006. A Scientist's Guide to Talking with the Media. Practical Advice from the Union of Concerned Scientists. Rutgers University Press, New Brunswick, NJ.

Hunter Jr., M., Dinerstein, E., Hoekstra, J., Lindenmayer, D., 2010. A call to action for conserving biological diversity in the race of climate change. Conserv. Biol. 24, 1169–1171. Available from: https://doi.org/10.1111/j.1523-1739.2010.01569.x.

Intergovernmental Panel on Climate Change (IPCC), Sixth assessment. 2019. https://www.ipcc.ch/2019/.

Intergovernmental Science-Policy Platform on Biodiversity and Ecosystem Services (IPBES), A million threatened species? 2019. https://ipbes.net/search?search_api_fulltext = million + species.

Moomaw, S.A., Massino, Faison, E.K., 2019. Intact forests in the United States: proforestation mitigates climate change and serves the greatest good. Front. For. Glob. Change 11 June 2019. Available from: https://doi.org/10.3389/ffgc.2019.00027.

Noss, R.F., Dobson, A.P., Baldwin, R., Beier, P., Davis, C.R., DellaSala, D.A., et al., 2012. Bolder thinking for conservation. Conserv. Biol. 26, 1–4.

Olson, R., 2009. Don't Be Such a Scientist. Island Press, Washington, D.C.

Parsons, E.C.M., 2016. "Advocacy" and "Activism" are not dirty words-how activists can better help conservation scientists. Front. Mar. Sci. 7 November 2016. Available from: https://doi.org/10.3389/fmars.2016.00229.

Ripple, W.J., Wolf, C., Newsome, T.M., Galetti, M., Alamgir, M., Crist, E., et al., 2017. World scientists' warning to humanity: a second notice. Bioscience 67, 1026–1028. Available from: https://doi.org/10.1093/biosci/bix125.

Ripple, W.J., Wolf, C., Newsome, T.M., Barnard, P., Moomaw, W.R., 11,258 scientist signatories from 153 countries, 2019. World scientists' warning of a climate emergency. Bioscience 70, 8–12. Available from: https://doi.org/10.1093/biosci/biz088.

Sagan, C., 1980. Cosmos. Penguin Random House, New York, NY.

Schneider, S.H., 2009. Science as a Contact Sport. National Geographic Society, Washington, D.C.

Index

Index

T

Y

Z

www.ingramcontent.com/pod-product-compliance
Lightning Source LLC
Chambersburg PA
CBHW081759200326
41597CB00023B/4081

* 9 7 8 0 1 2 8 1 2 9 8 8 3 *